2007 IEEE/LEOS International Conference on Optical MEMS and Nanophotonics

Hualien, Taiwan
12-16 August 2007

IEEE Catalog Number:	CFP07MOE-PRT
ISBN 10:	1-4244-0641-2
ISBN 13:	978-1-4244-0641-8

**Copyright © 2007 by The Institute of Electrical and Electronics Engineers, Inc.
All Rights Reserved**

Copyright and Reprint Permissions: Abstracting is permitted with credit to the source. Libraries are permitted to photocopy beyond the limit of U.S. copyright law for private use of patrons those articles in this volume that carry a code at the bottom of the first page, provided the per-copy fee indicated in the code is paid through Copyright Clearance Center, 222 Rosewood Drive, Danvers, MA 01923.

For other copying, reprint or republications permission, write to IEEE Copyrights Manager, IEEE Operations Center, 445 Hoes Lane, Piscataway, New Jersey USA 08854. All rights reserved.

IEEE Catalog Number:	CFP07MOE-PRT
ISBN 10:	1-4244-0641-2
ISBN 13:	978-1-4244-0641-8
LOC:	2006932761

Additional Copies of This Publication Are Available from:

IEEE Service Center
445 Hoes Lane
Piscataway, NJ 08854

Phone:	(800) 678-IEEE
	(732) 981-1393
Fax:	(732) 981-9667
E-mail:	customer-service@ieee.org

PREFACE

Welcome to Optical MEMS 2007. I wish to express my appreciation for those who had traveled a long distance to join us for the conference.

The conference began as an IEEE/LEOS Topical Meeting held in Keystone, Colorado, in 1996. The second meeting was held in Nara, Japan in 1997 as an independent international conference. Optical MEMS and Nanophotonics 2007 is the 12th in the series, following the conferences held in Oulu, Finland and Montana, USA. This time we have selected Farglory Hotel in Hualien, Taiwan. We believed Hualien would be the best venue for our conference.

To embrace the trend of submitted manuscripts shifting toward nano-related topics, our conference will be renamed as IEEE/LEOS Optical MEMS and Nanophotonics Conference from year 2007 on. The decision on the conference rename was made at the steering committee meeting last year; the technical program committee was also re-organized to reflect the expansion of our conference community. The committee looks forward to running a parallel-session conference in the near future, hopefully in 2008.

Best Wishes,

J. Andrew Yeh

General Chair of IEEE/LEOS Optical MEMS and Nanophotonics Conference 2007

TOPICS

Topics in the area of Nanophotonics
- Quantum Dots
- Nanowires
- Photonic Crystals
- Plasmonics
- Metamaterials
- Molecular Materials
- Nanocavities
- Nanoapertures
- Spectroscopy of Nanostructures
- Nanophotonics for Medicine and Biology
- Nanoscale Imaging
- Near Field Optical Microscopy
- Nanolithography
- Nanoimprint Lithography
- Self Assembly
- Laser Direct Writing
- Holographic Lithography
- Optical Storage
- Nanostructures for Displays
- Organic Emitters
- Light in Confined Space
- Modeling and Simulation

Topics of "Micro- and Nano-Optics" for the marriage devices
- Adaptive Optics
- Biological Applications
- Device Fabrication Technologies
- Integrated Microsystems
- Micro- and Nano-Optics
- Microscopies
- Modeling and Characterization
- Optical Scanners and Micromirrors
- Optical Sensors
- Optical and Fluidic Devices
- Packaging
- Surface Plasmon
- Telecommunications
- Tunable Devices

CONFERENCE OUTLINE

Date
August 12–16, 2007

Venue
Farglory Hotel, Hualien, Taiwan

Language
English

Sponsored by
- IEEE/LEOS (Laser and Electro Optical Society)
- National Science Council
- Ministry of Econoic Affairs
- National Tsing Hua University
- K-12 Education Center for Nanotechnology
- Sino American Silicon Products, Inc.
- Instrument Technology Research Center, National Applied Research Laboratories

ACKNOWLEDGEMENT

The organizers would like to thank the following organizations for helpful cooperation:
- Hualien County Government
- Taroko National Park
- Tourism Bureau, Ministry of Transportation and Communications
- National Dong Hwa University
- Fa-Naw Tribe

Optical MEMS & Nanophotonics 2007 Committees

Conference Chair

J. Andrew Yeh
National Tsing Hua University,
Hsinchu, Taiwan

Nanophotonics Committee	Optical MEMS Committee
Program Chair: Chennupati Jagadish *Australian National University, Canberra, Australia*	**Program Chair:** Minoru Sasaki *Toyota Technological Institute, Nagoya, Aichi, Japan*
Sailing He, *Zhejiang University, Hangzhou, China*	George Barbastathis, *Massachusetts Institute of Technology, Cambridge, MA, USA*
Bert Hecht, *University of Wuerzburg, Wuerzburg, Germany*	David L. Dickensheets, *Montana State University, Bozeman, MT, USA*
Hans-Peter Herzig, *University of Neuchatel, Neuchatel, Switzerland*	Lorenzo Faraone, *University of Western Australia, Crawley, Australia*
Hartmut H. Hillmer, *University of Kassel, Kassel, Germany*	Ekaterina Golovchenko, *Tyco Telecommunications Laboratories, Eatontown, NJ, USA*
Wonho Jhe, *Seoul National University, Seoul, Korea*	Jong-Hyun Lee, *GIST, Gwangju, Korea*
Sanjay Krishna, *University of New Mexico, Albuquerque, NM, USA*	Daniel Lopez, *Lucent Technologies, Murray Hill, NJ, USA*
Yong-Hee Lee, *KAIST, Taejon, Korea*	Hiroshi Miyajima, *Olympus Corporation, Tokyo, Japan*
Gong-Ru Lin, *National Taiwan University, Taipei, Taiwan*	Yael Nemirovsky, *Israel Institute of Technology, Haifa, Israel*
Lih Y. Lin, *University of Washington, Seattle, WA, USA*	Wilfried Noell, *University of Neuchatel, Neuchatel, Switzerland*
Michal Lipson, *Cornell University, Ithaca, NY, USA*	Fumikazu Oohira, *Kagawa University, Kanagawa, Japan*
Yoshiaki Nakano, *University of Tokyo, Tokyo, Japan*	Yves-Alain Peter, *Ecole Polytechnique de Montreal, Montreal, QC, Canada*
John D. O'Brien, *University of Southern California, Los Angeles, CA, USA*	Renshi Sawada, *Kyushu University, Fukuoka, Japan*
Toshiharu Saiki, *Keio University, Yokohama, Japan*	Richard R.A. Syms, *Imperial College London, London, UK*
Vahid Sandoghdar, *ETH Zurich, Zurich, Switzerland*	Joesph J. Talghader, *University of Minnesota, Minneapolis, MN, USA*
Olav Solgaard, *Stanford University, Stanford, CA, USA*	Hakan Urey, *Koc University, Istanbul, Turkey*
Elias Towe, *Carnegie Mellon University, Pittsburgh, PA, USA*	Ming C. Wu, *University of California – Berkeley, CA, USA*
M. Selim Unlu, *Boston University, Boston, MA, USA*	Hans Zappe, *University of Freiburg, Freiburg, Germany*

TABLE OF CONTENTS

Monday, 13 August 2007

PLE   **Plenary Session**
PLE1   Concepts of Nanophotonic Devices and Fabrications 1
PLE2   Combination of MEMS and Microoptics for New Application.....................................N/A
PLE3   Photonic Band Gap Materials: Engineering Light-Matter Interactions 3

MA   **Medical and Bio Sensing**
MA1   Wearable Laser Blood Flowmeter for Ubiquitous Healthcare Service 4
MA2   A Portable Two-Photon Fluorescence Microendoscope based on a Two-Dimensional
Scanning Mirro...6
MA3   Tunable Endoscopic MEMS-Probe for Optical Coherence Tomography.....................8
MA4   Forward-Imaging Swept Source Optical Coherence Tomography using Silicon MEMS
Scanner for High-Speed 3-D Volumetric Imaging 10
MA5   Miniaturized Optical Viscosity Sensor based on a Laser-induced Capillary Wave...............12

MB   **Active Nano Devices**
MB1   High-Q Photonic Crystal Nanocavities 14
MB2   Modeling of Slow Light in Vertical Cavity Surface Emission Lasers 15
MB3   A Nano-scale Nanocrystal Photodetector with High Sensitivity 17
MB4   Waveforms of Terahertz Radiation Emitted from Superconducting Dipole Antenna 19
MB5   Loss and Crosstalk in Quantum Dot Waveguides...21

MC   **Novel Fabrication Techniques**
MC1   Towards the Fabrications Platforms for MOEMS..23
MC2   FR-4 as a New MOEMS Platform 25
MC3   Optically Flat Micromirror Using Stretched Membrane with
Crystallization-Induced Stress 27
MC4   Polarization-Transmissive Thin-Film Solar Cell with Photodiode Nanowires.................29
MC5   Micro Knife-Edge Optical Measurement Devices Fabricated by SOI and CMOS
MEMS Processes 31

Tuesday, 14 August 2007

TuA   **Bio Nano Devices**
TuA1   Linear Tactile Nanodevice with Resolution on Par with Human Finger 35
TuA2   Optical NEMS Based Force Sensor Using Silicon Nanophotonics.............................37
TuA3   Photostable Single $KTiOPO_4$ Nanocrystals for Second-Harmonic
Generation Microscopy 39
TuA4   Iridescent Photonic Nano-Silica for Chemical and Biological Sensing 41
TuA5   A Novel Single-Cell Surgery Tool Using Photothermal Effects of Metal Nanoparticles.................43

TuB   **Actuation**
TuB1   Optically Controlled, Holographic Micro-Hand 45
TuB2   A Thermo-Pneumatically Actuated Tip-Tilt-Piston Mirror..47
TuB3   Bi-directionally Driven Metal Cantilevers Developed for Optical Actuation.................49
TuB4   Reconfigurable Nanophotonic Systems by Tunable Alignment between
Nanomagnet Arrays 51
TuB5   High-Accuracy Digital-to-Analog Actuators Using Parallel Spring Array....................53

TuP   **Poster Session**
TuP1   Novel Large Area Applications Using Optical MEMS...55
TuP2   Characterization of an Improved, Real-Time MEMS-Based
Phase-Shifting Interferometer..57
TuP3   Fast Tracking of Light Source with Micromirror and Associated Feedback Circuit.................59
TuP4   Parallel and Selective Trapping in a Patterned Plasmonic Landscape........................61
TuP5   Laser Doppler Vibrometer Using a 45°-Angled Optical Fiber for In-Plane Dynamic
Measurement of MEMS Actuators..63

TuP6	A Dielectrically Driven Liquid Lens with Optical Packaging	65
TuP7	Fully-Integrated Optofluidic Trap with Linear Microsphere Array	67
TuP8	Electrowetting-Based Total Internal Reflection Chip for Optical Switch and Display	69
TuP9	Self-Alignment Micro-Lens by Gradient of Surface Tension	71
TuP10	Design of a Holding System for Micro-Coil based MRI	73
TuP11	Glass Reflowed Microlens Array and its Optical Characteristics	75
TuP12	Photothermally Actuated Microcantilever Beams Using Nanoparticles	77
TuP13	Vertical Comb-Drive MEMS Mirror for Optical Spectrum Sensing	79
TuP14	Drift-Free Single Crystalline Silicon Micromirror with Floating Field Limiting Shields	81
TuP15	A Two-Axis MEMS Scanner Driven by Radial Vertical Combdrive Actuators	83
TuP16	Pull-in Analysis of Scanners Actuated by Electrostatic Vertical Combdrives	85
TuP17	Micromirrors for Multiobject Spectroscopy: Large Array Actuation and Cryogenic Compatibility	87
TuP18	Improved Control of the Vertical Axis Scan for MEMS Projection Displays	89
TuP19	High Temperature Operation of Gimbal-less Two Axis Micromirrors	91
TuP20	Mechanical-Contact-Based Submicron-Si-Waveguide Optical Microswitch at Telecommunication Wavelengths	93
TuP21	Two-Wavelength Grating Interferometry for Extended Range MEMS Metrology	95
TuP22	A New Fast Infrared Tracking System with Thermopile Array Implementation	97
TuP23	Self-Supported Pitch-Variable Guided-Mode Resonant Grating Filters at Telecom Wavelengths	99
TuP24	An Optical MEMS Pressure Sensor based on Phase Demodulation	101
TuP25	A MEMS-based Organic Deformable Mirror with Tunable Focal Length	103
TuP26	Novel Adaptive Optics System with an Electrostatically-Driven Deformable Mirror and Wavefront Compensation Algorithm	105
TuP27	Reflectance Study of Nano-Scaled Textured Surfaces	107
TuP28	Experimental Observation of Self-Propelled Cavity Soliton-like Evolutions in VCSELs with Photonic-Crystal Micro-Structures	109
TuP29	Elastic-like Collision of Gap Solitons in Nonlinear Nonlocal Photonic Crystals	111
TuP30	Wide-Angle Low-Loss 1×2 Multimode Interference Optical Power Divider with Tilted Input and Output Waveguides	113
TuP31	Double Reflection in the Blazed Grating	115
TuP32	Effect of a Vertical Stack of Aligned Subwavelength Metal Hole Arrays on Extraordinary Transmission Spectra	117
TuP33	The Measurement of Liquid Refractive Index by D-Shaped Fiber Bragg Grating	119
TuP34	A Compact Silicon-on-Insulator MMI-based Polarization Splitter	121
TuP35	Near-Field Images of Surface Plasmon Eigenmodes in Gold Nanogratings	123
TuP36	Surface Plasmon Leakage in Its Coupling with an InGaN/GaN Quantum Well through an Ohmic Contact	125
TuP37	Temperature-dependent Behaviors of the Surface Plasmon Coupling with an InGaN/GaN Quantum Well	127
TuP38	The Role of the Quantum-Confined Stark Effect in an InGaN/GaN Quantum Well During Its Coupling with Surface Plasmon for Light Emission Enhancement	129
TuP39	Passivation of Silicon Wafer Patterned by Aluminum for Micromachining	131

Wednesday, 15 August 2007

WA	**Microlenses**	
WA1	MEMS-based Microspectrometers for Infrared Sensing	137
WA2	Fabrication and Characterization of a Repositionable Liquid Micro Lens System	139
WA3	A Lateral-shift-free LVD Microlens Scanner for Confocal Microscopy	141
WA4	Implementation of CMOS-MEMS Compound Lens	143
WA5	Low Cost Adaptive Silicone Membrane Lens	145

WB	**Micro and Nano Lithography**	
WB1	Vortex Generation and Pixel Calibration Using a Spatial Light Modulator for Maskless Lithography	147
WB2	The Study on 3D Electron Beam Lithography for Sub-Micrometer Diffractive Optics	149

WB3	Fabrication of a Multi-Level Lens Using Independent-Exposure Lithography and FAB Plasma Etching	151
WB4	Self-Assembled Two-Dimensional Block Copolymers on Pre-Patterned Templates with Laser Interference Lithography	153
WB5	Fabrication of Large Size Photonic Crystal Templates by Holographic Lithography Technique	155
WB6	Extraordinary Transmission Through A Poly-SiC Membrane with Subwavelength Hole Arrays	157

WC   **Spectroscopy**

WC1	Compact Spectroscopic Sensor Using an Arrayed Waveguide Grating	159
WC2	A Micro-Optic-Fluidic Spectrometer with Integrated 3D Liquid-Liquid Waveguide	161
WC3	A Disposable Grating-integrated Multi-channel SPR Sensor Chip for Real-time Monitoring of Biomolecule Binding	163
WC4	Tunable Resonant Cavity Enhanced Detectors using Vertical MEMS Mirrors	165
WC5	Two-state Optical Filter Based on Micromechanical Diffractive Elements	167

WD   **Tunable Devices**

WD1	MEMS Tunable Filters for LWIR Spectral Imaging	169
WD2	Tunable Erbium Doped Fiber Laser Using a Silicon Micro-Electro-Mechanical Fabry-Perot Cavity	171
WD3	Design and Fabrication of Photonic MEMS Waveguide Modulators	173
WD4	A Study on Optical Diffraction Characteristics of Skewed MEMS Pitch Tunable Gratings	175
WD5	Tunable MEMS Actuated Microring Resonators	177

Thursday, 16 August 2007

ThA   **Nanofabrication**

ThA1	One-Way Waveguide and Strong Photon-Photon Interaction in Nanophotonic Structures	181
ThA2	Flower-Structured InGaN/GaN Quantum-Well Nanodisk Crystals on Micromachined Si Pillars	183
ThA3	Fabrication of Wafer-level Antireflective Structures in Optoelectronic Applications	185
ThA4	Structural and Optical Properties of III-V Nanowires and Nanowire Heterostructures Grown by Metalorganic Chemical Vapour Deposition	187
ThA5	Magnetic Alignment of Carbon Nanotube Interconnects	189

ThB   **Micromirrors**

ThB1	Optical MEMS for Future Instruments in Astronomy	191
ThB2	Passivated Piezoresistive Rotation Angle Sensor Integrated in Micromirror	193
ThB3	Integrated Piezoresistive Positionssensor for Microscanning Mirrors	195
ThB4	Ultra Flat High Resolution Microscanners	197
ThB5	Combined Device of Optical Microdisplacement Sensor and PZT-Actuated Micromirror	199

Author Index 201

MONDAY, 13 AUGUST 2007

PLE   Plenary Session

MA   Medical and Bio Sensing

MB   Active Nano Devices

MC   Novel Fabrication Techniques

Monday Missing Paper

PLE2   "Combination of MEMS and Microoptics for New Applications,"
A. Bräuer, *et. al., Fraunhofer-Institut, Jena, Germany.*

PLE1 (Plenary)
08:30 – 09:00

Concepts of nanophotonic devices and fabrications

Motoichi Ohtsu

Department of Electronics Engineering, The University of Tokyo
2-11-16 Yayoi, Bunkyo-ku, Toko 113-8656, Japan. http://uuu.t.u-tokyo.ac.jp

Nanophotonics utilizes the optical near field that mediates the interaction between nanometric particles. To realize qualitative innovation for future optical technology, this presentation reviews novel devices, fabrications, and systems.

I. Introduction

Quantitative innovation of optical technology is required for future information transmission systems, for example, increasing the integration of photonic devices by reducing their size and heat generation [1]. Furthermore, novel applications such as information-processing systems are expected by realizing "qualitative innovation," that is, by realizing novel functions and operations in photonic devices that differ from those of conventional photonic devices such as lasers, modulators, and waveguides.

To reduce the size of photonic devices beyond the diffraction limit, the local interaction between the nanometric particles and photons is required. Furthermore, the energy transferred through this interaction must be dissipated in the nanometric particles or the adjacent macroscopic materials to determine the magnitude of the energy transfer. These local energy transfers and their subsequent dissipation are possible only by using optical near fields, which are the elementary surface excitations on nanometric particles. Novel optical technology utilizing the local interaction between nanometric particles *via* optical near fields is called "nanophotonics," which was proposed by the author [2]. Nanophotonics enables quantum mechanical light emission by controlling the layout of the position and electronic energy state of the nanometric particles. As a result, novel photonic devices, fabrications, and systems are realized for qualitative innovation that meets the requirements of future optical technology. As evidence of qualitative innovation, this presentation reviews novel nanophotonic devices, nanophotonic fabrications, and nanophotonic systems.

II. Theoretical models, devices, and fabrications

Our group developed a novel theory of nanophotonics based on the energy transfer between nanometric particles *via* an optical near field [3]. Two nanometric particles and optical near fields, that is, the "nanometric subsystem" under study, are always buried in the "macroscopic subsystem," which consists of the macroscopic substrate material and the macroscopic electromagnetic fields of the incident and scattered light. The macroscopic subsystem is expressed as an exciton–polariton, which is a mixed state of material excitation and electromagnetic fields. Since the nanometric subsystem is excited by an interaction with the macroscopic subsystem, the projection operator method is effective in describing the quantum mechanical states of these systems. Under this treatment, two nanometric particles can be considered as being isolated from the surrounding macroscopic systems and as interacting by exchanging exciton–polariton energies. Since the time required for this local electromagnetic interaction is very short, the uncertainty principle allows the exchange of a virtual exciton–polariton between the two nanometric particles, as well as that of a real exciton–polariton. The former exchange corresponds to the nonresonant interaction between the two particles. The optical near field mediates this interaction, which is illustrated by a Yukawa function representing the localization of the optical near-field energy around the nanometric particles, like an electron cloud around an atomic nucleus whose decay length is equivalent to the material size.

As described above, the optical near field is an electromagnetic field that mediates the interaction between nanometric particles located in close proximity to each other. Nanophotonics is the technology utilizing this field to realize novel devices, fabrications, and systems. A nanophotonic device with a novel function can be operated by transferring the optical near-field energy between nanometric particles. In such a device, the optical near field transfers a signal and carries the information. Novel photonic systems become possible by using these novel photonic devices. Furthermore, if the magnitude of the transferred optical near-field energy is sufficiently large, structures or conformations of nanometric particles can be modified, which suggests the feasibility of novel photonic fabrications. Note that the true nature of nanophotonics is to realize "qualitative innovation" in photonic devices, fabrications, and systems by utilizing novel functions and phenomena caused by optical near-field interactions, which are impossible as long as conventional propagating light is used. The advantage of going

1-4244-0641-2/07/$20.00 ©2007 IEEE

beyond the diffraction limit, that is, "quantitative innovation," is no longer essential, and only a secondary nature of nanophotonics.

As key nanophotonic devices, we present a nanophotonic switch, logic gates including a not-gate, and an optical nano-fountain based on optical near-field energy transfer between quantum dots. A single photon operation capability is demonstrated. An architectural approach to nanophotonic information and communication systems is also discussed. One is a memory-based architecture, which is founded on a lookup table using the optical near-field interaction between quantum dots. In addition, content-addressable memories, digital logic, and matrix–vector multiplication can be implemented in this architecture. As fundamental functional elements, a data-summation mechanism and digital-to-analog conversion are presented, and their proof-of-principle experiments are outlined. Owing to its high spatial density and low power consumption, a massive array of such functional components will be useful in applications such as massive lookup table operations in networking and information processing systems.

For nanophotonic fabrications, this presentation demonstrates photochemical vapor deposition of nanometric particles based on the photodissociation of gas-phase metal–organic molecules using optical near fields under the nonresonant condition, which is possible by multiple-step excitation *via* molecular vibration modes. Such a nonadiabatic process violates the Franck–Condon principle, and can be applied to other photochemical phenomena to open a new field of nanofabrication. As an example, we demonstrate photolithography patterning using visible light for UV photoresist. This enables the fabrication of a replica of an electronic circuit pattern, multiple exposures, nanophotonic devices, and so on. Fabrication of a diffraction grating for soft X-rays is also demonstrated.

III. Summary

The term "nanophotonics" is occasionally used for plasmonics, and for photonic crystals, silicon photonics, and quantum dot lasers, although they are not based on optical near-field interactions. For the true development of nanophotonics, one needs deep physical insights into the virtual exciton–polariton and the nanometric subsystem composed of electrons and photons.

References

[1] M. Ohtsu, K. Kobayashi, T. Kawazoe, S. Sangu, and T. Yatsui: IEEE J. Selected Topics in Quantum Electronics, **8** (2002) pp.839-862.

[2] M. Ohtsu(ed): *Progress in Nano-Electro-Optics V* (Spinger-Verlag, Berlin, 2006), Preface to Volume V: Based on "nanophotonics" proposed by Ohtsu in 1993, OITDA (Optical Industry Technology Development Association, Japan) organized the nanophotonics technical group in 1994, and discussions on the future direction of nanophotonics were started in collaboration with academia and industry.

[3] M. Ohtsu, K. Kobayashi: *Optical Near Fields* (Springer-Verlag, Berlin, 2004), pp.109-120.

PLE3 (Plenary)
09:30 – 10:00

Photonic Band Gap Materials: Engineering Light-Matter Interactions

Sajeev John
Department of Physics
University of Toronto
Toronto, Ontario, Canada

Abstract

Photonic Band Gap (PBG) materials are 3D periodic dielectrics that enable engineering of the most fundamental properties of electromagnetic waves [1,2]. Most strikingly, they enable the trapping and localization of light [1,3]. Unlike electronic micro-circuitry, optical wave-guides in a PBG micro-chip can simultaneously conduct hundreds of wavelength channels of information in a three-dimensional circuit path without loss [4,5,6,7].

I review recent approaches to micro-fabrication of PBG materials, including single-beam optical lithography using a novel optical phase mask [8]. I describe all-optical switching in a PBG wave-guide through coherent resonant interaction with an inhomogeneously broadened distribution of quantum dots [9]. I discuss the possibility of electromagnetically induced exciton mobility in a PBG quantum well hetero-structure [10].

1. S. John, Phys. Rev. Lett. 58, 2486 (1987)
2. E. Yablonovitch, Phys. Rev. Lett. 58, 2059 (1987)
3. S. John, Phys. Rev. Lett. 53, 2169 (1984)
4. A. Chutinan, S. John, O. Toader, Phys. Rev. Lett. 90, 123901 (2003)
5. A. Chutinan and S. John, Phys. Rev. E 71, 026605 (2005)
6. A. Chutinan and S. John, Phys. Rev. B 72, 16, 161316 (2005)
7. A. Chutinan and S. John, Optics Express 14 (3), 1266 (2006)
8. T. Chan, O. Toader, S. John, Phys. Rev. E 73, 046610 (2006)
9. D. Vujic and S. John (to be published)
10. S. John and S. J. Yang (to be published)

MA1 (Invited)
10:30 – 11:00

Wearable Laser Blood Flowmeter for Ubiquitous Healthcare Service

T. Kiyokura, N. Tatara, J. Shimada and T. Haga
NTT Microsystem Integration Laboratories
3-1 Morinosato-Wakamiya, Atsugi, Kanagawa, 243-0198, Japan
Tel +81-46-240-2095, Fax +81-46-270-2323, E-mail kiyokura@aecl.ntt.co.jp

Abstract

We have fabricated a wearable laser blood flowmeter that is functionally the same as a conventional desktop laser blood flowmeter. This innovation was achieved as a result of the miniaturization of the optical flow sensor, which was fabricated by using surface-micromachining and surface-mounting techniques. The wearable laser blood flowmeter is battery-powered and equipped with a wireless communications device. Its ability to continuously monitor changes in blood flow makes it very useful for preventive medical care.

Keywords: blood flowmeter, integrated optics, optical sensor, silicon micromachining, telemedicine

1 INTRODUCTION

Recently, Importance of medical services change from treatment to prevention is proposed for the improvement of the quality of life and the containment of the medical cost inflation. For preventive medical care, wearable medical sensors that can monitor a person's physical condition regardless of time and place would be very useful. Moreover, the ability to use such sensors in combination with telecommunications systems would be very effective for remote health management.

Blood flow data is of special interest because it can reveal the physiological state of microcirculation and, when used in conjunction with blood pressure data, can elucidate the peripheral vascular resistance related to circulatory diseases. However, conventional blood flowmeters are optical fiber-based and desktop-sized and weigh several kilograms, so they are not portable. Although it is technically possible to reduce the size of the electronic parts, reducing the size of the optical parts -- lenses, fibers, and the canned laser diode and canned photodiode -- has been an obstacle to reducing the total device size. Recently, NTT has invented a laser blood flow sensor that is fingertip sized [1, 2]. This innovation motivated us to reduce the size of the electronic parts, which led to success in fabricating a wearable laser blood flowmeter [3]. This paper describes this device and presents some experimental data.

2 DEVICE DESCRIPTION

NTT's blood flow sensor chip [1, 2] is shown in detail in Fig. 1. To minimize the size of the laser blood flow sensor, we built a surface-integrated sensor chip that contains no discrete components. It was fabricated as follows: First, Au/Pt/Ti electrodes, alignment markers, AuSn solder film, and a fluorinated polyimide [4] optical waveguide were patterned onto a silicon substrate chip, which works as a

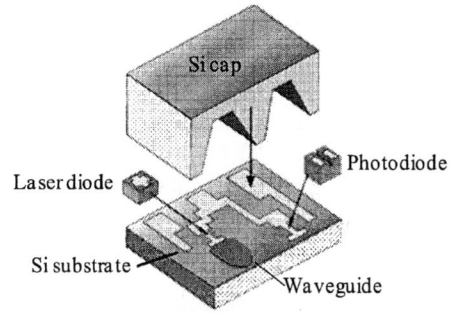

Fig.1. Schematic diagram showing how the sensor chip is integrated. The Si substrate is 2 mm × 3 mm.

heat sink. Then, the laser diode and photodiode chips with alignment markers were integrated onto the chip surface with high precision (±1 μm) [5]. After that, a silicon cap for light-shading was bonded to the chip.

The laser diode chip is a single-mode InGaAsP-InP distributed-feedback semiconductor laser diode. Whereas a conventional laser blood flowmeter operate at wavelength of 0.6–0.8 μm, our device operates at 1.3 μm, which enables better transmittance through skin because the absorbance of melanin and hemoglobin is very low in this wavelength region[6]. The convex-shaped optical waveguide works as a collimating lens. The photodiode is an edge-illuminated refracting-facet type [7], which works as a detector and as a spatial filter as well because of the small light receiving area (15 μm × 65 μm). To obtain a higher signal-to-noise ratio, a conventional blood flowmeter uses an optical fiber as a spatial filter for incoming light. Owing to the use of the special photodiode, we were able to eliminate this optical fiber. The light-shading silicon cap is covered with Ti, which prevents stray light from the laser diode reaching the photodiode without passing through tissue, which decreases the background noise.

1-4244-0641-2/07/$20.00 ©2007 IEEE

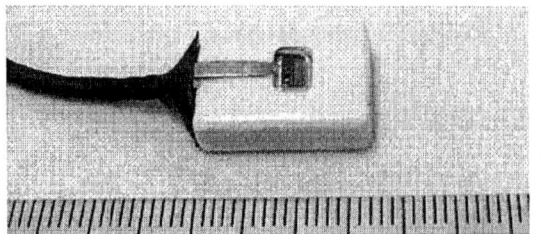

Fig. 2. The sensor head of the laser blood flowmeter (weight: 9.6 g) [3].

The sensor chip was mounted on a printed-circuit board with a preamplifier. Then, the sensor chip was packaged and fitted with a sensor head (Fig. 2), which makes contact with the skin. The sensor head is $17 \times 12 \times 6$ mm^3.

Figure 3 shows a photograph of the blood flowmeter. The main unit on the arm is $63 \times 45 \times 20$ mm^3. A liquid crystal display on the unit's case indicates the blood flow. The main unit contains the electronic parts: the main amplifier, an A/D converter, a DSP, a laser driving circuit, a display panel controller circuit, a flash memory, a secondary cell, a USB interface, and a Bluetooth wireless communication circuit.

3 BLOODFLOW MEASUREMENTS

We performed some measurements of blood perfusion in a fingertip. Figure 4 shows an example of blood flow data obtained simultaneously with the wearable blood flowmeter via wireless communication (lower curve) and a conventional one via a cable (upper curve). Data were acquired every 20 msec. To see the dynamic response of microcirculation, the wrist was occluded with a pressurized cuff and the pressure released gradually. The signal curve from the wearable laser blood flowmeter agrees well with the that from the conventional one. In these data, the correlation coefficient is 0.96. The wearable blood flowmeter is not inferior to the conventional one.

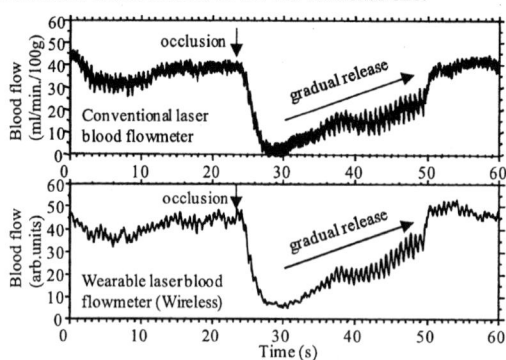

Fig. 4. Blood flow data obtained with the wearable laser blood flowmeter and with a conventional one [3].

4 CONCLUSIONS

This blood flowmeter will make it possible to monitor changes in blood flow regardless of time and place. Bedside monitoring in hospitals and home health monitoring with

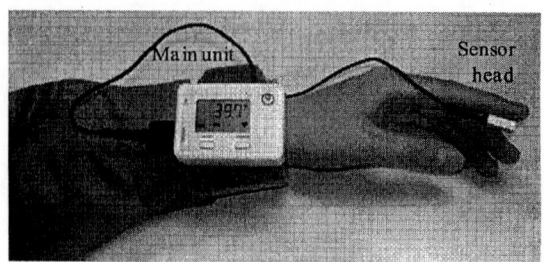

Fig. 3. Wearable laser blood flowmeter. The main unit weighs 55.8 g [3].

this device will allow medical staff to monitor physical conditions in real time. The wireless measurement function decreases the mental distress and annoyance of the patient. Devices like this one that exploit the capabilities of ubiquitous networks will cause a drastic change in medical services from hospital-centric ones to home-centric ones and shift the focus from treatment to prevention.

By miniaturizing the optical parts, we were able to reduce the total device size to wearable. Thus, new applications that have hardly been imagined before are now possible. Optical MEMS technology is expected to produce new instruments with new applications in the biomedical field.

REFERENCES

[1] E. Higurashi, R. Sawada, and T. Ito, "An integrated laser Doppler blood flowmeter for a wearable health monitoring system" Proc. of the Optical MEMS 2001, Okinawa, Japan, August 25-28, 2001, pp.49-50.

[2] E. Higurashi, R. Sawada, and T. Ito, "An integrated laser blood flowmeter" J. Lightwave Tech., Vol. 21, No. 3, pp. 591-595, 2003.

[3] T. Kiyokura, S. Mino, and J. Shimada, "Wearable Laser Blood Flowmeter" NTT Technical Review, Vol.4, No.1, pp. 38-43, 2006.

[4] T. Matsuura, S. Ando, S. Sasaki, and F. Yamamoto, "Low-loss, heat resistant optical waveguide using new fluorinated polyimide" Electron. Lett., Vol. 29, No. 3, pp. 269-270, 1993.

[5] R. Sawada, E. Higurashi, and T. Ito, "Highly accurate and quick bonding of a laser-diode chip onto a planar lightwave circuit" Precision Eng., Vol. 25, No. 4, pp. 293-300, 2001.

[6] R. R. Anderson and J. A. Parrish, "The optics of human skin"J. Investigative Dermatol., Vol.77, No.1, pp.13-19, 1981.

[7] H. Fukano and Y. Matsuoka, "A low-cost edge-illuminated refracting-facet photodiode module with large bandwidth and high responsivity" J. Lightwave Technol., Vol. 18, No. 1, pp. 79-83, 2000.

MA2
11:00 – 11:15

A Portable Two-photon Fluorescence Microendoscope Based on a Two-dimensional Scanning Mirror

Wibool Piyawattanametha[1,2], Eric D. Cocker[1], Robert P. J. Barretto[1], Juergen C. Jung[1], Benjamin A. Flusberg[1]
[1]James H. Clark Center for Biomedical Engineering & Sciences, Stanford University, Stanford, California 94305
Hyejun Ra[2] and Olav Solgaard[2]
[2]Edward L. Ginzton Laboratory, Stanford University, Stanford, California 94305
Mark J. Schnitzer
James H. Clark Center for Biomedical Engineering & Sciences, Stanford University, Stanford, California 94305
Email: wibool@stanford.edu

Abstract

Towards overcoming the size limitations of conventional two-photon fluorescence microscopy for brain imaging in freely moving mice, we introduce a portable laser-scanning microendoscope based on a microelectromechanical systems (MEMS) two-dimensional (2-D) scanning mirror, compound gradient refractive index (GRIN) micro-lenses, and a photonic bandgap fiber (PBF). The microendoscope achieves fast line scanning acquisition rates up to 3.5 kHz and micron-scale imaging resolution.

Keywords: scanning mirror, scanner, two-photon, 3-D image, fluorescence image

INTRODUCTION

Two-photon fluorescence microscopy has become a widely used technique for three-dimensional imaging of biological specimens due to proven advantages over conventional epi-fluorescence and confocal fluorescence microscopy. Among the most significant of these are inherent optical sectioning without the use of a pinhole due to the non-linear excitation process, reduced photobleaching and phototoxicity, and greater penetration depth into highly scattering tissue. To date, use of two-photon fluorescence microscopy in live subjects has generally been limited to anesthetized animals. Integration of fiber optics [1] and microlens based imaging probes [2], or the two in combination [3, 4] into an imaging system may help to overcome these limitations and provide the ability to image in freely moving subjects. Our goal is to develop a portable two-photon microendoscopy system for brain imaging in freely moving mice. We have chosen to focus on the use of mice, because of the wide availability of transgenic mouse lines with genetically targeted alterations to cellular processes and animal behaviors. Prior miniaturized scanning mechanism for two-photon fluorescence imaging have typically involved resonant scanning of optical fibers [3,4], which limit size reduction, restrict the choice of scanning rates, and are unsuitable for batch fabrication. Recently, we have demonstrated the use of a MEMS scanning mirror in a tabletop two-photon microscope [5]. In this paper we introduce a portable two-photon fluorescence microendoscope based on a MEMS scanning mirror, compound GRIN micro-optics, and a PBF.

MICROENDOSCOPE DESIGN

The 2-D gimbal scanning mirrors are batch fabricated on silicon-on-insulator wafers that have two single-crystalline silicon device layers (30 μm each). The mirror, movable comb teeth, and inner torsional springs reside in the upper device layer. The frame, outer torsional springs, and fixed comb teeth are fabricated within both device layers. Fabrication involves four deep-reactive-ion-etching steps. The first three steps self-align the comb fingers in the device layers by transferring mask features sequentially from upper to lower layers. The last step removes the backside of the substrate behind the mirror, releasing the gimbal for rotation [6]. The scanning mirror is actuated by 6 banks of vertical comb actuators. The total die is 3.2 × 3.0 mm^2 in size and the mirror is 760 × 760 μm^2.

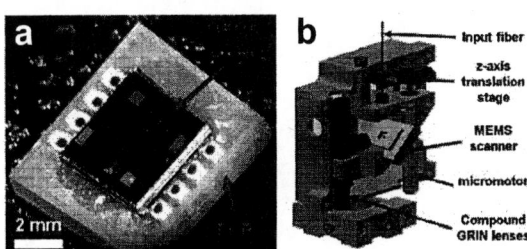

Fig. 1: a) Photograph of a MEMS scanning mirror. b) Schematic drawing of the microendoscope.

Figure 1a and 1b show a photograph of a scanning mirror mounted on a printed circuit board (size = 6.0 × 5.0 mm^2) and a schematic drawing of a microendoscope (size = 2.0 × 1.9 × 1.1 cm^3), respectively. Figure 2 shows a photograph of the microendoscope with overlay drawing of optical paths. The optical circuit can be described as follows. An ultrashort pulsed Ti:sapphire laser (λ_0=800 nm) provides an excitation

1-4244-0641-2/07/$20.00 ©2007 IEEE

beam (red arrow) through a PBF. The beam is collimated by a collimator before it reflects off the scanning mirror. The beam is then re-expanded, passes through a dichroic prism, and fills the back aperture of a GRIN objective, which focuses the light at the specimen plane. Fluorescence (green arrow) returns back through the GRIN objective, reflects off the dichroic prism, and is detected by a photomultiplier tube. The excitation and collection numerical aperture (NA) values of the system are 0.46 and 0.61, respectively. The microendoscope working distance from the GRIN objective is 400 μm. The power output at the specimen can reach up to 60 mW.

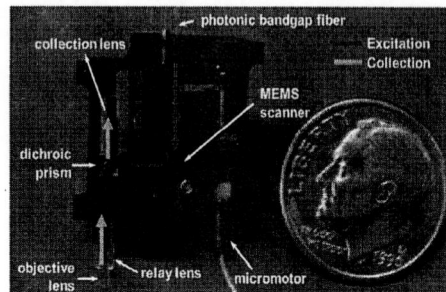

Fig. 2: Photograph of the microendoscope.

RESULTS & DISCUSSIONS

Table 1: Summary of scanning mirror performance

	Inner-axis	Outer-axis
Mirror radius of curvature	> 1 m	
Mirror surface avg. roughness (*RMS*)	< 20 nm	
Resonant freq. (kHz)	1.6 to 1.8	0.6 to 0.8
Max. optical DC scan angle (°)	± 4.0° to ± 8.0°	± 1° to ± 3°
Voltage at Max. DC angle (V)	70 V to 100 V	120 V to 180 V

The general performance range of non-metalized 2-D scanning mirrors is summarized in Table 1. To image with the scanning mirror, both inner- (fast-axis) and outer- (slow-axis) axes have their opposing comb actuator banks driven 180° out of phase with unipolar sinusoidal waveforms. This is done to maximize the linear region of the angular deflection [5]. The fast-axis is driven at resonance while the slow-axis is driven in DC mode (5 Hz). All images are 402 pixels × 162 pixels taken at 5 frames per second with double-sided fast-axis acquisition. The maximum field of view is 80 × 20 μm^2. No frame averaging is performed on the images. Figure 3a and 3b show parts of a spiky-shape

and a clover-leaf shape pollen grains, respectively. The maximum power to obtain these auto-fluorescence images is 30-40 mW. These power levels were needed because of the fast acquisition rate. The scale bars are 10 μm.

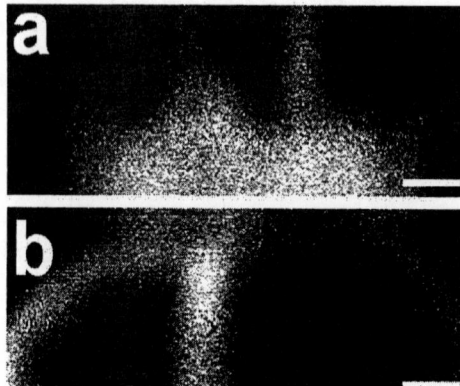

Fig.3: Fluorescence images of pollen grains (a) spiky shape (b) clover-leaf shape.

CONCLUSION

We have built the first MEMS based miniaturized and portable two-photon fluorescence microendoscope with micrometer-scale resolution. We anticipate a broad set of future imaging applications for this microendoscope including freely moving mice brain imaging.

REFERENCES

[1] S. Kimura and T. Wilson, "Confocal scanning optical microscope using single-mode fiber for signal detection," Appl. Opt. 30, 2143 (1991).

[2] J. C. Jung, A. D. Mehta, E. Aksay, R. Stepnoski, and M. J. Schnitzer, "In vivo mammalian brain imaging using one- and two-photon fluorescence microendoscopy," J. Neurophysiology, 92:3121-3133 (2004).

[3] B. A. Flusberg, E. D. Cocker, W. Piyawattanametha, J. C. Jung, E. L. M. Cheung, and M. J. Schnitzer, "Fiber optic fluorescence imaging," Nature Methods, 2, 941-950 (2005).

[4] F. Helmchen, M. S. Fee, D. W. Tank, and W. Denk, "A miniature head-mounted two-photon microscope. high-resolution brain imaging in freely moving animals," Neuron, 31, 903-912 (2001).

[5] W. Piyawattanametha, R. P. J. Barretto, T. H. Ko, B. A. Flusberg, E. D. Cocker, H. Ra, D. Lee, O. Solgaard, and M. J. Schnitzer, "Fast-scanning two-photon fluorescence imaging based on a microelectromechanical systems two-dimensional scanning mirror," *Opt. Lett.*, Vol. 31, No. 13, July 1, 2006.

[6] H. Ra, W. Piyawattanametha, Y. Taguchi, D. Lee, M. J. Mandella, and O. Solgaard, "Two-dimensional MEMS scanner for Dual-Axes confocal microscopy," J. Microelectromech. Syst., in print, 2007.

MA3
11:15 – 11:30

Tunable Endoscopic MEMS-Probe for Optical Coherence Tomography

K. Aljasem, A. Werber, and H. Zappe

Laboratory for Micro-optics, Department of Microsystems Engineering (IMTEK), University of Freiburg,
Georges-Koehler-Allee 102, 79110 Freiburg, Germany
Tel +49-761-203-7519, Fax +49-761-203-7562, E-mail aljasem@imtek.uni-freiburg.de

Abstract

A novel miniaturized tunable endoscopic MEMS probe employable in an endoscopic optical coherence tomography (OCT) system is presented. The probe consists of a pneumatically-actuated micro-lens and a GRIN lens coupled to an optical fiber. Silicon micromachining using polydimethylsiloxane was used to fabricate the lens. The probe has a diameter of 5 mm. The successful integration of the tunable endoscopic probe with an OCT imager promises to be of value in medical diagnostics, particularly for the early detection of internal tumors and cancers.

Keywords: Tunable MEMS probe, Endoscopic optical coherence tomography, polydimethylsiloxane (PDMS), micro-lens

1 INTRODUCTION

Optical coherence tomography (OCT) is a non-destructive, sub-surface imaging technique for cross-sectional in-vivo analysis of biological tissue [1]. OCT coupled with endoscopy enables high resolution, minimally-invasive diagnostics inside the human body, and is thus at present a topic of active research. In endoscopic OCT, the use of an objective lens with a fixed focal length has the following drawbacks:

1- The difficulty of manually adjusting the focal point at the desired position, particularly for irregular tissues;
2- The inverse relation between the lateral resolution and the depth of focus (DOF) in OCT;
3- The low intensity of the backscattered light from layers in the out-of-focus regions, particularly for in-vivo applications.

An approach to circumvent these drawbacks while maintaining a high lateral resolution to vary the focus of the measuring beam with the axial scan [2]. We demonstrate here a novel endoscopic probe with a dynamic focusing capability using pneumatically-actuated tunable lenses for OCT applications.

2 DESIGN AND CHARACTERIZATION

2.1 Design and Components of the Probe

The probe consists of a tunable micro-lens, a mechanical holder and a GRIN-lens collimator connected to a fiber optic cable. Figure 1 shows the schematic view of the design of the MEMS probe.

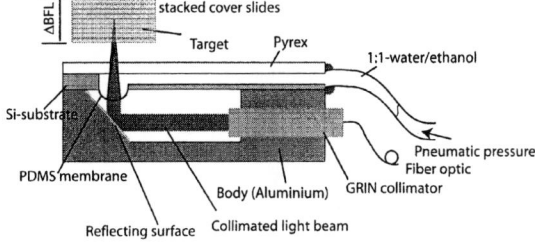

Figure 1: Principle design of the tunable probe

As can be seen in Figure 1, the tunable micro-lens is fixed on top of the mechanical holder made of aluminum. The holder has a 1.8 mm opening for a gradient-index (GRIN) lens collimator. The collimated light beam of 1 mm diameter is reflected by a thin mirror placed onto the mechanical holder with an angle of 45°. The collimated light beam is then focused using the tunable micro-lens. By tuning the micro-lens, a dynamic focus change can be achieved. This arrangement of the elements, particularly the micro-lens, can offer a side-imaging. The outer diameter of the probe is 5 mm.

2.2 Design and Characterization of the Micro-lens

The lens system consists of a silicon chip bonded onto a Pyrex chip. A 50 µm thick polydimethylsiloxane (PDMS) layer is spun on the front side of the silicon substrate [3] spanning a micro-fluidic cavity. Changing the pressure of the fluid causes a distension of the membrane, whose convex surface forms a lens with pressure-dependent curvature and thus focal length.

A 200 µm wide channel is created on the backside of the

1-4244-0641-2/07/$20.00 ©2007 IEEE

silicon substrate to enable the connection between a pressure source and the cavity of the lens. Deep-reactive ion-etching (DRIE) process is used to create the channel and cavity which is then filled by a 1:1 ethanol/water mixture. Figure 2 shows a photograph of the micro-lens integrated into the probe. The circular opening shown in Figure 2 is the lens aperture with a diameter of 1.5 mm.

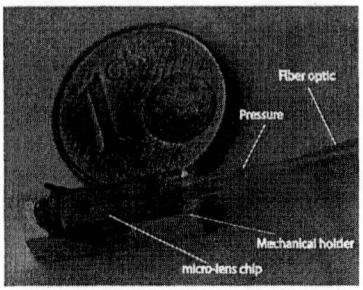

Figure 2 Photograph of the micro-lens chip integrated into the probe.

Figure 3 shows the back focal length (BFL) of the tunable lens as a function of applied pressure.

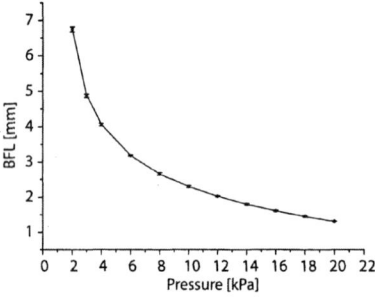

Figure 3 BFL as a function of applied pressure for the tunable micro-lens.

3 MEASUREMENS

To demonstrate the functionality of the tunable probe, a time-domain OCT system was assembled.

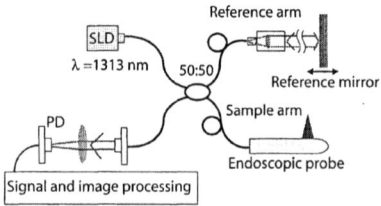

Figure 4 Endoscopic probe integrated into the OCT system.

Figure 4 shows the OCT system schematically, where the output is coupled to the tunable probe.

To demonstrate the functionality of the adaptive lens employed in the endoscopic probe in an OCT configuration, we used a test sample for imaging consisting of eight cover glass slides, each 140 µm thick. The measured intensity peaks correspond to reflections from the slide surfaces. It is clearly seen in Figure 5a, taken with a fixed focal length, that the interference intensity decreases with depth. However, as the tunable lens moves the focus of the measurement signal deeper below the surface (i.e., to the right) (Figure 5b), the envelope of maximum intensity also moves to the right. By tuning the focal length concomitantly with the A-scan (the depth scan), a large effective DOF may be achieved. The axial scanning range is thereby markedly increased.

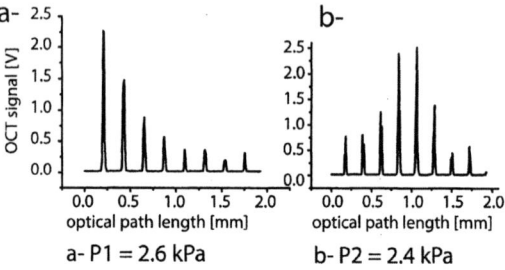

Figure 5a- and 5b- Intensity of backscattered interference signals corresponding to different focal lengths.

4 CONCLUSION

A novel design for an endoscopic probe using an integrated tunable micro-lens for use in an endoscopic OCT system has been presented. The increase in depth of field and resulting enhanced depth scanning range will allow flexible application of this system for new applications, particularly for more effective detection of early-stage tumors and cancers.

REFERENCES

[1] A.F. Fercher, W. Drexler, C.K. Hitzenberger and T. Lasser, "Optical coherence tomography - principles and applications", *Reports on Progress in Physics*, vol. 66, pp 239-303, 2003.

[2] N. Iftimia, B. Bouma, J. de Boer, B. Park, B. Cense, and G. Tearney, "Adaptive ranging for optical coherence tomography", *Opt. Express,* vol. 12, pp 4025-4034, 2004.

[3] A. Werber and H. Zappe, "Tunable microfluidic microlenses", *Appl. Opt.* vol. 44, pp. 3238-3245, 2005.

MA4
11:30 – 11:45

Forward-Imaging Swept Source Optical Coherence Tomography using Silicon MEMS Scanner for High-Speed 3-D Volumetric Imaging

Karthik Kumar*[1], Jonathan C. Condit[1,2], Austin McElroy[2], Nate J. Kemp[2],
Kazunori Hoshino[1], Thomas E. Milner[1], and Xiaojing Zhang[1]

* Telephone: +1-512-232-4275, Fax: +1-512-471-0616, E-mail: kkumar@mail.utexas.edu

[1]Department of Biomedical Engineering, The University of Texas at Austin, Austin, TX 78712 USA

[2]CardioSpectra, Inc., San Antonio, TX 78229 USA

ABSTRACT

We demonstrate a fiber-based forward-imaging swept source OCT (SS-OCT) system using a two-axis silicon micromachined vertical comb-drive microscanner for high-speed, 3-D imaging of biological specimens. Higher signal-to-noise ratio of SS-OCT over traditional time-domain techniques, combined with low beam-steering loss of silver-coated scanning micromirrors with over 90% reflectivity, provide good imaging performance. Fast wavelength scanning of the laser source (scan rate: 20kHz) over 110nm spectral bandwidth enabled image acquisition at 8 million voxels/sec (3-D imaging) or 40fps (2-D imaging, 500 transverse pixels per image). We successfully acquired *en face* and tomographic *in vitro* images of rigid structures (microscanner), soft materials (onion and pickle slices), and *in vivo* images of epidermis. Lateral resolution of 12.5μm and axial resolution of 10μm over a $2\times1\times4mm^3$ imaging volume has been demonstrated. The compact forward-imaging OCT probe may be suitable for image-guided minimal-invasive examination of various diseased tissues.

Keywords: optical coherence tomography (OCT), swept source, forward-imaging, MEMS, scanner

I. INTRODUCTION

Optical coherence tomography (OCT) has emerged as a high-resolution diagnostic imaging tool in cases where biopsy is difficult, for image-guided microsurgery, and for three-dimensional pathology reconstruction [1]. Three-dimensional OCT enhances a physician's visualization of morphology by providing tomographic and microscopic views simultaneously [2]. Miniaturization of optical diagnostic equipment is critical for translation of OCT techniques from research laboratories to clinical medicine. Micro-electro-mechanical system (MEMS) technologies offer the unique capability to package micro-optical elements with actuators for imaging in *in vivo* environments. Previous studies have utilized one-dimensional [3-5] and two-dimensional [6-8] MEMS scanning systems to perform Time Domain OCT. MEMS elements have also been used for dynamic beam focus control in Doppler OCT [9] and grating spectrometer resolution enhancement in Fourier Domain OCT [10]. Spectral Domain (i.e., Fourier Domain/ Swept Source) OCT has been shown to provide signal-to-noise ratio advantages over Time Domain OCT [11-12]. Swept Source OCT (SS-OCT) has potential for real-time high-resolution imaging due to continual improvements in laser wavelength scan range, linewidth, and scan speed, some of which are enabled by MEMS techniques [13-14]. Here we demonstrate a miniaturizable forward-looking Swept Source OCT system incorporating a silver-coated silicon MEMS scanner for high-speed 3-D volumetric imaging. The silicon MEMS scanner reported here provides for the first time, two-dimensional angular scanning of incident broadband light using silver-coated surfaces in a common plane.

II. EXPERIMENTAL METHODS

We adopted a forward-imaging configuration for our fiber-based system (Fig. 1A) that can be miniaturized into an OCT endoscope. Forward-imaging OCT endoscopes are technically challenging to build but much-needed for image-guided surgery in sensitive tissues including the gastrointestinal (GI) tract, breast, liver, and ovaries, cardiovascular system, brain, urinary, and reproductive systems [15]. Our system utilizes a tunable laser (10mW, 1310nm λ_{center}, 110nm range, 20,000 A-scans/sec, Santec), and a Steinheil triplet lens (JML Optical TRP14340/100,

0.6 NA, 7.9mm EFL) to provide aberration-free focus of broadband illumination. Optical design software (ZEMAX) simulation (Fig. 1B) shows that a 500μm diameter beam

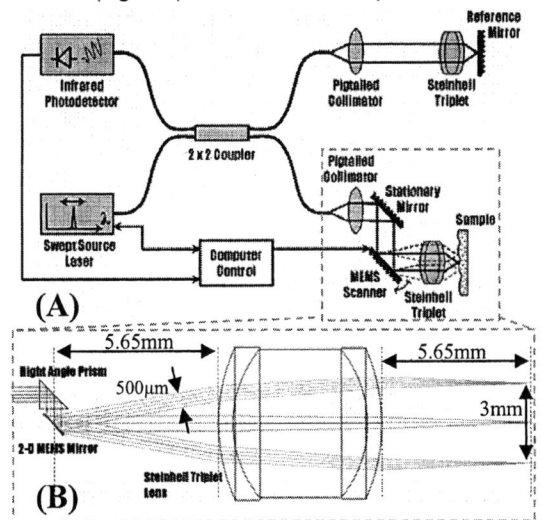

Fig. 1: Illustration of forward-imaging SS-OCT system. (A) Experiment setup (B) Optical characteristics of scanning system - ZEMAX simulation

deflected by ±10° (optical) by a microscanner placed at the back focal plane of the triplet objective lens produces an approximately 3mm linear scan. Therefore, adopting fiber-fused GRIN lenses and right-angle micro-prisms with micro-objectives enables OCT endoscopes for 3-D imaging in many typically-inaccessible human organs. The two-axis microscanners (Fig. 2) fabricated by a process described previously [16] have 500μm×700μm mirror dimensions for 45° incident illumination. 125nm-thick silver coating results in ~95% uniform reflectance over the source spectrum. An important advantage of the scanner reported here is that two-dimensional scanning is obtained with a single plane device. Incorporating two scan directions in a single plane reduces optical field distortions. We observed inner and outer axis resonance at 2.28kHz and 385Hz (Fig. 3A), respectively, and 9° optical deflection on both axes for single-sided voltage input at low frequencies (Fig. 3B).

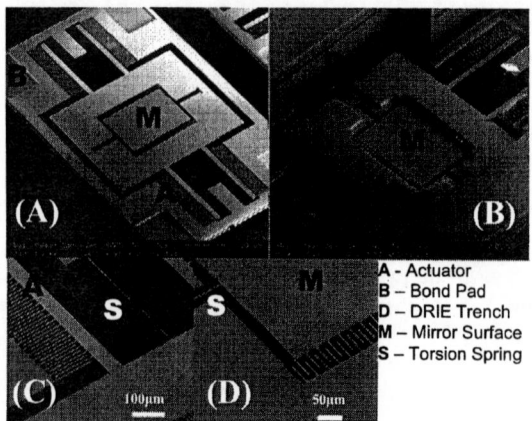

A - Actuator
B – Bond Pad
D – DRIE Trench
M – Mirror Surface
S – Torsion Spring

Fig. 2: SEM images of MEMS scanners. (A) Top view (B) Bottom view (C) Outer spring, comb bank (D) Inner spring, comb bank.

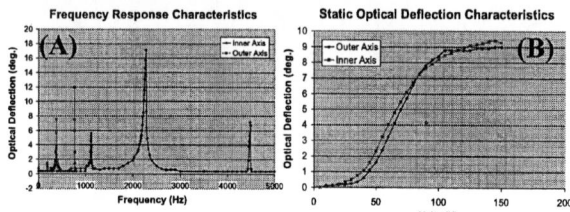

Fig. 3: MEMS microscanner operating characteristics. (A) Frequency response curve. (B) Static single-sided voltage deflection curve.

III. RESULTS AND DISCUSSION

We used our system to simultaneously reconstruct *en face* and tomographic images of rigid structures (our microscanner), soft materials (onion and pickle slices), and epidermis *in vivo*. The scanning micro-optical system and high-speed broad-spectrum swept laser allows imaging of $2\times1\times4mm^3$ volume with $12.5\times12.5\times10\mu m^3$ resolution at 8million voxels/sec.

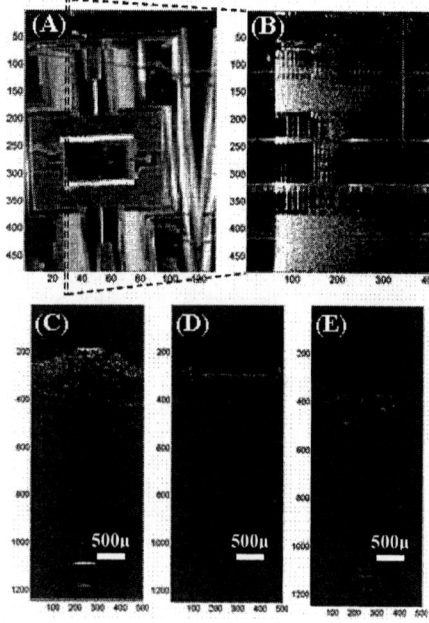

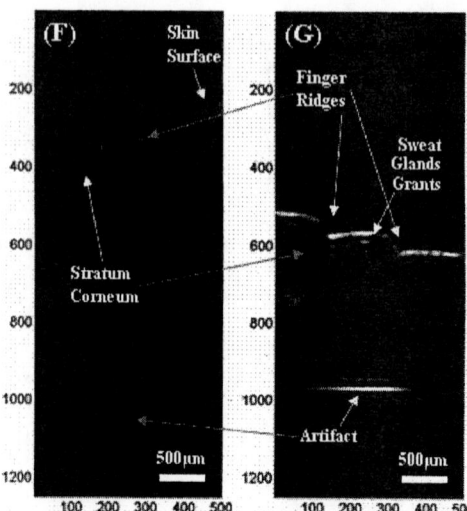

Fig. 4: Forward-looking SS-OCT Images using MEMS Scanner. (A) *En face* image of scanner acquired at 8million voxels/sec (B) Tomographic view (B-scan #31) of the microscanner (C-D) *In vitro* tomographic images of pickle slices obtained at 40fps using microscanner, galvanometer respectively (500 transverse pixels/image). (E) *In vitro* tomographic image of onion peel obtained at 40fps using microscanner. (F-G) *In vivo* images of human epidermis (finger) obtained at 20fps, 40 fps (500 transverse pixels/image).

Microscanner structure and biological samples are clearly visible from the tomographic and *en face* views. Higher-than-video-rate (40fps) *in vivo* acquisition of B-scans at micron resolution will allow real-time monitoring of sub-surface morphology for disease diagnostics and image-guided biopsy and therapy.

IV. CONCLUSION

Swept Source three-dimensional OCT is demonstrated using a miniaturized forward-imaging probe incorporating a two-axis silicon microscanner. Tomographic, *en face* views of $2\times1\times4mm^3$ volume with $12.5\times12.5\times10\mu m^3$ resolution are acquired at 8million voxels/sec. Further miniaturization of the MEMS scanning optics, with the use of fiber-fused GRIN lens collimators, will enable clinical applications of these catheters for diagnosis of the cardiovascular stenosis.

REFERENCES

[1] M.E. Brezinski, ed. *Optical coherence tomography: principles and applications*, (Academic Press, Boston, 2006)
[2] R. Huber *et al.*, Optics Express **13**, p10523-10538 (2005)
[3] T. Xie *et al.*, Applied Optics **42**, p6422-6426 (2003)
[4] P.H. Tran *et al.*, Optics Letters **29**, p1236-1238 (2004)
[5] C. Chong *et al.*, IEEE Photo. Tech. Lett. **18**, p133-135 (2006)
[6] A.D.Aguirre *et al.*, Optics Express **15**, p2445-2453 (2007)
[7] Z. Chen *et al.*, Proc. SPIE **6466**, 64660H (January 2007)
[8] W. Jung et al., IEEE JSTQE **11**, p806-810 (2005)
[9] V.X.D. Yang et al., Optics Letters **31**, p1262-1264 (2006)
[10] K. Yu et al., Intl. Conf. Optical MEMS, p42-43 (2006)
[11] R. Leitgeb et al., Optics Express **11**, p889-894 (2003)
[12] M. Choma et al., Optics Express **11**, p2183-2189 (2003)
[13] D. Anthon et al., Proc. Opt. Fiber Conf., p97-98 (2002)
[14] Q. Chen et al., IEEE Photo. Tech Lett. **16**, p1438-1440 (2004)
[15] Z. Yaqoob *et al.*, Journal of Biomed. Opt. **11**, p063001 (2006)
[16] K. Kumar et al., Intl. Conf. Optical MEMS, p120-121 (2006)

We gratefully acknowledge the support from Wallace H. Coulter Foundation Early Career Award 2006-08 (Zhang) and grants from the NIH (R01 EY 016462) and Veterans Administration (Milner). Microscanners were fabricated at Stanford and UT-Austin NNIN Facilities supported by NSF Grants 9731293 and 0335765 respectively.

MA5
11:45 – 12:00

Miniaturized Optical Viscosity Sensor based on a Laser-induced Capillary Wave

Y. Taguchi, A. Ebisui and Y. Nagasaka

Department of System Design Engineering, Keio University

3-14-1 Hiyoshi, Yokohama, Japan

Tel +81-45-566-1809, Fax +81-45-566-1720, E-mail tag@sd.keio.ac.jp

Abstract

A novel micro optical viscosity sensor (MOVS), by laser-induced capillary wave method enabling us non-contact, short-time (several hundreds of nano seconds), and small sample volume (several tens of micro litters) *in situ / in vivo* measurement, is reported in this paper. The microfabricated MOVS chip consists of two deep trenches holding photonic crystal fibers for excitation laser, and two shallow trenches holding the lensed-fibers for probing laser. The optical interference fringe excited by two pulsed laser beams heats the sample surface, and the temporal behavior of surface geometry is detected as a first-order diffracted beam, which contains the information of liquid properties (viscosity and surface tension). The preliminary measurements using distilled water and sulfuric acid with dye of carbon black are demonstrated. The high-speed damped oscillation signals are successfully detected by MOVS.

Keywords: Laser-induced capillary wave, Measurement technique, Optical Interference, Surface tension, Viscosity

1 INTRODUCTION

The viscosity is one of the most essential parameters for the diagnosis and the design of fluid dynamics in a broad field of science and engineering such as clinical research[1], food industry[2], petroleum industry[3], and polymer material science[4]. We have developed a novel micro optical viscosity sensor (MOVS) based on a laser-induced capillary wave method[5] enabling optical, non-contact, high-speed, and small sample volume measurement, which is applicable to *in vivo/in vitro* and *in situ*. This paper reports the principle and the fabrication of MOVS, and the feasibility of MOVS through experiments using distilled water and sulfuric acid with a small amount of dye are preliminarily demonstrated.

2 PRINCIPLE OF MEASUREMENT

The schematic diagram of MOVS is shown in Fig. 1. The sample surface is interferometrically heated by two pulsed laser beams generating optical grating. Therefore, the temperature distribution of the thermal grating creates a capillary wave on the liquid surface driven by the thermal expansion and the temperature dependence of the surface tension. The sinusoidal surface geometry induced by the grating pattern of temperature distribution is described as

$$u(x) = u_0 + \Delta u \cos\left\{ \frac{4\pi x}{\lambda_h} \times \sin\left(\frac{\theta}{2}\right) \right\}, \qquad (1)$$

where u_0 and Δu is the mean and the amplitude of capillary wave, respectively. λ_h is the wavelength of heating laser beam. In the present setup, the actual mean temperature rise and the amplitude of capillary wave are several milli-kelvin and less than 10 nm, respectively. The motion of the laser-induced capillary wave after the instantaneous heating

is described by the Navier-Stokes equation, the continuity equation, and the heat conduction equation. By solving three fundamental phenomenological equations under several boundary conditions, the geometric behavior of sample surface is written by the material properties (i.e. viscosity, surface tension, thermal diffusivity, thermal conductivity, temperature dependence of surface tension, sound velocity, and absorption coefficient of laser light). Therefore, observing the motion of the laser-induced capillary wave, the viscosity and surface tension is numerically analyzed from the decay time and the frequency of damped oscillating behavior.

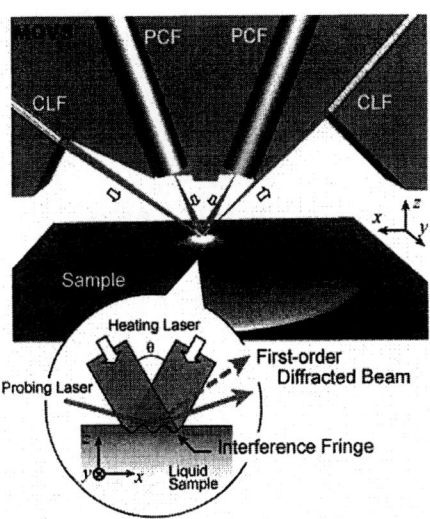

Fig. 1 Principle of MOVS

1-4244-0641-2/07/$20.00 ©2007 IEEE

When the probing laser is irradiated onto the heated area, the interference fringe of laser-induced capillary wave diffracts the beam as the temporal diffraction grating corresponding to the motion of the wave. Hence, by detecting the first-order diffracted beam from the interference fringe of laser-induced capillary wave, the surface geometric behavior of laser-induced capillary wave is optically monitored. The characteristics of MOVS are summarized as follows: (1) completely non-contact method, (2) high-speed measurement within micro second order, and (3) requirement of small sample volume within micro litter order.

3 DEVICE DESIGN AND PRELIMINARY RESULTS

The four trenches with springs holding fibers are fabricated on a silicon-on-insulator (SOI) wafer (thickness: device layer of 269 ± 2 μm, buried oxide layer of 4 μm, and substrate of 525 μm) by deep reactive ion etching (DRIE) using positive photo resist and Al mask (Fig. 2). The photonic crystal fibers (PCFs) guide the pulsed laser beam (Nd:YAG, wavelength of 1064 nm and pulse width of 6 ns) to the sample surface at an angle of $\theta = 15.3°$. The collimated lensed fibers (CLFs) with fusion-spliced micro gradient index (GRIN) lenses condense the probing laser (wavelength of 658 nm) on the laser-induced capillary wave, and collect the first-order diffracted beam. The diameter of PCF is approximately twice the size of that of CLF, and thus, the depth-control of the trenches for CLF by a timed-etch of DRIE is indispensable in order for the center of fiber-cores of PCF and CLF to be aligned at the same height (Fig. 2 (b and c)).

The preliminary measurements to verify the applicability of MOVS were demonstrated using distilled water and sulfuric acid (29.6 wt%) with black carbon dye (0.1 wt%). In the design of the first prototype of MOVS, the fringe space of laser-induced capillary wave, which is equal to that of optical interference, was set to 4 μm. The high-speed damped oscillating signals of first-order diffracted light are shown in Fig. 3. The fast relaxation phenomena of laser-induced capillary wave corresponding to the theory have been successfully observed, thus, the feasibility of MOVS has been confirmed.

ACKNOWLEDGEMENT

This work described in this paper was partially supported by the Keio Leading-edge Laboratory of Science and Technology (KLL) specified research projects, and the Ministry of Education, Culture, Sports, Science and Technology, Grant-in-Aid for No. 19106004(S).

REFERENCES

[1] L. Dintenfass, Blood Microrheology, Butterworth & Co., London, 1971.
[2] M. Bourne, Food Texture and Viscosity – Concept and Measurement, Academic Press, New York, 2002

[3] K. A. Miller, L. A. Nelson, and R. M. Almond, "Should you trust your heavy oil viscosity measurement?" Journal of Canadian Petroleum Technology, vol. 45 no. 4, pp. 42-48, 2006.

[4] S. C. Tjong, "Structure, morphology, mechanical and thermal characteristics of the in situ composites based on liquid crystalline polymers and thermoplastics," Materials Science and Engineering, vol. R41, pp. 1-60, 2003.

[5] T. Oba, Y. Kido, and Y. Nagasaka, "Development of Laser-Induced Capillary Wave Method for Viscosity Measurement Using Pulsed Carbon Dioxide Laser," International Journal of Thermophysics, vol. 25 no. 5, pp. 1461-1474, 2004.

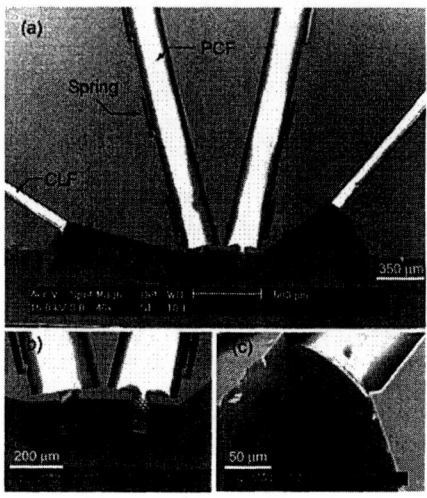

Fig. 2 The scanning electron microscope (SEM) image of MOVS. (a) four trenches and springs are monolithically fabricated on a silicon-on-insulator (SOI) wafer by microfabrication technique. (b) Deep trenches for PCFs, and (c) shallow trenches for CLFs are successfully fabricated.

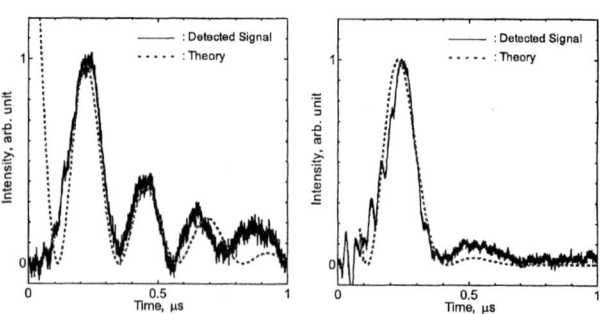

(a) distilled water　　　　(b) sulfuric acid (29.6% wt)

Fig. 3 Preliminary results detected by prototype of MOVS (Λ=4 μm). The nano-second order dumped oscillating signals have been observed.

MB1 (Invited)
13:30 – 14:00

High-Q Photonic Crystal Nanocavities

Susumu Noda
Department of Electronic Science and Engineering,
Kyoto University,
Katsura, Nishikyo-ku, Kyoto 615-8510, Japan
E-mail: snoda@kuee.kyoto-u.ac.jp

Abstract: Recent progress of high-Q nanocavities is reviewed, where Q-factors more than 2million have been successfully achieved while keeping small modal volume of $1.2(\lambda/n)^3$. New design and dynamic tuning of Q-factor of nanocavities are also discussed.

A high-Q photonic nanocavity, which can confine photons strongly in volumes of optical-wavelength size, is attracting much attention in various fields, including photonics, telecommunications, quantum information and cavity quantum electrodynamics, as a strong light–matter interaction is obtained. Recently, an important concept to achieve a high-Q nanocavity in a two-dimensional photonic crystal slab has been proposed and demonstrated [1, 2]: the cavity should have a gentle electric field distribution—like a Gaussian function—to suppress the out-of-slab leakage of photons while maintaining ultrasmall modal volume. Tuning of air holes in a cavity [1] and/or utilization of photonic double heterostructure [2, 3] have been fount to be a key to satisfy the concept. The Q factors more than 2 million, with the modal volume of ~$1.2(\lambda/n)^3$, have been successfully achieved [4].

In this presentation, details of the progress of high Q nanocavities are explained at first. Then, time domain measurement results for such high Q nanocavities are described, where photon decay time longer than 1.7ns is successfully observed [4]. Then, an advanced cavity design with Q-factor of 10^9 with a modal value of less than the previous design will be described. Finally, dynamic tuning of Q-factor of nanocavities aiming at slow light application [5] and/or cavity quantum electrodynamics [6] will be described.

References

[1] Y. Akahane, T. Asano, B. S. Song, and S. Noda, *Nature*, 425 (2003) 944.

[2] B. S. Song, S. Noda, T. Asano, and Y. Akahane, *Nature Materials* 4 (2005) 207.

[3] T. Asano, B. S. Song, and S. Noda, *Optics Express* 14 (2006) 1996, and T. Asano and S. Noda, *IEEE Selected Topics on Quantum Electronics*, 12, issue 6 (Nov.-Dec. 2006).

[4] S. Noda, M. Fujita, and T. Asano, *Nature Photonics* (to be published).

[5] For example, Y. Tanaka, T. Asano, and S. Noda, IQEC-CLEO/PR, CWE4-3, Tokyo, Japan, Jul. 11-15 (2005);

[6] For example, S. Noda, *Science*, 314 (2006) 260.

1-4244-0641-2/07/$20.00 ©2007 IEEE

MB2
14:00 – 14:15

Modeling of Slow Light in Vertical Cavity Surface Emission Lasers

C.-S. Chou[1], R.-K. Lee[1*], P. C. Peng[2], H. C. Kuo[3], G. Lin[4], H. P. Yang[5], and J. Y. Chi[5]

[1]Institute of Photonics Technologies, National Tsing-Hua University, Hsinchu, 300 Taiwan

[2]Department of Applied Materials and Optoelectronic Engineering, National Chi Nan University, Nantou, 545 Taiwan

[3]Department of Photonics and Institute of Electro-Optical Engineering, National Chiao-Tung University, Hsinchu, 300 Taiwan

[4]Department of Electronics Engineering, NCTU, Hsinchu, 300 Taiwan

[5]Opto-Electronics and System Laboratory, Industrial Technology Research Institute, Hsinchu, 300 Taiwan

Keywords: VCSEL, slow-light

Abstract

We develop a model for the slow light in the Vertical Cavity Surface Emission Lasers (VCSELs), with the combinations of cavity and the population pulsation effects. The dependences of pumping power, injection current and wavelength detuning for the group delays are demonstrated theoretically and experimentally. Up to *65 ps* group delays and up to *10 GHz* modulation frequency can be achieved in the room temperature at the wavelength of *1.3 μ m*. Based on the experimental parameters of quantum dot VCSEL structures, we show that the resonance effect of laser cavity plays a significant role to enhance the group delays.

1 INTRODUCTION

Slow lights using electromagnetically induced transparency (EIT) and coherent population oscillation (CPO) have been under intensive studies due to their significant applications in optical communication, optical memories, signal processing, and phase-array antenna systems [1]. Unlike EIT in the cryogenic systems, slow light in semiconductor optoelectronic devices based on CPO is more promising due to its inherent compactness, direct electrical controllability, and room temperature operation. CPO is the effect that the population of the ground state of the material will oscillate in time at the beat frequency of the two input waves. With state of art fabrication technologies, quantum well and quantum dot semiconductor optical amplifiers (SOAs) have been demonstrated as a flexible platform for studying slow light phenomenon as well as its applications in room temperature [2].

Recently, we demonstrate a tunable optical group delay in monolithically single-mode quantum dot (QD) Vertical-Cavity Surface-Emitting Laser (VCSEL) at *10 GHz* experimentally [3]. Compared to SOA devices with the active region for the gain medium about several *mm*, instead the active region of VCSEL is typically only several μ m long. In this scenario, the common adopted population pulsation model of traveling waves induced dynamic carrier index grating can not directly applied to semiconductor lasers. In this work we develop a model for the slow light in the VCSELs with the combinations of Fabry-Perot filter for the cavity effect and the rate equation for carrier undulation. Experimental data of up to *65 ps* group delays and up to *10 GHz* modulation

frequency operated in the room temperature at the wavelength of *1.3 μ m* are in agreement with proposed theoretical results. Based on the experimental parameters of quantum dot VCSEL structures, we show that the resonance effect of laser cavity plays a significant role to enhance the group delays.

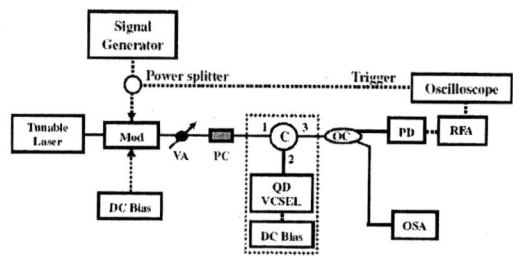

Figure 1: Experimental setup for measuring the slow light in quantum dot VCSELs.

2 EXPERIMENTAL SETUP FOR SLOW-LIGHT in VCSELS

The experiment setup for the slow-light in VCSELs is illustrated in the Figure 1. The key component in our experiment is a monolithically single-mode GaAs based QD VCSEL, grown by molecular beam epitaxy (MBE) with fully doped n- and p-doped AlGaAs distributed Bragg reflectors (DBRs). A probe signal is generated by a tunable laser modulated with an electro-optical modulator and its power is controlled by a variable optical attenuator. Then incident probe signal is tuned to the resonance of the QD VCSEL cavity with different

1-4244-0641-2/07/$20.00 ©2007 IEEE   15

detuning in the modulation frequency. The time delay of the reflected probe signal is measured by a circulator and compared to the reference one, see Ref. [4] for more details about our experimental results.

3 MODELING AND SIMULATION RESULTS

To model the population oscillation in the semiconductor lasers, our theoretical starting point is based on carrier undulation induced by the frequency beating between two optical waves [2]. The probe signal experiences gain and refractive index changes by the pump wave through the carrier index and gain grating. The dynamics of the carrier density, N, at an injected current, I, can be derived from the carrier rate equation,

$$\frac{dN}{dt} = \frac{I}{qV} - \frac{N}{\tau_s} - \frac{g(N)}{\hbar w}|E|^2 + D\nabla^2 N(z) \quad (1)$$

where q is the unit electron charge, V is the active region volume, g is the model gain, t_s is the carrier lifetime, and D is the diffusion coefficient. In addition to the carrier rate equation, we simplify the DBR reflectors in the VCSELs by a Fabry-Perot filter with the response of the cavity gain described by [5]

$$Gr = \frac{(\sqrt{R_t} - \sqrt{R_b}g_s)^2 + 4\sqrt{R_t}\sqrt{R_b}g_s \sin^2\phi}{(1 - \sqrt{R_t R_b}g_s)^2 + 4\sqrt{R_t}\sqrt{R_b}g_s \sin^2\phi} \quad (2)$$

where R_t is the top mirror reflectance, R_b is the bottom mirror reflectance, g_s is the single-pass gain, and ϕ_s is the single-pass phase detuning. With Equations (1-2), we assume that the probe signal is much weaker than the pump wave, and obtain the index change and gain of the probe beam by

$$\Delta n = \gamma g(\bar{N})\frac{c}{2\omega}[1 - \frac{p_0(1 + p_0 - \frac{\Omega t_s}{\gamma})}{(1 + p_0)^2 + (\Omega t_s)^2}]$$

$$g = g(\bar{N})[1 - \frac{p_0(1 + p_0 + \Omega t_s \gamma)}{(1 + p_0)^2 + (\Omega t_s)^2}]$$

where γ is the line-width enhancement factor, P_0 is the normalized pump power with respect to the saturation power, Ω is the detuning modulation frequency of the probe signal.

The simulation results of the group delays for different signal powers at different bias current, and for different modulation frequency detuning are shown in Figures 2 and 3, respectively. The most important

signature of our modeling is that the length of the active region is only 1.13 μm. Without imbedded the cavity effect in our modeling, there is no possibility to have large group delays up to 65 ps in semiconductor devices with such a short active length.

In conclusion, we develop a model for the slow light in semiconductor lasers with consideration of the cavity effect. The simulation results of our proposed model agree well with experimental data for different operations of signal power, bias current and modulation frequency detuning. Based on the experimental parameters of quantum dot VCSEL structures, we show that it is possible to have 65 ps within a compact active region as short as 1.13 μm.

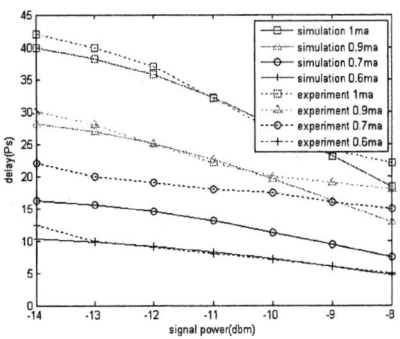

Figure 2: Group delay of QD VCSELs is shown as a function of the optical power of probe signal beam at different bias currents, where solid lines are simulation results while the dashed lines are experimental data.

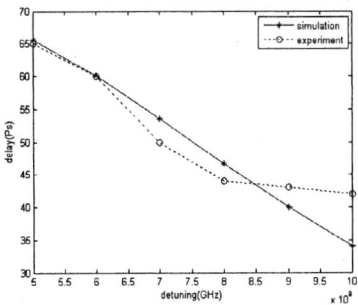

Figure 4: Group delay of QD VCSELs is shown for different modulation frequency detuning, where solid lines are simulation results while the dashed lines are experimental data.

REFERENCES

[1] R. W. Boyd, D. J. Gauthier, and A. L. Gaeta, *Optics & Photonics News* **19**, 18-23 (2006).
[2] H. Su, and S. L. Chuang, *Appl. Phys. Lett.* **88**, 061102 (2006).
[3] P. C. Peng *et al.*, *Opt. Express* **14**, 12880 (2006).
[4] P. C. Peng *et al.*, *Electron. Lett.* **42**, 1036 (2006).
[5] N. Laurand *et al.*, Opt. Express **14**, 6858 (2006).

MB3
14:15 – 14:30

A Nano-scale Nanocrystal Photodetector with High Sensitivity

Michael C. Hegg, Lih Y. Lin

Department of Electrical Engineering, University of Washington, Box 352500, Seattle, WA 98195
heggm@u.washington.edu

Abstract

We present the design, fabrication and testing results of a nano-scale quantum dot photodetector composed of quantum dots that are positioned between a nano-gap in electrodes for high sensitivity and high-resolution photodetection.

Keywords: Nanophotonics, photodetectors, nanocrystals, quantum dots

1 Introduction

Nano-scale photonic integrated circuit design is an active field of research where nano-scale photonic components with high integration densities may have the potential to outperform traditional VLSI technology in many areas, including bandwidth, speed, integration density, and power consumption [1]. Colloidal QDs have unique optoelectronic properties such as sharp absorption, emission and gain spectra. They are nanometer in size, and have flexible surface chemistry, which make them ideal candidates for building a nano-photonic integrated circuit. Previously, we have demonstrated sub-diffraction quantum dot (QD) cascade waveguides for high-density on-chip interconnection [2]. Another critical component of the QD nano-photonic integrated circuit is a nano-scale photodetector with high sensitivity that can be fabricated in an integrated process. This can be achieved by fabricating a nano-sized gap between electrodes and depositing QDs in the gap, so that electrons can easily tunnel from the source electrode to the drain electrode through the QDs. The tunneling current is enhanced by photo-excited carriers. We present the design, fabrication, and experimental results of a new nano-scale QD photodetector that employs such principles.

2 Design and Fabrication

Figure 1 shows a schematic and the fabrication steps involved in constructing a nano-scale quantum dot photodetector.

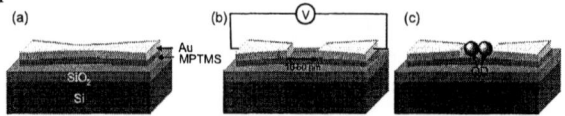

Figure 1. a) EBL and MPTMS deposition, b) break-junction, and c) drop-casting QD deposition

A Si wafer with a 1 µm surface layer of SiO_2 is used as the substrate. The wafer is spin-coated with ~100 nm layer of PMMA and patterned by electron-beam lithography (EBL) to make electrode patterns with line widths of 50 nm. After developing the EBL pattern, the exposed SiO_2 is silanized with a monolayer of (3-Mercaptopropyl)trimethoxysilane

(MPTMS) purchased from Aldrich and Co., USA. To perform silanization, the wafer is oxygen-plasma cleaned and then exposed to the MPTMS gaseous molecules inside a vacuum dessicator for two hours. The sample is then immediately transferred to a thermal evaporator for deposition of a 300-Å Au layer, followed by lift-off of PMMA to form the Au electrodes (Fig. 1(a)). The MPTMS monolayer acts as an insulating adhesion layer for the Au film [3]. A nanometer-sized gap is then created by applying voltage across the electrode (Fig. 1(b)). A force proportional to the current density causes the Au atoms to migrate, a process known as electromigration. After confirmation of the break using SEM, a 100 µM solution of CdSe nanocrystals in toluene is drop-cast onto the wafer (Fig. 1(c)). The nanocrystals have a nominal core/shell diameter of 5.2 nm and an emission wavelength of 620 nm.

Electrodes having widths of 50 nm were broken with gap sizes ranging from 30 nm to 1 nm, in order to demonstrate the flexibility of the procedure. Figure 2(a) shows an SEM micrograph of a narrow break-junction with ~25 nm gap size.

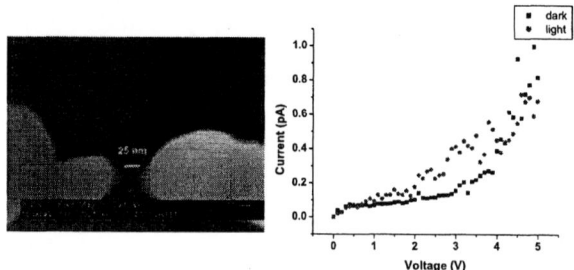

Figure 2. Scanning electron micrograph of (left) pre-QD deposition break-junction, (right) I-V characteristics of the junction in dark (squares) and illuminated (circles) conditions

The core/shell diameter of the quantum dots is nominally 5.2 nm, so at most an array of 5x10 quantum dots can fill the gap. Although QDs are deposited in a large area, only tunneling through the narrow break-junction gap can contribute significantly to the measured current. Figure 2(b) shows control measurements performed prior to quantum dot deposition, with and without optical excitation.

1-4244-0641-2/07/$20.00 ©2007 IEEE   17

3 Experimental Results

Optical light used for excitation is from a 405 nm laser source coupled to the device through a tapered fiber probe. The measured tunneling current is less than 1 pA without the QDs throughout a bias range of 0-5V as shown in Figure 2(b). The same measurement is then repeated for the nano-scale photodetector after QD deposition. Figure 3 shows the current-voltage characteristics of the device under illumination and dark conditions. The illumination intensity corresponds to an optical power of 1.25 nW over the device area of 25 nm x 50 nm. The tunneling current increases by nearly 18 times after the QDs are attached, confirming the placement of QDs between the electrodes. The photocurrent at 4 V is over 3 times larger than the dark current, confirming photosensitivity of the device.

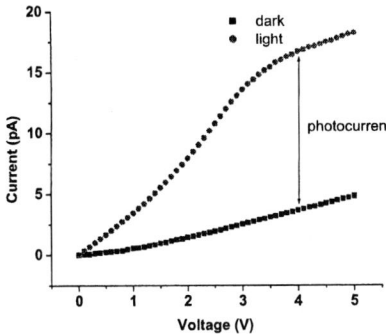

Figure 3. I-V characteristics of the photodetector under dark (squares) and illuminated (circles) conditions

To characterize the sensitivity and responsivity of the device, photocurrent measurements were performed under various illumination intensities, with the device biased at 4 V. The results are shown in figure 4

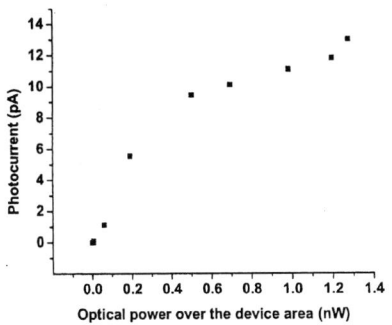

Figure 4. Responsivity measured at room temperature using a calibrated 405 nm laser and biased at 4.0 V

The device starts to show measurable photocurrent under 5 pW illumination over the device area. The device has a responsivity of 0.029 A/W under low-intensity illumination, and starts to exhibit saturation effects when the optical power increases beyond 0.4 nW.

Several devices with a range of gap sizes from 1 – 30 nm were fabricated and tested. Figure 5 shows the signal-to-noise (SNR) and responsivity of these devices as a function of gap size. The largest measured SNR is 3.5 for a gap of 25 nm and decreases with smaller gap size due to an increase in the dark current as the number of tunneling barriers between source and drain electrodes is reduced. The largest device responsivity is nearly 0.15 A/W over the device area for a gap size of 1.5 nm and decreases with increasing gap size for gap sizes less than 10 nm.

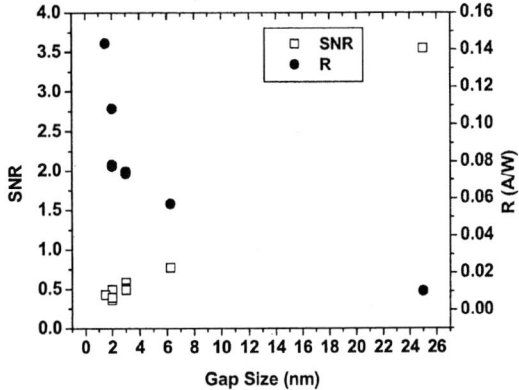

Figure 5. Signal-to-noise ratio (squares) and responsivity at 4V bias (circles) for devices of variable gap size

4 Conclusions

We present design, fabrication, and characterization results of a new nano-scale quantum dot photodetector. The device consists of QDs positioned by drop-casting into a nano-gap between two closely spaced electrodes. The gap is fabricated using a break-junction technique. The device is able to detect optical power as low as 5 pW. SNR and responsivity are characterized for devices with various gap sizes. The SNR increases linearly with gap-size due to a reduction of dark current with increasing tunneling length. The responsivity is largest for small device area, as the input optical power increases linearly with area, but not the tunneling photocurrent. The nano-scale QD photodetector can be integrated into a QD nano-photonic integrated circuit for detection of on-chip optical signals.

5 References

[1] M. Ohtsu, K. Kobayashi, T. Kawazoe, S. Sangu, and T. Yatsui, "Nanophotonics: Design, Fabrication, and Operation of Nanometric Devices Using Optical Near Fields," *Journal of Selected Topics in Quantum Electronics*, vol. 8, no. 4, pp. 839-862, July 2002.

[2] C. J. Wang, L. Huang, B. A. Parviz, and L. Y. Lin, "Subdiffraction Photon Guidance by Quantum-Dot Cascades," *Nano Letters*, vol. 6, no. 11, pp. 2549-2553, Nov. 2006.

[3] A. K. Mahapatro, S. Ghosh, and D. B. Janes, "Nanometer Scale Electrode Separation (Nanogap) Using Electromigration at Room Temperature," *IEEE Transactions on Nanotechnology*, vol. 5, no. 3, pp. 232-236, May 2006.

MB4
14:30 – 14:45

Waveforms of Terahertz Radiation Emitted from Superconducting Dipole Antenna

Shyh-Shii Pai* and Cheng-Chung Chi

Department of Physics, National Tsing Hua University, Hsinchu, Taiwan, ROC.
* Nano and Micro Technology Division, Instrument Technology Research Center, National Applied Research Laboratories, Hsinchu, Taiwan, ROC.

Tel +886-3-577-9911, Fax +886-3-577-3947, E-mail paiss@itrc.org.tw
* Instrument Technology Research Center, National Applied Research Laboratories,
20, R&D Rd. VI, Hsinchu Science Park, Hsinchu, Taiwan

Abstract

We have observed the ultrashort electromagnetic pulse radiation from a current-biased bow-tie structure of $YBa_2Cu_3O_{7-\delta}$ thin film dipole antenna on MgO using 100 fs, 750 nm laser pulses. With the electro-optic detection, we obtained the THz pulses with 1.0 ps full width at half maximum, containing frequency components up to 1.0 THz. The THz peak amplitude dependence shows the saturation and a nonlinear behavior with a higher excitation pumping power and with the applied bias currents. The saturation on the dependence with the excitation powers exhibits the bolometric heating in nature. However, the nonlinear characterization of the THz radiation from the superconductive thin film antenna revealed that the inadequacy of pure supercarrier approximation on a two-fluid model. The ultrashort transient response and the deviation from the classical theory are discussed in relation to the quasiparticle dynamics of the nonequilibrium mechanism.

Keywords: ultrashort, THz radiation, superconductor, dipole antenna, quasiparticle

1 INTRODUCTION

Using a femtosecond (fs) laser, pulsed electromagnetic (EM) radiation from several kinds of materials within a terahertz (THz) spectral bandwidth has been extensively studied by many groups recently. Although, the radiated mechanism of THz pulses emitted from semiconductors is well established and the observed waveforms can be explained by the photoexcitation of carriers followed by ultrafast modulation of photocurrent and locally induced polarization [1, 2]. The ultrafast EM properties of radiated THz pulses from superconductors (SCs) still are quite complicated and not well understood [3]. This is particularly true to consider the total contributions of time-varying suppercarriers and normal carriers in the nonequilibrium state under a constant biased current density.

2 EXPERIMENTAL DETAILS

2.1 Sample Preparation

Superconducting terahertz radiations emitted from an $YBa_2Cu_3O_{7-\delta}$ (YBCO) thin film antenna were carried out

with a broadband scheme of free space electro-optic sampling detection (EOSD) for a typical THz radiation setup [4].

The c-axis oriented film with a thickness of 100 nm was deposited on the highest transmission in THz regime, 0.5 mm thick MgO substrates. Its critical temperature T_c was 85 K and the critical current density at 77 K was about 1.9×10^6 A/cm². A coplanar transmission line with a 60° angled bow-tie structure at the center was patterned into the films using standard photolithographic techniques and wet chemical etch. After the etching process was finished, silver paste is glued on those pads with platinum wires and the samples are subsequently annealed under 1 atm oxygen at about 500°C for 1 hour to supply the samples with oxygen and solidify the contacts between the silver paste and thin films. In order to disperse the high applied current, a couple of platinum wires are glued on one pad at the same time.

2.2 Experimental Setup

About 5.5 nJ, 100 fs pump pulses derived from a Ti:Sapphire laser were focused onto the center bridge (beamsize was slightly larger the line width) of constant current biased YBCO bow-tie antenna to generate THz radiation. THz radiation then concentrates by an MgO hemispherical lens with a diameter of 5 mm attached directly to the substrate upon a pair of parabolic mirrors and the EOSD system.

3 RESULTS

3.1 Typical Waveform

Figure 1 (a) shows the typical waveform of the EM pulses radiated from the bow-tie antenna made of YBCO thin film. The representative measurement was performed under a pump power of 190 mW, and a bias current of 750 mA at 10 K. For this bow-tie antenna, a slightly wider but smaller negative pulse followed by a larger positive pulse, was observed at around a 1.3 ps delay. This waveform is quite different from the empirical result of semiconductor emitters for the alternate of a fast component consequently with a slow one.

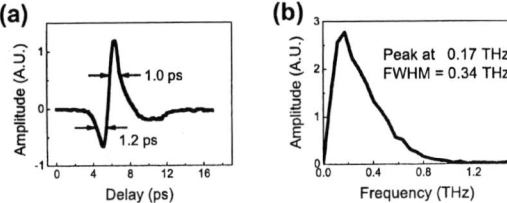

Figure 1. (a) The typical waveform of the EM pulses radiated from the bow-tie YBCO antenna. (b) The corresponding FFT spectrum of the measured EM pulses.

By Fourier transforming to the pulse shown in Fig. 1 (a), we obtain the corresponding amplitude spectrum of the EM pulse, as shown in Fig. 1 (b). The amplitude spectrum of the measured waveform has the FWHM of 0.3 THz with peak amplitude at around 0.17 THz.

3.2 Polarity

Figure 2 shows the electrical pulses of the freely propagating THz beam generated by YBCO thin film measured at 11 K. The pump beam power is 105 mW. The polarity of the pulse shape is changed by reversing the bias current direction. No signal is observed when the bias current is not applied. This result indicates that the observed phenomena are directly the EM pulse radiated from the YBCO bow-tie antenna and relate to the polarity of bias current.

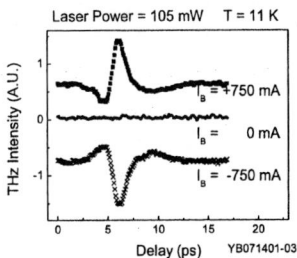

Figure 2. The polarity of the THz radiation waveforms measured with the bias current of +750, -750, and 0 mA at 11 K.

3.3 Polarization

Figure 3 shows the polarization of THz radiation emitted from a SC antenna. Using a wire grid polarizer, we observe that the polarization of EM waves radiated by a current biased emitter is parallel to the direction of the bias current. This parallel polarization exhibits that the THz radiation is due to the straight current flowing along the bridge, not the current loop at the illuminated area of the bridge.

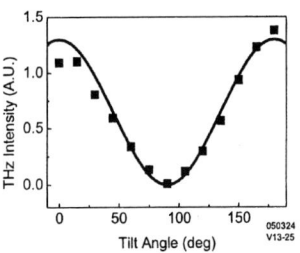

Figure 3. The polarization of THz radiation emitted from a SC antenna, which sinusoidal curve is only a guide to eyes.

REFERENCES

[1] D. H. Auston, K. P. Cheung and P. R. Smith, "Picosecond photoconducting Hertzian dipoles", Appl. Phys. Lett. 45, pp. 284-286, 1984.

[2] X.-C. Zhang *et al.*, "Generation of femtosecond electromagnetic pulses from semiconductor surfaces", Appl. Phys. Lett. 56, pp. 1011-1013, 1990.

[3] Masahiko Tani *et al.*, "Emission properties of $YBa_2Cu_3O_{7-\delta}$-film photoswitches as terahertz radiation sources", Jpn. J. Appl. Phys. Part 1, 36, pp. 1984-1989, 1997.

[4] Y. Cai *et al.*, "Coherent terahertz radiation detection: Direct comparison between free-space electro-optic sampling and antenna detection", Appl. Phys. Lett. 73, pp. 444-446, 1998.

MB5
14:45 – 15:00

Loss and Crosstalk in Quantum Dot Nanophotonic Waveguides

C.-J. Wang and L. Y. Lin

Department of Electrical Engineering, University of Washington, Box 352500, Seattle, Washington 98195
Tel +1-206-616-4785, Fax +1-206-543-3842, E-mail jeanwang@u.washington.edu

Abstract

We measure loss and crosstalk for optically pumped nanophotonic quantum dot waveguides for sub-diffraction transmission. The structures are composed of an array of colloidal semiconductor nanoparticles self-assembled onto a silicon-silicon dioxide substrate through a simple two-layer process. We find ~3dB/3μm loss for throughput in 500 nm wide devices, which is lower than plasmonic waveguides. The crosstalk between waveguides with 500 nm separation is not measurable.

Keywords: Quantum dots, waveguides, sub-diffraction limit

1 INTRODUCTION

Quantum dots (QD) have found a niche in several areas of applications, such as biological tagging [1], lasers [2] and solar cells [3]. Usage of the nanoparticles hinges on exploiting the size-dependent emission, photostability property, the Stokes shift from the absorption spectrum [4], potential for multiparticle ejection [3], as well as gain and stimulated emission criteria [5]. To harness the useful characteristics and further extend QD implementation in optics, we proposed the nanophotonic waveguide as a method for sub-diffraction energy transfer [6,7]. While nanoscale guiding commonly relies upon negative dielectric material or metals to generate surface plasmon polaritons by strip [8], slot [9] or nanoparticle [10] formations, we aim to improve upon the propagation length through semiconductor gain materials and facilitate ease-of-fabrication through self-assembly processes. In this paper, we present the measured loss figure and crosstalk suppression from quantum dot waveguides of 500 nm width.

2 WAVEGUIDE OPERATION

The nanophotonic quantum dot waveguide, depicted in Fig. 1a, utilizes a pump light to excite electron hole (e-h) pairs and activate the quantum dots while a signal light introduced at the device edge prompts recombination and stimulated emission. Through near field interactions, a cascade of photons as modified by the quantum dot gain and inter-dot coupling propagate downstream to the output side. We assume a three level system to model the QD optical response such that the pump raises an electron from the valence band to the second excited level in the conduction band. After fast relaxation to the first excited state, the nominal radiative relaxation process, which is ~25 ns for CdSe/ZnS nanocrystals is circumvented through the femtosecond time scale signal induced stimulated emission.

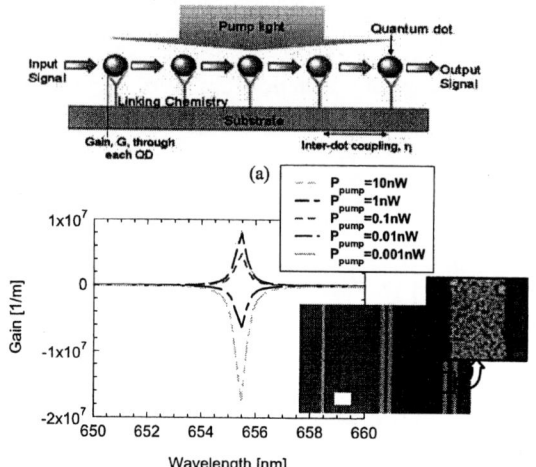

Fig. 1. (a) QD waveguide diagram. (b) Theoretical gain spectra of a 7.4 nm × 7.4 nm × 7.4 nm CdSe/ZnS (shell thickness = 1.1 nm) quantum cube with CW excitation, inset: AFM image of 655 nm emission QDs in 500 nm wide structures, scale bar is 1 μm [8].

Fabrication of the waveguide is accomplished with a two-layer self-assembly technique [7] and the atomic force micrograph for 500 nm wide 655 nm emission QD arrays in single and pair formation show well-defined packing. In addition, the gain profile [6] as a function of wavelength and pump power, in Fig. 1b, indicates that the negative to positive gain swing takes place between 0.01 to 0.1 nW for a CdSe/ZnS core/shell cubic particle with 7.4 nm side length and 1.1 nm shell thickness. Saturation occurs around 1 nW due to the limited number of e-h pairs a quantum dot may support. However, the Auger process raises the gain threshold [5] and is observed in our results. Propagation and crosstalk for 1D arrays modeled through ABCD matrix [6] and FDTD [11] approaches, respectively, reveal a tradeoff between gain and inter-dot coupling as well as suppressed crosstalk to adjacent waveguides compared to conventional dielectric structures of same waveguide width.

1-4244-0641-2/07/$20.00 ©2007 IEEE

3 TEST RESULTS

The efficiency of transfer through the waveguide is measured by contrasting the output signal at various pump power to the output under no pump. As the quantum dots are under optical excitation, the absorption reduces and we expect a rise in net throughput with respect to increase in pump intensity (in Fig. 2a). Specifically, the setup consists of a collimated 405 nm laser placed overhead to illuminate the sample and two fiber probes aligned directly to the waveguide edge for delivery and detection of the 639 nm signal light. The input is modulated at 470 Hz and the detector connects to a lock-in amplifier, which selectively monitors the signal contribution over CW pump and fluorescence baseline.

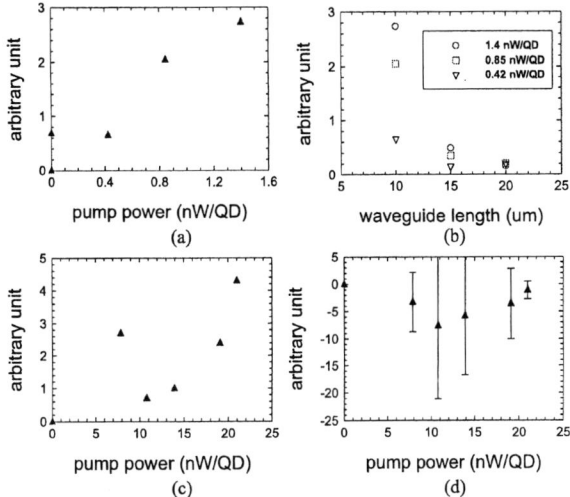

Fig. 2. For 500 nm waveguides: (a) net signal vs. pump power for 10 μm length, (b) transmission vs. length, pump powers noted, (c) relative power for signal and (d) adjacent waveguide at 500 nm separation vs. pump.

For loss, the probes are first optimally positioned with respect to the waveguide and to one another and signal is measured over a ramp profile in pump light. The transmitted power measured under zero pump power is used as the baseline. The net pump induced signal is calculated from the readings. We repeat the process for different waveguide lengths. Using Fig. 2b as a guide, the pump induced transmission decreases at a rate of ~3dB/3μm. Low pump power was used for this measurement to reduce photobleaching of QDs, which would result in higher loss. Crosstalk measurement was done with different waveguides at pair formation with 500 nm spacing. The probes are aligned to the same device first and then the output probe is shifted downward to the adjacent waveguide edge. As crosstalk is the output power ratio between the two waveguides and the photobleaching effect would cancel, high pump power was used to ensure the QDs are in the positive gain regime. Figure 2c and 2d provide results for the transmitted signal and the crosstalk, with zero pump power data as the baseline. The transmitted signal shows an overall positive trend, while the crosstalk is basically not measurable, dominated by large standard deviation.

To draw a comparison, with metal nanoparticle plasmonic waveguides, there is a minimum of 3 dB/500 nm theoretical loss. Hence, the quantum dot waveguide loss may be reduced by the gain property of the semiconductor.

4 CONCLUSIONS

Using self-assembly, we show that quantum dots may be formed as a waveguide for energy transfer with relative low loss and crosstalk. While the current process requires e-beam lithography to define the patterns, nanoimprint techniques represent a way to enable mass production of devices and possible integration with other photonic components. Overall, the quantum dot waveguide shows promise for sub-diffraction optical propagation with low loss and crosstalk.

REFERENCES

[1] W. C. W. Chan and S. Nie, "Quantum dot bioconjugates for ultrasensitive nonisotopic detection," Science, vol. 281, pp. 2016-2018, 1998.

[2] V. M. Ustinov, A. E. Zhukov, A. Y. Egorov, and N. A. Maleev, Quantum Dot Lasers, Oxford University Press, 2003.

[3] V. I. Klimov, "Detailed-balance power conversion limits of nanocrystal-quantum-dot solar cells in the presence of carrier multiplication," Appl. Phys. Lett., vol. 89, pp. 123118-123121, 2006.

[4] C. B. Murray, C. R. Kagan, and M. G. Bawendi, "Synthesis and characterization of monodisperse nanocrystals and close-packed nanocrystal assemblies," Annu. Rev. Mater. Sci., vol. 30, pp. 545-610, 2000.

[5] V. I. Klimov, A. A. Mikhailovsky, S. Xu, A. Malko, J. A. Hollingsworth, C. A. Leatherdale, H.-J. Eisler, and M. G. Bawendi, "Optical gain and stimulated emission in nanocrystal quantum dots," Science, vol. 90, pp. 314-317, 2000.

[6] C.-J. Wang, L. Y. Lin, and B. A. Parviz, "Modeling and simulation for a nano-photonic quantum dot waveguide fabricated by DNA-directed self-assembly," J. of Select. Topics in Quantum Electron., vol. 11 no. 2, pp. 500-509, 2005.

[7] C.-J. Wang, L. Huang, B. A. Parviz, and L. Y. Lin, "Sub-diffraction photon guidance by quantum dot cascades," Nano Lett., vol. 6 no. 11, pp. 2549-2553, 2006.

[8] J. R. Krenn, B. Lamprecht, H. Ditlbacher, G. Schider, M. Salerno, A. Leitner, and F. R. Aussenegg, "Non–diffraction-limited light transport by gold nanowires," Europhys. Lett., vol. 60 no. 5, pp. 663-669, 2002.

[9] L. Chen, J. Shakya, and M. Lipson, "Subwavelength confinement in an integrated metal slot waveguide on silicon," Opt. Lett., vol. 31 no. 14, pp. 2133-2135, 2006.

[10] M. L. Brongersma, J. W. Hartman, and H. A. Atwater, "Electromagnetic energy transfer and switching in nanoparticle chain arrays below the diffraction limit," Phys. Rev. B, vol. 63 no. 24, pp. R16356-R16359, 2000.

[11] L. Huang, C.-J. Wang, and L. Y. Lin, "A comparison of crosstalk effects between colloidal quantum dot waveguides and conventional waveguides," Optics Lett., vol. 32 no. 3, pp. 235-237, 2007.

MC1 (Invited)
15:30 – 16:00

Towards the Fabrication Platforms for MOEMS

[1,2]Weileun Fang, [1]Mingching Wu, [1]Hung-Yi Lin, and [1]Jerwei Hsieh

[1]Power Mechanical Eng. Department, [2]MEMS Inst., National Tsing Hua University, Hsinshu, Taiwan

Abstract

Fabrication platform is one of the key factors to accelerate the progress of MOEMS (Micro opto-electro-mechanical systems). The two poly-Si MUMPs process is recognized as one of the most popular fabrication platforms for MOEMS. This study intends to introduce three fabrication platforms evolved from MUMPs to improve the performances of MOEMS devices.

Keywords: Fabrication platform, MUMPs process, MOSBE process.

1. INTRODUCTION

The two poly-Si MUMPs process is recognized as one of the most popular fabrication platforms for MOEMS. Many promising MOEMS devices have been realized through the MUMPs platform. However, the MUMPs platform is unable to fulfill the requirements of some MOEMS applications, for instance, the stiffness and moving space issues. Various approaches have been employed to improve the performances of MUMPs devices. For instance, the thin film residual stresses and some special micro mechanisms have been developed to increase the moving space of MUMPs structures [1-2]. This study intends to introduce three fabrication platforms evolved from MUMPs, named (100) MOSBE [3-4], (111) MOSBE [5], and SOI MOSBE [6], to improve the performances of MOEMS devices. In summary, this study shows the importance and potential of fabrication platforms for different MOEMS applications.

2. (100) MOSBE PLATFORM

The MOSBE process is illustrated in Fig.1 [4]. The DRIE is used to etch trenches around ~20µm deep (Fig.1a). The two poly-Si MUMPs process is exploited to fabricate structures around ~2µm thick, meanwhile the trench-refilled molding technique is used to realize structures of ~20µm thick (Fig.1b). The trenched-refilled poly-Si is trimmed by DRIE to fabricate structures of different vertical positions (e.g. vertical comb electrodes, Fig.1c). Finally, the bulk silicon etching is employed to provide a large free space of ~100µm under the structure (Fig.1d).

Some typical MOSBE components are demonstrated in Fig.2, for instance, the stiff rib-reinforced supporting frame and mirror plate in Fig.2a and the flexible thin torsional spring in Fig.2b. In Fig.2c, the vertical comb electrodes located at different vertical positions are demonstrated. Since these poly-Si components are fabricated using the same process, they are monolithically integrated on wafer.

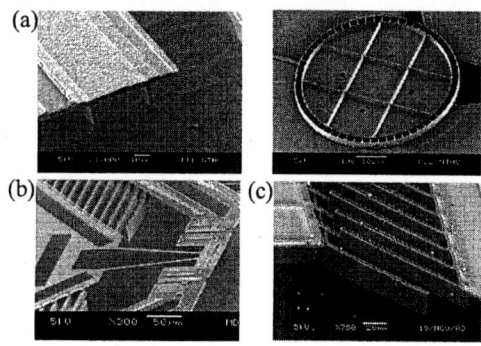

Fig.2 MOSBE structures, (a) Stiff rib reinforced structure, (b) flexible spring, (c) comb electrodes at different vertical positions for VCA (vertical comb actuator).

In applications, the 1D and 2D optical scanners are shown in Fig.3 [4,7]. These scanners, which suspended above a bulk micromachined cavity, consist of the vertical comb actuator (VCA), the flexible torsional spring, and the stiff mirror plate and supporting frame. It is easy to improve the linearity and maximum deflection of MOSBE scanner using the concept of sequential-engagement comb electrodes [8].

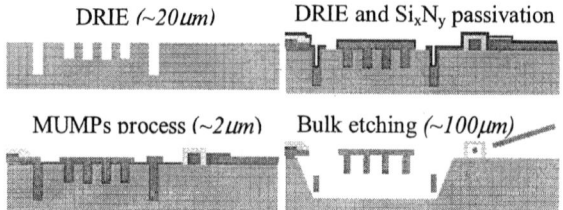

Fig.1 The fabrication process steps of MOSBE

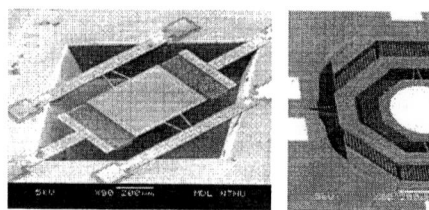

Fig.3 The 1D and 2D (100) MOSBE scanners.

1-4244-0641-2/07/$20.00 ©2007 IEEE

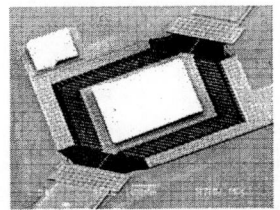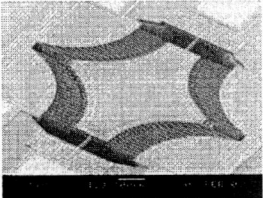

Fig.4 The scanners with sequential-engagement VCAs.

As in Fig.4, by varying the in-plane distribution of comb fingers, the engagement as well as the electrostatic forces of electrodes can be easily tuned. Fig.5 shows the MOEMS lens positioner monolithically integrates bidirectional focusing (out-of-plane) VCA, V-beam tracking (in-plane) actuators, in-plane and out-of-plane springs, supporting frame, and lens holder [9]. The UV-cured polymer lens can be dispensed on lens holder after releasing of device, so that the incident light beam is focused by polymer lens and modulated by focusing and tracking actuators.

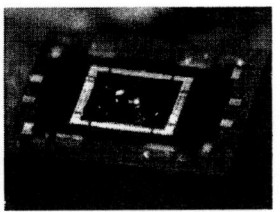

Fig.5 The (100) MOSBE 3-axis lens positioner for tracking and focusing of light spot.

3. (111) MOSBE PLATFORM

Fig.6 illustrates the (111) MOSBE platform which can further integrate thick SCS (single crystal Si) structures with multi-thickness and multi-depth freely suspended thin film structures [5]. The characteristic of selective lateral wet anisotropic etching on (111) wafer is used to provide a large moving space and SCS structures for the device. Therefore, the variety of components with higher stiffness and larger inertia can be significantly increased by adding the SCS structures. The 1-axis scanning mirror shown in Fig.7a is a typical fabrication result realized by the (111) MOSBE process [5]. This vertical-comb-drive scanner had a thick SCS mirror plate and suspended by flexible thin springs, as shown in Fig.7b.

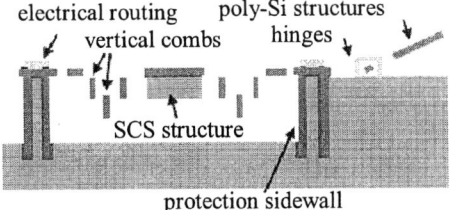

Fig.6 The (111) MOSBE platform.

Fig.7 The (111) MOSBE scanner with thin spring, VCAs, and thick SCS mirror plate.

4. (111) MOSBE PLATFORM

Fig.8 shows the SOI MOSBE platform of integrating polymer, two poly-Si MOSBE, and HARM structures on SOI wafer to form a SiOB (silicon optical bench) [6]. In general, the poly-Si can be employed to implement thin film structures, such as hinges and flexible springs. The device layer of SOI wafer are exploited to realize various HARM structures, for instance, the thick fiber housing of good clamping ability, and the thick SCS actuator with larger output force and displacement. Finally, the thin poly-Si components are anchored to the thick device Si layer of SOI to establish the SiOB, as indicated in Fig.9 [6].

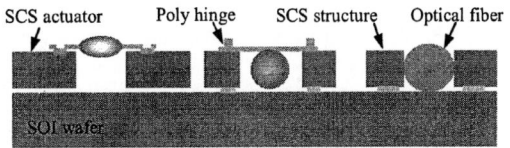

Fig.8 The SOI MOSBE platform.

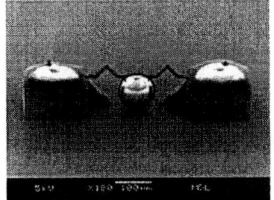

Fig.9 A typical SOI MOSBE SiOB consists of an in-plane polymer lens, fiber housing, and a thermal actuated shutter.

REFERENCES

[1] V A Aksyuk, et.al. *Proc. of SPIE*, p.984, 1999.
[2] Y-P Ho, et.al., *Microsys. Techn.*, 11, p.214, 2004.
[3] H-Y Lin, et.al., *J. Micromech. Microeng*, 10, p.93, 2000.
[4] M Wu, et.al., *J. Micromech. Microeng*, 15, p.535, 2005.
[5] M Wu, et.al., *J. Micromech. Microeng*, 16, p.260, 2006.
[6] S-Y Hsiao, et.al., *Proc. of APCOT*, 2006.
[7] M Wu, et.al., *IEEE J. of Photo. Tech. Lett.*, p.2111, 2007.
[8] M Wu, et.al., *IEEE MEMS'07*, p.655, 2007.
[9] M Wu, et.al., *J. Micromech. Microeng*, 16, p.1290, 2006.

FR-4 as a New MOEMS Platform

S. Holmstrom[1], A.D. Yalcinkaya[1,2], S. Isikman[1], C. Ataman[1], H. Urey[1]

[1] Koç University, College of Engineering, Istanbul, TURKEY

Tel +90-212-338-1474, Fax +90-212-338-1548, E-mail: hurey@ku.edu.tr

[2] Boğaziçi University, College of Engineering, Istanbul, TURKEY

Abstract

FR4 is a well-engineered material and widely used in the PCB industry and lend itself to high degree of integration of optoelectronic, micro-optic, and electronic devices. FR4 is used for the first time as an actuated mechanical device that integrates several functions on the same device. Two different approaches to 2D laser scanning using a single electromagnetic actuation coil and application to Fourier Transform spectroscopy are presented; many other applications can be envisioned.

Keywords: electromagnetic actuators, scanners, interferometers

1 INTRODUCTION

This paper introduces FR4, most commonly used material for printed circuit boards (PCB), as an actuated micromechanical platform. Integration of optoelectronic, micro-optic and electronic devices on FR4 is widely explored; however, the mechanical functionality brings about a new found flexibility and higher degree of integration. PCB fabrication technology using lithographic techniques and microvia technology is now able to produce devices with micron precision and nearly as small feature sizes. PCB machining technology is low-cost, require short design and fabrication cycles, and produce more robust devices compared to silicon MEMS devices.

2 LORENTZ FORCE ACTUATOR

The Lorentz type FR4 actuator consists of a structural layer made of an epoxy-fiber layer, copper conductor lines on both sides of the 130 μm thick FR4 substrate as seen in Fig. 1. A 100μm thick aluminum coated silicon die is attached as a mirror. Exemplary scanner designs are shown in Fig. 2. When the current and the magnetic field, produced by an external magnet placed underneath, are directed as in Fig. 1, the forces marked as F_1 and F_2 exert a net torque, thus a rotation about x-axis (torsion mode). Similarly, forces shown as F_3 and F_4 create a rotation about the y-axis (rocking mode). Fig. 3 shows the FEM simulation results for two orthogonal rotation modes and the frequency response of the torsion mode. Oscillation modes are well separated and the mechanical quality-factors (Q) are between 50 and 100. Fig. 4 illustrates the large angle scanning capability, which exceeds 180° in some designs. Scan angle (θ) increase linearly at low currents and the rate of increase is lower at high drive current levels. The coil power level is smaller than 100mW.

For 2D actuation, magnet is oriented off-axis to energize both modes (preferably at 45°) and the drive current includes two superimposed sinusoidal signals at frequencies f_1 and f_2 corresponding to two orthogonal modes [1]. The scanner mechanical modes separate the drive torque and create a 2D sinusoidal raster, as seen in Fig. 5. The Lorentz force actuated scanners presented here can achieve $f^2\theta$ product of up to 50 KHz^2·deg and a θD-product of 700deg•mm with a Total Optical Scan Angle (TOSA) of 140° (operating at 250mA). These figures enable a number of scanning applications in both DC and resonant mode. Reliability tests are performed and the scanner resonant frequency fluctuated <1% during >250Million cycles of operation at 20° TOSA.

3 SOFT MAGNETIC FILM ACTUATOR

The soft magnetic actuator consists of a mirror suspended by a polymer cantilever beam, a permalloy sheet plated on the mirror, and an external coil to generate the driving magnetic field (fig. 6) [3]. For permalloy actuators, due to thin film anisotropy the force is unidirectional, but can be attractive or repulsive depending on the electromagnet configuration [2]. A 2D sinusoidal raster can be achieved by superposing two sinusoidal drive signals corresponding to two orthogonal scan modes (Fig. 7).

4 FR4 BASED FT SPECTROMETER

With the low frequency operation, long and controllable travel range and the easy fabrication and integration methods available for FR4 based scanners give a great potential for development of novel and compact spectrometers with good performance. One example is seen in Fig. 8 where an FR4 scanner, vibrating out-of-plane translation mode, is used in a Michelson Interferometer setup with integrated position feedback. The crucial requirement here is to accurately measure the position of the moving mirror, which is realized through a feedback system consisting of a secondary micromachined grating coupled with a monochromatic source [4]. This method eliminates interpolation and resampling of the spatially non-uniform sampled interferogram.

In summary, several devices and applications using FR4 MOEMS platform are demonstrated. FR4 has attractive engineering material properties and has proven to be a reliable mechanical material for electromagnetic actuated devices. We expect to see many other systems and devices enabled using FR4 in the future.

1-4244-0641-2/07/$20.00 ©2007 IEEE

ACKNOWLEDGEMENT

Authors would like to thank Aselsan Inc. and Ümit Tümkaya for help with the PCB prototypes and Microvision for partial funding of this project.

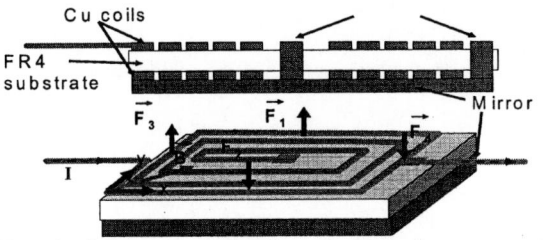

Fig. 1: Schematic of the Lorentz type electromagnetic actuator fabricated on FR4.

Fig. 2: Three scanner designs to obtain a range of resonant frequencies. Substrate is 130μm FR4. Square mirror dimensions are D=5mm or D=8mm. Double-sided coil trace-width is 100μm.

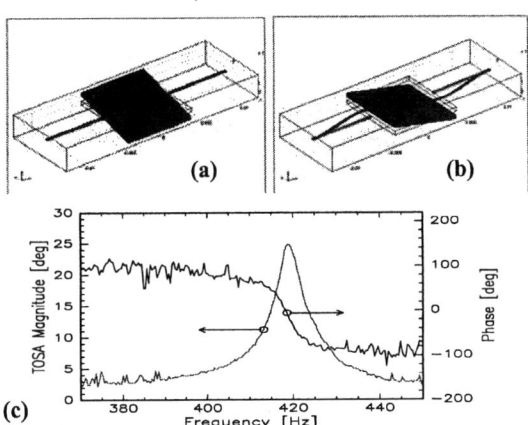

Fig. 3: FEM simulations for (a) torsion and (b) rocking mode. (c) Frequency response at low (10 mA) drive current for Design (B)

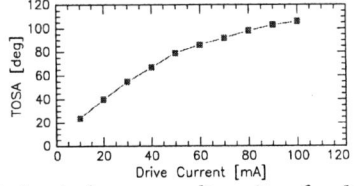

Fig. 4: Mechanical response linearity of a 130μm thick design (B).

Fig. 5: 2D-sinusoidal pattern created by Design B.

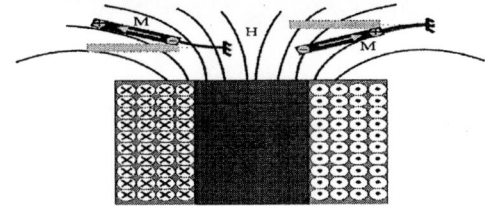

Fig. 6: Permalloy-plated scanner in push or pull modes.

Fig. 7: (a) Test setup for permalloy actuator and (b) the resulting 2D-sinusoidal pattern [3]

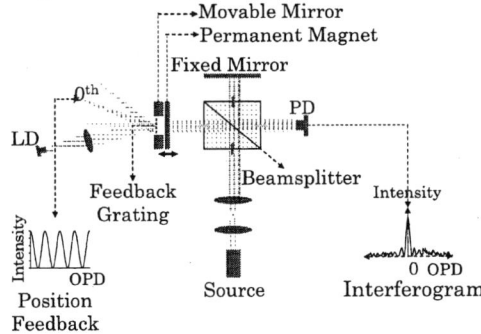

Fig. 8: FR4 Scanner as a Fourier transform spectrometer in the Michelson interferometer configuration.

REFERENCES

[1] A. D. Yalçınkaya, H. Urey, D. Brown, T. Montague, R. Sprague, "Two-axis Electromagnetic Microscanner for High Resolution Displays," *IEEE J. MEMS*, Vol. 15, p. 786-794, 2006.

[2] S.O. Isikman, O. Ergeneman, A.D. Yalcinkaya and H. Urey, "Modeling and Characterization of Soft Magnetic Film Actuated 2-D Scanners", *IEEE J. Sel. Top. Quantum Electron.*, vol. 12, pp. 283-289, 2007.

[3] A. D. Yalcinkaya, O. Ergeneman, H. Urey, "Polymer Magnetic Scanners for Bar Code Applications," *Sens. Act. A*, Vol. 135, pp.236-243, 2007.

[4] C. Ataman, H. Urey, A. Wolter, "MEMS-based Fourier Transform Spectrometer," *J. Micromech. and Microeng.* Vol.: 16, Pages: 2516-2523, 2006

MC3
16:15 – 16:30

Optically Flat Micromirror Using Stretched Membrane with Crystallization-Induced Stress

Minoru Sasaki[1], Takashi Sasaki[2], Kazuhiro Hane[2], and Hideo Miura[2]

[1]Dept. of Advanced Science and Technology, Toyota Technological Institute
Hisakata 2-12-1, Tenpaku-ku, Nagoya 468-8511, Japan
Phone: +81-52-809-1840, E-mail: mnr-sasaki@toyota-ti.ac.jp
[2]Dept.of Nanomechanics Eng., Tohoku University, Aramaki 6-6-01 Aoba-ku, Sendai, 980-8579, Japan

Abstract

The flat and light-weighted micromirror is realized using the tense poly-Si film across a rigid c-Si drum. The tensile stress of ~600 MPa is obtained using the crystallization of a-Si film. Compared to the research carried out by Nee et al., the initial film has the purer amorphous phase and generates larger stress. The mirror satisfies the better optical flatness $<\lambda/10$ for the visible light. The peak-to-valley distance of the micromirror is ~20 nm. The membrane profile is found to depend on the process sequence relating to the timing of the annealing. The mirror shape is stable against the high temperature.

Keywords: optical flatness, micromirror, tensile stress, amorphous, crystallization

1 INTRODUCTION

Generally the mirror is required to be flat. As the criterion of the flatness, the peak-to-valley distance of $<\lambda/10$ is frequently used. λ is the wavelength of the light used. Supposing the blue light, this criterion becomes <40 nm. Satisfying this criterion is not easy when the thin film is used, because the film bends due to the stress and its distribution inside. Dividing one mirror into small mirror array does not become the solution when the single mirror with the large diameter (>500 μm) is necessary. Although the use of the thicker plate (>30 μm) is a simple solution, there is a trade-off between the resonant frequency. The increase of the inertia decreases the resonant frequency. The scanner needs the fast response especially for the high-resolution application. The flat and light-weighted mirror is ideal.

The light-weighted flat mirror is prepared by thin film with tensile stress (said to be 338 MPa) [1]. Figure 1 shows the illustration of the mirror, which consists of a rigid c-Si drum and the thin film. The tension applied in the film suppresses the bending of the membrane [2]. The tensile stress increases the stiffness without increasing the mass. Another trial is preparing the honeycomb core inside the mirror plate [3]. As a simpler method, the timed Si etching is applied to the micromirror [4]. The etching for the mass-reduction is carried out from the back side.

Although the mechanical structure proposed by Nee et al. is ideal [1], their mirror has 100 nm deflection in peak-to-valley. If the tensile stress can be larger, the larger flattening effect will be obtained. In this study, the crystallization-induced stress is proposed.

2 PRINCIPLE

The film prepared by Nee et al. is deposited at 590 °C and considered to be the mixture of amorphous (a-) and polycrystalline (poly-) Si phases. In this study, the deposition is carried out at a-Si phase. After the annealing, the crystal matrix formation generates the strong tensile stress shrinking the volume and removing hydrogen. The reported stress is ~800 MPa [5]. The temperature performance is considered to be stabilized, since the crystallization-induced stress is temperature independent. Even after the metal deposition generating the layered structure, the bending will be suppressed.

3 FABRICATION

Figure 2 shows the fabrication sequence. The mirror consists of the drum and the membrane. The crystalline (c-) Si substrate is thermally oxidized (step 1), and LPCVD grown a-Si film is deposited by ~500 nm (step 2). After

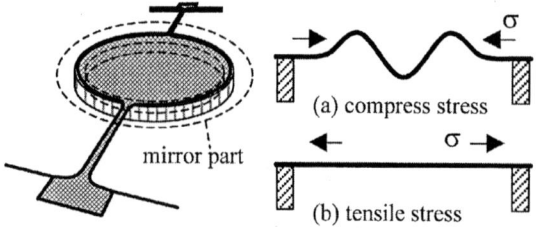

Figure 1 Schematic drawing of the mirror, which consists of a rigid c-Si drum and the thin film.

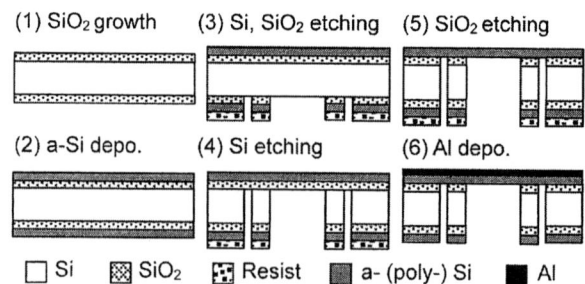

Figure 2 Fabrication sequence of the mirror structure.

1-4244-0641-2/07/$20.00 ©2007 IEEE

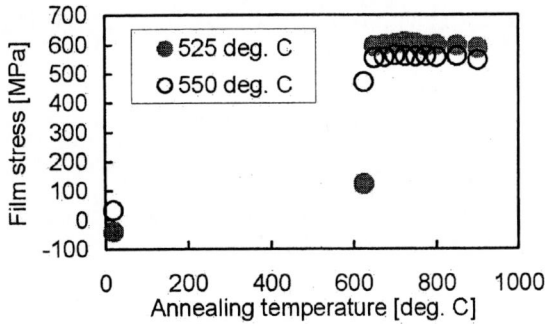

Figure 3 Film stress as the function of the annealing temperature.

patterning the wafer backside (step 3), the drum is prepared using Si etching (step 4). The etching stops at the front SiO_2 film. The membrane has the bi-layer structure. The bending occurs at this stage, since the thermally grown SiO_2 film has the large compress stress. After SiO_2 etching (step 5), the Si membrane is obtained. Al is deposited on the structure for increasing the reflectivity (step 6).

4 RESULTS

Figure 3 shows the film stress plotted against the annealing temperature. After the deposition, the film at the single side is remained and annealed. The stress is evaluated from the radius of curvature of the wafer. The film deposition is carried out at 550 and 525 °C using SiH_4 gas. As deposited a-Si film show the slight stress. The film deposited at the lower temperature generates the larger tensile stress. After the annealing at >650 °C, the tensile stress of ~600 MPa occurs.

Figure 4 shows the optical image and the cross-section of the mirror. The outer and inner diameters of the drum are 500 and 380 µm, respectively. Figures 4(a) and 4(b) are the mirrors annealed after the process steps 5 and 2, respectively. The a-Si films are same. Figure 4(a) shows the significant subsidence. The bending of the membrane before the crystallization was 8.5 µm and it decreases to 136 nm in the peak-to-valley distance as seen from Fig. 4(a). This remained value does not satisfy <λ/10. Figure 4(b) shows the case when a-Si film is annealed before the diaphragm structure is prepared. The mirror shows the smaller bending of 20 nm. Even though the bi-layer at the step 4 has the combination of large tensile and compressive films, the membrane remains without breaking.

Figure 5 shows the shape when the mirror shown in Fig. 4(b) is heated to 110 °C. The peak-to-valley distance is 15 nm. The almost structural material is Si and the thermal expansion ratio is almost same between c-Si and poly-Si. Figure 6 shows the mirror after Al deposition by 100 nm. Figures 6(a) and 6(b) show the peak-to-valley distances of 16 and 18 nm observed at 20 and 110 °C, respectively. This satisfies <λ/10, supposing the visible light. Although Al/Si

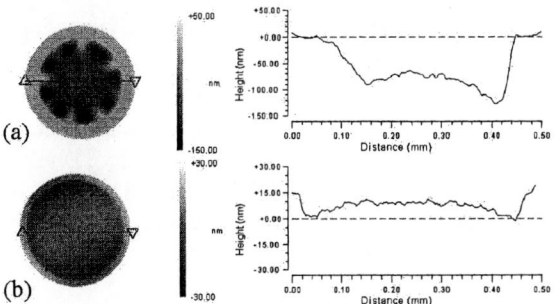

Figure 4 Micromirror with the film crystallized after steps (a) 5 and (b) 2.

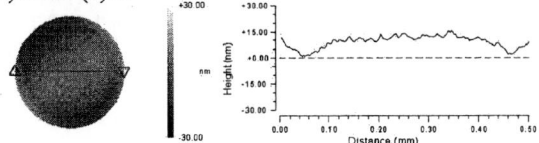

Figure 5 Micromirror heated at 110 °C.

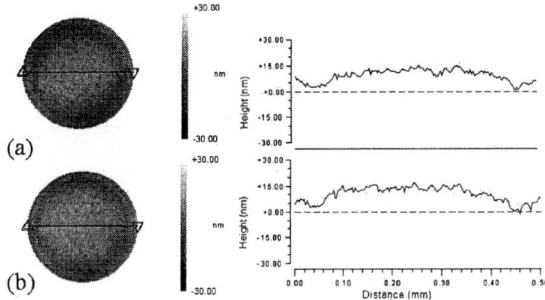

Figure 6 Al metalized micromirror at (a) 20 and (b) 110 °C.

bi-layer generates the bending moment with the different thermal expansion, the optical flatness is maintained.

This research was supported by a grant-in-aid for scientific research on priority areas (no. 19016003). The facilities used for this research include the micro/nano-machining research and education center, at Tohoku University.

REFERENCES

[1] J. T. Nee, R. A. Conant, K. Y. Lau, R. S. Muller, Proc. IEEE 13th Annual Int. Conf. Micro-Electro-Mechanical Systems, 2000, 704-709.
[2] S. D. Senturia, "Microsystem Design", Chapter 9, Springer Science+Business Media, Inc., New York.
[3] P. R. Patterson, G.-D. J. Su, H. Toshiyoshi, M. C. Wu, Proc. Int. Workshop Solid-State Sensors and Actuators, Hilton Head, 2000, 17–18.
[4] V. Milanovic, M. Last, K. S. J. Pister, Proc. 11th International Conference on Solid-State Sensors and Actuators 2001, 1298-1301.
[5] H. Miura, H. Ohta, N. Okamoto, T. Kaga, Trans. Japan Society Mechanical Eng. A, vol. 58 1992, 1960-1965.

MC4
16:30 – 16:45

Polarization-Transmissive Thin-Film Solar Cell with Photodiode Nanowires

Kenichiro Hirose, Yoshio Mita*, and Shuichi Sakai

Graduate School of Information Science and Technology, The University of Tokyo

7-3-1, Hongo, Bunkyo-ku, Tokyo, 113-8656, Japan

Tel +81-3-5841-6763, Fax +81-3-5841-6763, E-mail {hirken, sakai}@mtl.t.u-tokyo.ac.jp

*Graduate School of Electrical Engineering, The University of Tokyo   E-mail MEMS@else.k.u-tokyo.ac.jp

Abstract

The Polarization-Transmissive Thin-Film Solar Cell, which consists of a 400nm-wide silicon photodiode-nanowire grid, transmits the light polarized in one direction and generates photocurrent from the light polarized in the other direction. It can efficiently use light in the system which uses polarization. The fabricated device generated 52nA photocurrent from the 250μW incident light by a fluorescent light and achieved the extinction-ratio of 4 for the incident light whose wavelength was 675nm.

Keywords: polarization, thin-film solar cell, DRIE with controlled vertical profile, high aspect-ratio trench, energy scavenging

1 INTRODUCTION

Polarization is widely used in optical systems such as Liquid Crystal Displays (LCDs). In such systems, polarization plates are usually used to extract the light polarized in one direction by shutting the light polarized in the other directions. Therefore, half of the light energy is wasted. This paper proposes the polarization-transmissive thin-film solar cell (PTTF solar cell) that can recover light energy by means of photocurrent as shown in Fig. 1.

It is widely known that the metal wire grid structure, which consists of the wires narrower than wave length of incident light, reflects the light polarized in parallel to the grid and transmits the light polarized perpendicularly to the grid [1]. In place of the metal wire grid, PTTF solar cells consist of the silicon photodiode-nanowire grid. They can generate photocurrents from the incident light polarized in parallel to the grid and transmit the incident light polarized perpendicularly. The fabricated PTTF solar cell, which consists of the 400nm wide photodiode wires, achieved the extinction ratio of 4 and generated photocurrent from the 675nm-laser.

2 FABRICATION OF THE DEVICE

2.1 Dry Release Technique with Profile-Controlled DRIE

PTTF solar cells are fabricated by using a simple Deep Reactive Ion Etching (DRIE). After some cycles of the standard Bosch's process, which is a combination of sidewall passivation and etching [2], an intentionally long isotropic etching is carried out. As shown in Fig. 2, large undercuts are formed on the bottom of trenches [3]. Then, these undercuts can cut off the thin-film nano-grid structure from the substrate.

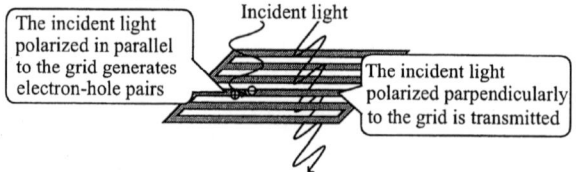

Figure 1. Concept of the PTTF solar cells.

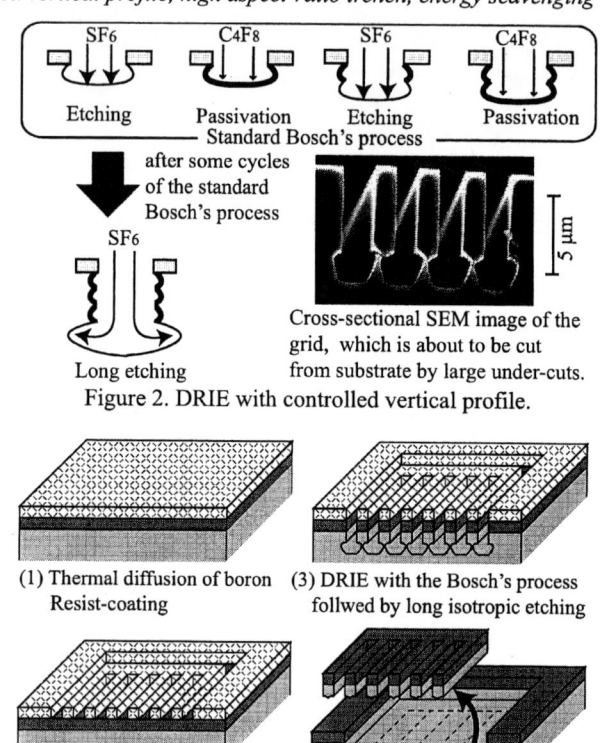

Figure 2. DRIE with controlled vertical profile.

(1) Thermal diffusion of boron Resist-coating

(2) Lithography Developmetnt

(3) DRIE with the Bosch's process follwed by long isotropic etching

(4) Removing resist and passivation Peeling off the device

▨ Silicon (n-type)   ▧ Doped silicon (p-type)   ▦ Resist

Figure 3. Fabrication process

2.2 Fabrication Process

As shown in Fig. 3, PTTF solar cells require only three fabrication steps: diffusion, lithography, and DRIE. An n-type 4" silicon wafer (thickness: 525μm, resistivity: approximately 5 Ω·cm) was used. First, thermal diffusion was carried out with boron nitride planar diffusion source (BN-975, Saint-Gobain) to form the p-n junction on the wafer surface. After thermal diffusion, the SiO_2 film was

1-4244-0641-2/07/$20.00 ©2007 IEEE   29

removed by 50% Hydro Fluoric acid (HF). Then, a 400nm-thick Electron Beam (EB) resist (ZEP-520A, ZEON Co.) was spun on the wafer. The EB resist was exposed by the EB-lithography machine (F5112+VD01, ADVANTEST) and developed by an organic developer (ZED-N50, ZEON Co.) to define the grid pattern. As mentioned in the section 2.1, the wafer was etched by using the standard Bosch's process followed by the long isotropic etching with an Inductively Coupled Plasma Reactive Ion Etching (ICP-RIE) apparatus（AMS-100, Alcatel). Then, the wafer was cleaned by O_2 plasma to remove EB resist and sidewall-passivation. The PTTF solar cell was peeled off the substrate.

3 MEASUREMENT RESULTS

PTTF solar cell was pinched between two transparent acrylic plates, both partly covered with aluminum film as shown in Fig. 4. These Al films work as electrodes and the generated photocurrent was measured by the semiconductor parameter analyzer (4156B, Agilent). There is a polarization plate and a half-wavelength plate between the measured device and light source. Power of the transmitted light was measured using an optical power meter (TQ8210, ADVANTEST).

The measured device is approximately 5mm² and consists of the 5µm thick, 400nm wide, and 500nm interval photodiode wires as shown in Fig. 5. Fig. 6 demonstrates the power of transmitted light, which is normalized by that of the incident light, when the polarizations of the incident lights vary. It achieved the extinction-ratio of 4 for the laser whose wavelength is 675nm. Fig. 7 indicates that the spectrums of a fluorescent light and its transmitted light. The light whose wave length is longer than 550 nm is strongly affected by the PTTF solar cell. It explains the higher extinction-ratio of the 675nm-LASER. Fig. 8 indicates that the generated

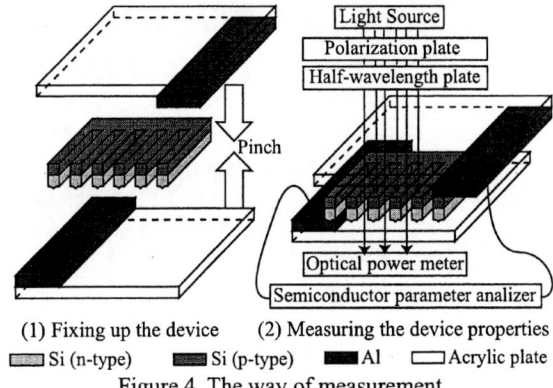

(1) Fixing up the device　　(2) Measuring the device properties

▨ Si (n-type)　▨ Si (p-type)　■ Al　▢ Acrylic plate

Figure 4. The way of measurement.

photocurrent increased proportionally to the power of the incident light. Judging from Fig. 6 and Fig. 9, the less incident light is transmitted, the more photocurrent is generated. Fig. 10 shows the p-n junction characteristics without any incident light. Dark current of the p-n junction of this device is approximately 70nA. Surface damage recovery as well as passivation is necessary to reduce the dark current.

4 CONCLUSIONS

A PTTF solar cell consisting of the 400nm×5µm×50µm silicon photodiode nanowires was fabricated by using DRIE with the controlled vertical profile. It achieved the extinction-ratio of 4 for the 675-nm laser and generated photocurrents from the incident lights.

REFERENCES

[1] Y. Pochi, Optics Communications, Vol. 26, pp. 289-292
[2] F. Lärmer and A. Schilp, US Patent 5501893, 1996.
[3] A. K. Shaw, et al., Proc. of MEMS `93, pp. 63-70, 1993

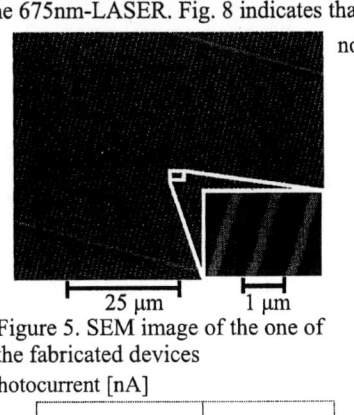

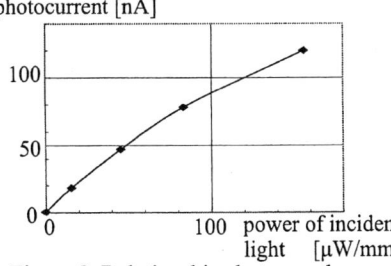

Figure 5. SEM image of the one of the fabricated devices

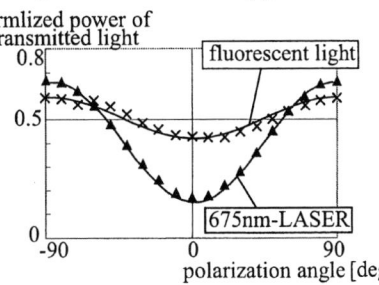

Figure 6. Transmission of the light polarized in various directions

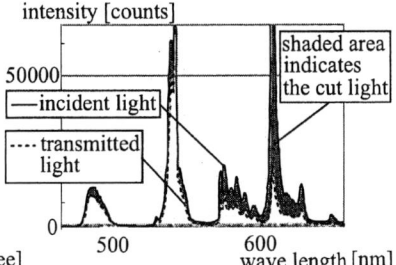

Figure 7. Spectrums of the transmitted and incident lights

Figure 8. Relationships between the photocurrent and the power of incident light

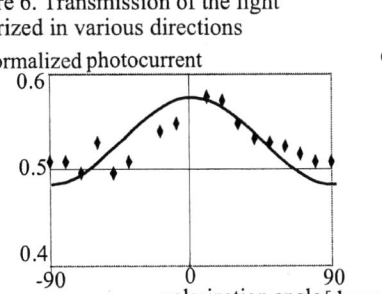

Figure 9. Relationships between the photocurrent and polarization of the light

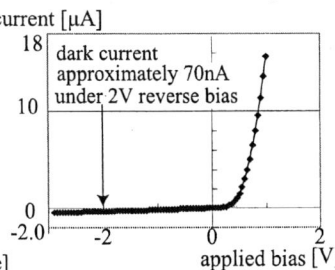

Figure 10. I-V characteristics of the p-n junction

MC5
16:45 – 17:00

Micro Knife-edge Optical Measurement Devices Fabricated by SOI and CMOS MEMS Processes

Tzu-Lin Chang, Victor Farm-Guoo Tseng, Yi Chiu

Department of Electrical and Control Engineering, National Chiao Tung University
1001 Ta Hsueh Road, Hsinchu, Taiwan 300, ROC
Tel: 03-5731838, Fax: 03-5715998, email: yichiu@mail.nctu.edu.tw

Abstract

The knife-edge method is a commonly used technique to characterize the optical profiles of laser beams or focused optical spots. In this paper, we present the design, fabrication, and test of a micro knife-edge scanner based on the micro-electromechanical-system (MEMS) technology. Silicon-on-insulator (SOI) processes are used to demonstrate the feasibility of the new device, whereas the CMOS-MEMS processes are used to enable the integration of the photo detector and on-chip signal conditioning circuitry. Focused optical spot size measured by the reflection-type SOI device is demonstrated to be c lose to the diffraction limit. Preliminary measurement results of the CMOS-MEMS device are also presented.

Keywords: knife edge, focused optical spot, diffraction limit, SOI, CMOS MEMS

1 INTRODUCTION

The knife-edge method is a commonly used technique to characterize the optical profiles of laser beams or focused optical spots with high spatial resolution [1-5]. By using MEMS technology, photo detectors can be integrated with on-chip signal conditioning circuitry. Such a micro measurement system can be used to measure tightly focused spots or near-field optical distributions, which are difficult to measure by using bulk optical systems.

A scanning knife-edge setup is schematically shown in Fig. 1(a). A sharp knife edge plate scans across the optical field distribution. The photo detector placed behind the plate detects the partial optical energy which is not blocked by the plate. When the plate scans across the distribution, the photo current $I(x)$ is measured as function of edge position x,

$$I(x) \propto \int_x^\infty P(x')dx' \qquad (1)$$

where $P(x)$ is the optical field distribution. The optical distribution can then be found from

$$P(x) \propto \frac{dI(x)}{dx} \qquad (2)$$

To implement the scanning knife edge system using MEMS technology, a comb drive actuator is used to scan knife edge plate (Fig. 1(b)). A right-angle triangular region in the movable part of the comb drive is used to scan the optical profile in two orthogonal directions. Three configurations can be implemented [6], namely the transmission type, the reflection type, and the absorption type. In this paper, a reflection-type device, where the knife-edge plate is a triangular reflective surface, is fabricated in an SOI substrate. An absorption-type device, where the knife-edge plate is a triangular photo detector, is also designed and fabricated using CMOS MEMS processes. Optical measurement using the SOI device is presented.

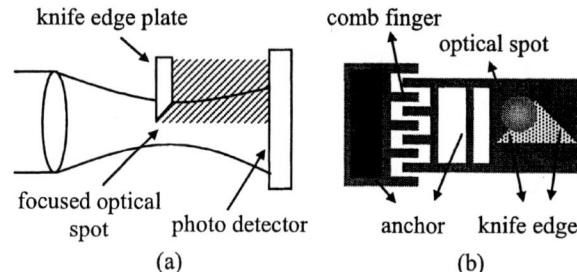

Fig. 1 (a) A scanning knife-edge setup, (b) system top view

2 DESIGN AND FABRICATION

If the maximum spot size to be measured is about 5 μm, a scanning range of ±10 μm should be achievable for reasonable applied voltage. To reduce the required voltage, the comb drive can be driven at resonance. In such a case, a static actuator displacement of 0.1 μm is needed for a typical quality factor $Q = 100$ for similar devices. SOI substrates are used in the demonstration since it is also possible to fabricate photo detectors therein. A typical folded-spring actuator is designed in the 20 μm-thick device layer to achieve the above requirement with an applied voltage of $V_{DC} = 15$ V and $V_{ACp-p} = 15$ V.

The fabrication process of the SOI device is shown in Fig. 2. It begins with a LPCVD nitride on a p-type wafer. Ion implantation of arsenic and boron is used to fabricate the photo detector (Fig. 2(a)). Then the aluminum layer is deposited and patterned for electrical connection (Fig. 2(b)). After the front silicon is etched by deep reactive ion etch (DRIE) to define the 20-μm-thick structure (Fig. 2(c)), a backside DRIE followed by HF vapor etch is used to release the structure (Fig. 2(d)).

1-4244-0641-2/07/$20.00 ©2007 IEEE   31

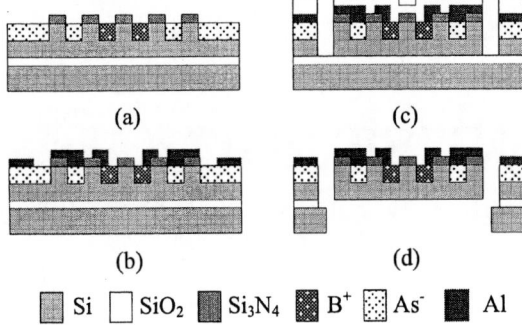

(a)   (c)

(b)   (d)

☐ Si ☐ SiO₂ ▨ Si₃N₄ ▨ B⁺ ▨ As⁻ ■ Al

Fig. 2 Fabrication processes of the SOI device

A similar absorption-type device is also designed for the 0.35-μm 4P2M TSMC CMOS MEMS process. The fingers are formed by 3 metal layers. The finger width is 4 μm and the gap spacing is 2 μm. The pn junction between the p+ region and the n well is used as the photo diode. The device resonance frequency is 12.4 kHz according to simulation. To convert the photo current into voltage signal, a trans-impedance amplifier with a gain of 400 kΩ and a 100 kHz bandwidth is integrated on the chip.

The post-processed CMOS-MEMS device is schematically shown in Fig. 3. The back side of the die is first etched by DRIE; the front side is dry-etched with M4 as the mask to define the structure. After the oxide is removed, isotropic Si etching is used to release the structure. During the processes, M4 is removed to reveal the photo diode.

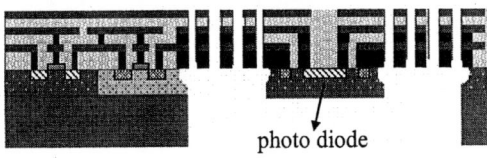

photo diode

Fig. 3 Post-processed CMOS device

3 MEASUREMENT

The reflection-type SOI device is used to measure focused optical spot size. The laser light is focused with a microscope objective lens on the triangular scanning plate of the device. The reflected light is collected by the same objective lens and detected by a remote amplified photo detector. Fig. 4(a) shows the detected knife-edge signal. The derived spot profiles in two orthogonal directions are plotted in Fig. 4(b), from which the full-width-at-half-maximum (FWHM) spot size can be calculated. Table 1 lists the measured spot size using various light sources and objective lenses. The measurement results are also compared to the theoretical diffraction-limited spot size $s = 0.5\lambda/NA$, where λ is wavelength and NA is the numerical aperture. It can be seen that the measured results are reasonably close to the theoretically values.

For CMOS-MEMS absorption-type device, the responsivity of photodiode is measured to be 0.1 A/W, while the post processing is in progress to release the structure.

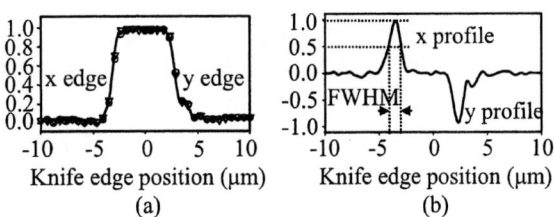

Fig. 4 (a) Measured knife edge signal, (b) derived spot profiles

Table 2 Measured and diffraction-limited focused spot size

Wave-length	Object-ive	NA	Diffraction limited spot size	Measured spot size
633 nm	20X	0.4	0.79 μm	0.90 μm
633 nm	40X	0.65	0.53 μm	0.57 μm
543 nm	20X	0.4	0.68 μm	0.82 μm
543 nm	40X	0.65	0.44 μm	0.50 μm

4 CONCLUSION

The reflection-type spot profile measurement using the SOI micro knife-edge scanner is demonstrated. The measured results are close to the diffraction limit and show a high resolution in the nanometer scale. The absorption-type device fabricated by the CMOS-MEMS process is currently being post-processed.

REFERENCES

[1] J. M. Khosrofian and B.A. Garetz, "Measurement of a Gaussian laser beam diameter through the direct inversion of knife-edge data," Appl. Opt., 22, 3406-3410 (1983).

[2] A. H. Firester, M. E. Heller, and P. Sheng, "Knife-edge scanning measurements of subwavelength focused light beams," Appl. Opt., 16, 1971-1974 (1976).

[3] F. Zamkotsian and K. Dohlen, "Surface characterization of micro-optical components by Foucault's knife-edge method: the case of a micromirror array," Appl. Opt., 38, 6532-6539 (1999).

[4] D. Karabacak, T. Kouha, C. C. Huang, and K. L. Ekinci, "Optical knife-edge technique for nanomechanical displacement detection," Appl. Phys. Lett., 88, 193122, 1-3 (2006).

[5] J. Murakowski, M. Cywiak, B. Rosner, and D. van der Weide, "Far field optical imaging with subwavelength resolution," Opt. Comm., 185, 295-303 (2000).

[6] Y. Chiu and J-H Pan, "Micro knife-edge optical measurement device in a silicon-on-insulator substrate," Opt. Express, 15, 6367-6373 (2007).

TUESDAY, 14 AUGUST 2007

TuA Bio Nano Devices

TuB Actuation

TuP Poster Session

TuA1 (Invited)
08:30 – 09:00

Linear Tactile Nanodevice with Resolution on Par with Human Finger

Ravi F. Saraf*, Vivek Maheshwari, Chieu Nguyen
Department of Chemical and Biochemical Engineering
University of Nebraska – Lincoln, Lincoln, NE 68516, USA
Tel 402 472 8284, FAX 402 472 6989, E-mail rsaraf@unlnotes.unl.edu

Abstract

A large area thin-film nanodevice made by self-assembly containing electroluminescent nanoparticles is reported. The ~100 nm thick device on application of potential across the top and bottom surface of the film converts local pressure to light. The intensity of the electroluminescent light is linearly proportional to the applied local compressive stress. By imaging the light, the stress distribution over the area of contact is obtained at resolution on par with human finger.

Keywords: Tactile sensor, Electron Tunneling, Thin film device, Nanoparticle self-assembly, Robotics

1. INTRODUCTION

Recently, Bill Gates speculated that after the Personal Computer, a humanoid robot in every home would be the next revolution that would affect our lives intrinsically (1). Excluding taste, among the four basic senses, touch at resolution on par with human finger remains a challenge. Without a high resolution and sensitivity large area tactile device, also sometimes referred to as electronic skin, the humaonoid robot may find it difficult to pour and deliver a glass of water to its master (2), or fold the laundry. The current tactile devices with active area of over 1 cm² have a resolution of ~1-2 mm (3) in contrast to human finger that has a resolution of ~40 µm (4) allowing us to feel a filament of hair on a smooth surface that has a typical diameter of ~100 µm. Recently, by self-assembling layers of nanoparticles by simple solution deposition process invented by Decher (5) we fabricated a large area device of thickness ~100 nm to convert compressive stress distribution or pressure on physical contact to light (6). By imaging the distribution of emitted light on a digital camera a "stress image" can be obtained that resolves the distribution of contact-pressure at a resolution of ~20 µm (6). Apart form the resolution on par with human sensitivity at contact pressures of 10-60 KPa that are comparable to the human touch, the approach has two practical advantages: (a) The fabrication process that is essentially sequential dip-coating followed by washing and drying will allow deposition of thin film device on large surface of area comparable to displays. (b) The response of the device, i.e., the intensity of light, is linearly proportional to applied stress. The linear relationship makes the grey-scale of the stress-image quantitatively comparable to the contact-pressure

distribution. In this report, we present our observations that indicate why an electron tunneling based device that is intrinsically non-linear leads to highly linear response.

2. DEVICE CHARACTERISTCS

The ~100 nm thick nanodevice is a film (Fig. 1) consisting of alternating layers of Au (10 nm diameter) and CdS (3 nm diameter) nanoparticles separated by dielectric layers (DL), composed of stacked alternating layers of poly(styrene sulfonate) (PSS) and poly(allylamine hydrochloride) (PAH) (6). The 2.5 by 2.5 cm device is made by layer-by-layer self-assembly process (5) on a sputter deposited ~400 nm thick transparent conductive electrode coating of indium-tin-oxide (ITO) on glass. Finally, a flexible electrode (Au on polymer film) is placed on the top of the multilayered to make a contact. The particular device described here has three Au layers and two CdS layers, with 4 layers each of PAH and PSS as the inter-lying DL.

The film is insulating in the in-plane direction (6). In the vertical direction, the film is conducting due to tunneling between the Au and CdS layers. The current density, J, through the film as a function of bias, V, between the ITO and Au electrode, under a uniform pressure (i.e., compressive stress) σ, is shown in Fig. 2(a). The uniform pressure is applied by placing an optically flat quartz disk on the flexible Au electrode. The nonlinear J-V curve is fit based on a model combining the field assisted electron tunneling current (7) through the nanoparticles and the ionic (leakage) current due to ions in the polyelectrolyte. The total current (solid curve in Fig. 2(a)) is therefore modeled as,

$$J = J_T + J_I = P \exp\left(-\frac{aK}{V}\right) + \frac{V}{R} \qquad (1)$$

1-4244-0641-2/07/$20.00 ©2007 IEEE

In the first term due to tunneling, "a" is the (vertical) inter-particle distance, "K" is a critical field for activated tunneling that depends on the work functions of the particles, and "P" at constant temperature is proportional to V^2 and the number density of carriers for conduction(7). The second term due to the mobile ions (H^+, Na^+ and OH^-, Cl^-) in the polyelectrolytes, "R", is the ohmic resistance that is proportional to the distance between the electrodes. The fits are performed as a two step process: (i) At low V, the current is dominated by ionic current. Thus, the slope of the linear region at low-V is 1/R. The estimated resistivity of ~ 1.6×10^8 Ω-m is reasonable for ion conductivity in PSS and PAH under ambient humidity. (ii) Thus, J-(V/R) is J_T that is fit by a single exponential to obtain (Ka) and P. Consistent with the theory on field-assisted tunneling, P~V^2 (Fig.2(b)). Assuming 'a' (at σ = 0) ~ 5 nm, the critical field K of ~ 10^9 V/m is reasonable. Because, electroluminescence will occur only when tunneling through CdS occurs, electroluminescence intensity, I_{EL}, should be measurable above the bias of ~8±1V when the total J begins to increase beyond the straight line (i.e., J_I) (Fig.2(a)). Fig. 2(c) is consistent with electroluminescence initiation at ~8±1V.

As the electroluminescence intensity from the device is virtually continuous (because inter-particle spacing is well below 25 nm), the resolution of the stress image is determined by the optics and the CCD camera and the "smearing" of the stress due to finite thickness of the plastic backing for the Au-electrode. Using the current device the stress distribution image due to the embossing of "Ashok Chinha" on 5 Rupee coin of India is clearly visible by pressing the coin at a reasonable load (Fig. 2(d))(6). For the CCD with a 512x512 array of ~16 μm pixels, the I_{EL} is well over noise level to achieve lateral spatial resolution of at least ~20 μm and the height modulation of ~10 μm is

measurable, indicating the device is fairly deformable (6). The linearity of the device is shown in Fig. 3(a). Both J and I_{EL} increases linearly with increasing magnitude of the compressive stress. The linearity of the device is due to the dominance of 'P' for modulation of tunneling current (Fig. 3(b)). Although, the tunneling process has an exponential dependence on 'a' that decreases with increasing σ (eq. (1)), the linear change in 'P' with load (=σ) dominates the modulation of tunneling process (i.e., current). Physically, 'P' is proportional to electron density of conduction electrons. At fixed temperature, the electron conduction density is constant. Thus, the increase in 'P' is proportional to the increase in number of percolation channels between the electrodes - as load increases, the number density of percolating channels of nanoparticles traversing through the thickness increases linearly causing a linear increase in 'P'.

3. SUMMARY AND CONCLUSION

We have briefly discussed the characteristics of a thin film tactile device based on electron tunneling phenomena that converts applied stress to electroluminescent light. By focusing the electroluminescent light on a camera a high resolution stress-image is obtained. The tunneling current is modulated by altering the percolation behavior of the nanoparticle clusters leading to linear response.

4. REFERENCES

1. B. Gates, *Scientific American*, p. 58, January 2007.
2. R. Crowder, Science, **312**, 1478 (2006).
3. T. Someya *et al.*, *Proc. Nat. Acad. Sci.* **101**, 9966 (2004).
4. J. W. Morley, et al., *Expt. Brain Research* **49**, 291 (1983).
5. G. Decher, *Science* **277**, 1232 (1997).
6. V. Maheshwari, R.F. Saraf, *Science*, **312**, 1501 (2006).
7. R. H. Fowler, et al., *Proc. Royal Soc. A.* **119**, 173 (1928).

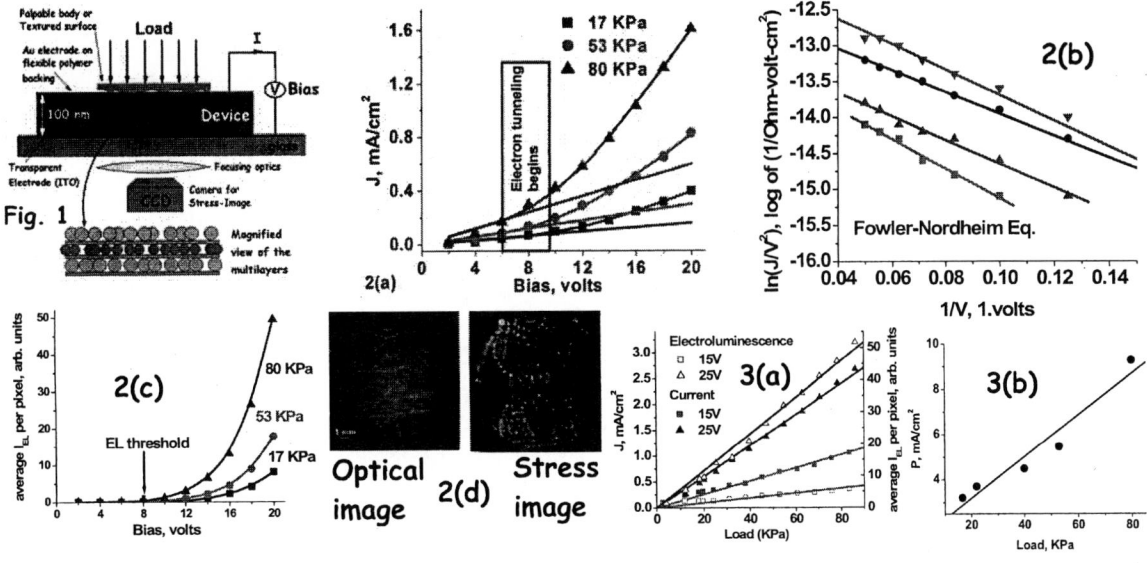

TuA2
09:00 – 09:15

Optical NEMS Based Force Sensor Using Silicon Nanophotonics

Chengkuo Lee[1,2,*], Rohit Radhakrishnan[1], Chii-Chang Chen[3], Jing Li[2] and Narayanan Balasubramanian[2]

1. Department of Electrical and Computer Engineering, National University of Singapore, Singapore 117576, Republic of Singapore

2. Institute of Microelectronics, A*Star, Singapore 117685, Republic of Singapore

3. Department of Optics and Photonics, National Central University, Jhong-Li, 320, Taiwan
E-mail elelc@nus.edu.sg; leeck@ime.a-star.edu.sg

Abstract

A line defect in a silicon two-dimensional (2-D) photonic crystal (PhC) is created as a waveguide for light propagation via the PhC. By introducing micro-cavities within the line defect so as to form the resonant band gap structure for PhC, we demonstrate a PhC waveguide (PhCWG) filter with clear resonant peak in output wavelength spectrum. We conceptualized a novel nanomechanical beam structure embedded with this PhCWG filter, i.e., a NEMS (Nanoelectromechanical system) based force sensor. Since the output resonant wavelength is sensitive to the shape of air holes and defect length of the micro-cavity. Shift of the output resonant wavelength is correlated with beam deformation or force loading for this free-standing PhCWG beam. Simply speaking, the induced strain modifies the shape of air holes and the spacing among them for micro-cavities along the silicon waveguide of PhCWG. For a silicon PhCWG beam structure with dimension of 340nm(thickness) x 5µm(width) x 20µm(length), the measurable vertical deformation of 20~25 nm at the center and detectable strain of defect length of 0.004% is derived according to simulation results.

Keywords: Optical NEMS, Nanophotonics, Sensor, Strain, Force

1 INTRODUCTION

Recently O. Levy et al. have proposed a novel displacement sensor comprising two planar photonic crystal waveguides (PhCWG) aligned along the same axis of light propagation [1]. They provided simulation results to prove this device concept. In this paper, the output light intensity is strongly dependent on the alignment accuracy, i.e., the coupling efficiency between input and output PhCWGs. Any deformation of structure will lead to misalignment so as to reduce the output light intensity. Later in 2005, O. L. J. Pursiainen et al., created a flexible three-dimnesional (3-D) PhC by using a self-assembly fabrication process. This flexible 3-D PhC contains multi-shelled polymer spheres of high-refractive-index and absorbing materials filled in the interstitial space surrounding said spheres. The dimension of this device is a 1-cm wide strip film. This device was placed on top of a sample holder and was uniformly stretched in micrometer scale while the optical property is measured in situ. It exhbited 50% reduction of transmission intensity at only 1% strain regarding to about 5nm wavelength shift of the resonant peak in the reflection spectrum [2]. However, good discrimination of resonant wavelength shift regarding to 1% strain may not be easily due to very low quality factor of the resonant wavelength peak.I. EI-Kady et al., proposed a new device concept of detecting submicron crack of substrates based on using a 3-D PhC structure. It reported that a 3-D PhC structure was attached on a polymer substrate to form a PhC sensor. According to the simulated results derived by using the FDTD (finite difference time domain) approach, this PhC sensor experienced changes in its band gap profile when micro-damage is induced in said substrate [3].

Interestingly either the intensity reduction of output peak or resonant wavelength shift has been deployed as the sensing scheme for PhC based physical sensors, like displacement and strain sensors. Thus we measure the change in optical signals and correlate such changes with physical parameters, like, displacements, strains and forces. For instance, 1% strain represents about 1nm~5nm deformation in hole diameter for silicon 2-D PhC devices in present study. It implies silicon PhC based physical sensors should exhibit outstanding performance intrinsically. In this paper, we proposed a free-standing beam structure comprising PhCWG as force and strain sensors.

2 MODELING METHOD

Sandoghdar et al. has proposed a PhCWG based filter structure comprising four air holes. It reported that strong optical resonant signals have been observed at the center of these four holes by using SNOM (scanning near-field optical microscopy). The peak intensity of resonance shown at 3.84µm has been observed [4, 5]. In current study, we revised such design and proposed a nanomechanical beam structure embedded with this PhCWG based filter, i.e., four holes.

1-4244-0641-2/07/$20.00 ©2007 IEEE

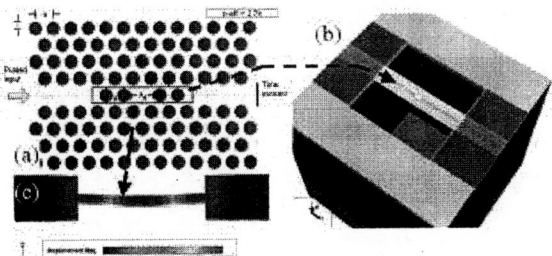

Fig. 1. (a) Schematic drawing of PhCWG filter comprising micro-cavity, i.e., four holes, on the line defect, i.e., silicon waveguide; (b) Nanomechanical suspended beam comprising PhCWG filter, (c) The deformed PhCWG beam under force load.

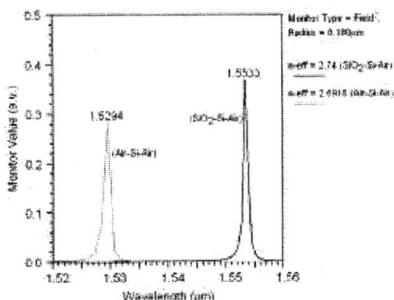

Fig. 2. The resonant wavelength peaks regarding to effective refractive indexes of typical SOI based PhCWG (1.5533μm) and suspended PhCWG beam (1.5294μm).

As shown in Fig. 1, the PhCWG is formed by an array of air holes in a silicon suspended beam with a hexagonal lattice constant of a = 500 nm, and the radius of all holes of r = 180 nm. A point defect is defined as a micro-cavity formed between the 2-holes pairs of reflectors with a defect length of A_d = 640 nm. The 2-D finite-difference time-domain (FDTD) method is performed to simulate the propagation of the electromagnetic waves in the waveguides. According a way proposed by K. Kawano and T. Kitoh [6], we derived the effective refractive index for the air holes and silicon portion of said suspended beam in Fig. 1 (b). The simulated results point out the resonant wavelength peak shifts to lower wavelength region (eg. 1.5294μm in current configuration) for suspended PhCWG beam device from the original peak (eg. 1.5533μm in the same layout) for the conventional configuration of a PhCWG filter made of silicon device layer from a SOI (silicon-on-insulator) substrate. We deploy effective refractive index (n-eff) of 2.6918 throughout this study regarding to air-Si-air configuration with silicon device layer thickness of 340nm. In the simulation of deformation of beam with these holes under force loads, it is conducted by using the CoventorWare FEM (Finite element method) modeling tool [7]. A typical simulated result is shown in Fig. 1 (c).

3 RESULTS AND DISCUSSION

Within the range of 0.7 to 1.7 μN loading force, rather linear behavior observed for the shift of resonant wavelength

regarding to the applied force (Fig. 3). For force loads between 2 uN and 4 uN, it is seen that the relationship is best described by a second order equation. A polynomial regression line of degree two fits well with the data points. In terms of the vertical deflection under the system measurement limitation of 0.1 nm resonant wavelength shift, the minimum detectable displacement is 20-25 nm for a beam of 340nm(thickness) x 5μm(width) x 20μm(length). We defined the strain in PhCWG as the percentage change in the defect length. Fig. 4 shows the relationship between the shift in resonant wavelength and the absolute value of strain. Although another polynomial regression line of degree two fits the overall data perfectly, it is also observed that the change in resonant wavelength fits in linear behavior for strain within 0.03%. Again the detectable smallest strain is derived as 0.004% regarding to a force load of 0.25 μN. It is a significant improvement with three orders of magnitude than previous strain sensing data reported by Ref. [2].

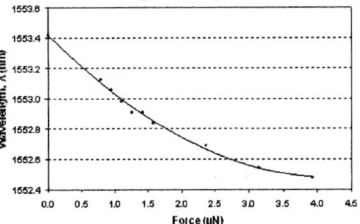

Fig. 3. Applied force versus resonant wavelength shift of deformed PhCWG beam.

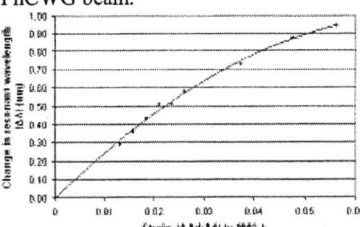

Fig. 4. Induced strain of defect length A_d versus resonant wavelength shift of deformed PhCWG beam.

REFERENCES

1. O. Levy, B. Z. Steinberg, N. Nathan and A.Boag, " Ultrasensitive displacement sensing using photonic crystal waveguides," Appl. Phys. Lett. 86, 104102 (2005).
2. O. L. J. Pursiainen, J. J. Baumberg, K. Ryan, J. Bauer, H. Winkler, B. Viel and T. Ruhl, "Compact strain-sensitive flexible photonic crystals for sensors," Appl. Phys. Lett. 87, 101902 (2005).
3. I. El-Kady, M. M. Reda Taha and M. F. Su, "Application of photonic crystals in submicron damage detection and quantification," Appl. Phys. Lett. 88, 253109 (2006).
4. P. Kramper, A. Birner, M. Agio, C. M. Soukoulis, F. Müller, U. Gösele, J. Mlynek and V. Sandoghdar, "Direct spectroscopy of a deep two-dimensional photonic crystal microresonator," Phys. Rev. B 64, 233102-1 (2001).
5. P. Kramper, M. Kafesaki, C. M. Soukoulis, A. Birner, F. Müller, R. Wehrspohn, U. Gösele, J. Mlynek and V. Sandoghdar, "Near-field visualization of light confinement in a photonic crystal microresonator," Opt. Lett., 29, 174–176 (2004).
6. K. Kawano and T. Kitoh, " Introduction to optical waveguide analysis: solving maxwell's equations and the schrödinger equation," John Wiley & Sons, Inc., ISBNs: 0-471-40634-1, (2001).
7. http://www.coventor.com/coventorware

TuA3
09:15 – 09:30

Photostable Single KTiOPO$_4$ Nanocrystals for Second-Harmonic Generation Microscopy

L. Le Xuan, C. Zhou, A. Slablab, D. Chauvat, N. Sandeau, S. Brasselet , J.-F. Roch

C. Tard*, S. Perruchas*, T. Gacoin*, P. Villeval**

Laboratoire de Photonique Quantique et Moléculaire, ENS Cachan, UMR CNRS 8537,

61, av. du Président Wilson, 94235 Cachan Cedex, France

*Laboratoire de Physique de la Matière Condensée, Ecole Polytechnique, UMR CNRS 7643,
91128 Palaiseau Cedex, France
**Cristal Laser S.A, 32, rue Robert Schumann, F-54850 Messein, France
Correspondent email: xuan-loc.le@m4x.org

Abstract

The finding of nonlinear nanometric-sized probes is of key importance for the development of nonlinear microscopy in nanosciences and biology. We isolate nonlinear KTiOPO$_4$ nanocrystals with remarkable photostability in second-harmonic generation under femtosecond infrared laser light excitation. Their size distribution is determined using dynamic light scattering and atomic force microscopy. With both polarization analysis and defocused imaging of the emitted second-harmonic field, we also extract the Euler angles of the crystalline axes of a single nanocrystal. These sub-wavelength particles can find application as near-field vectorial probes.

Keywords: nanocrystal, second harmonic generation, nonlinear microscopy

1 INTRODUCTION

The recent development of non-linear second-harmonic generation (SHG) microscopy has lead to the study of single nanoobjects like ZnO [1], GaAs [2], and CMONS [3]. Optical materials such as KTiOPO$_4$ (KTP) have been optimized for growing large efficient non-linear crystals with high damage threshold, and are widely used as frequency doublers or converters in laser systems. Here we isolate single KTP nanocrystals, and determine their complete properties in SHG microscopy.

2 SAMPLE PREPARATION AND MEASUREMENT

We obtain KTP nanocrystals by size-selection from a

powder of KTP, which remains at the end of the flux-growth process for large crystal. For that selection, the powder is diluted in isopropanol and mixed with a polymer (polyvynilpyrolidone, PVP). The solution is then centrifuged resulting in a colloidal solution which contains dispersed and size-selected KTP nanoparticles. Dynamic light scattering (DLS) analysis of these solutions show size distributions with a mean size ranging from 150 nm to a few tens of nanometers depending on centrifugation time and speed (Figure 1).

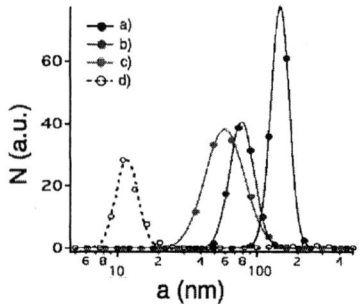

Figure 1. DLS spectra measured after successive centrifuged solutions. Mean size of (a) 150 nm, (b) 80 nm, (c) 60 nm are obtained. (d) is close to the detection sensitivity, the particle sizes being in fact (30±10) nm (cf. Figure 4).

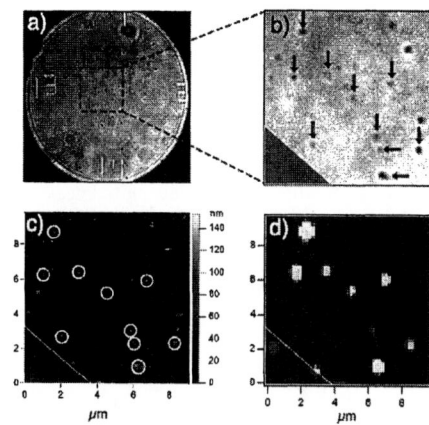

Figure 2. Observations of nanoparticles by different methods. a) Overview in white-light microscopy, b) 10×10 μm^2 zoom of a), c) corresponding AFM image, with circles pinpointing the SHG emitters, and d) corresponding SHG raster scan image.

1-4244-0641-2/07/$20.00 ©2007 IEEE   39

The sample is then prepared so that a single nanoparticle can be unambiguously isolated. A chosen colloidal solution is spin-coated on a plasma-cleaned glass coverslip, which results in a uniform 100-nm thin polymer coating embedding the nanocrystals. A transmission electron microscope copper grid with $25 \times 25 \ \mu m^2$ empty squared holes is deposited onto the polymer film, and the sample is etched for 15 minutes under oxygen plasma. The polymer is thus removed from the empty squares, leaving the KTP nanocrystals uncovered on the glass surface. We measure their height with an atomic force microscope (AFM). An SHG image which reveals nonlinear nano-emitters is obtained using a inverted microscope with high numerical aperture objective (x100, NA =1.4) and femtosecond pulse illumination (100 fs, 987 nm, 1-15 mW). Nearly all these emitters can be unambiguously found in the AFM image (Figure 2). Some objects revealed in the AFM image do not correspond to an SHG emission spot. We attribute them to a polymer residue, or a KTP polycrystalline nanoparticle.

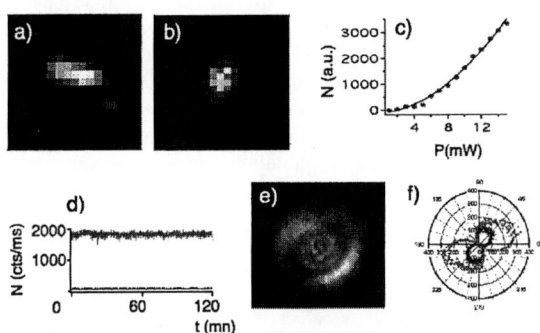

Figure 3. Investigation of a single KTP nanocrystal. a) AFM image of a nanocrystal: height = 59 nm, b) SHG scanning image (FWHM = 350 nm), (c) check of second-order nonlinearity, quadratic data fit, d) photostability of the x- and y-polarized emitted field components, (e) SHG defocused image and (f) polar graph of x- and y-polarized emitted field components, as a function of the excitation polarization angle.

We can now fully investigate a single nanocrystal. Figure 3a) shows a zoomed AFM image on one particle selected among those observed in figure 2, which measured height is 59 nm. The corresponding SHG image is well contrasted, with a signal-to-background ratio of 120. The number of photons emitted is proportional to the square of excitation power as expected (Figure 3c). Most remarkably, figure 3d) shows the very high photostability of the emitter, since perfectly constant signals are detected for both polarizations for more than 120 minutes, at a mean excitation power of 15 mW focused on a 350 nm diameter surface. We attribute this feature to the non-resonant character of the SHG emission in our experimental conditions. Finally, the emission pattern of this sub-wavelength emitter is accessed by a defocused imaging method [4] adapted to the nonlinear situation (Figure 3e). Combined with a polarization analysis technique [5] (Figure 3f), this method allows us to determine the three Euler angles that describe the nanocrystals

orientation, $(\theta, \varphi, \psi) = (85°\pm5° , 35°\pm5° , 70°\pm15°)$ for the described example. The absolute direction of the non centro-symmetric emitter could also be obtained using a phase-sensitive detection scheme that we have recently developed [6].

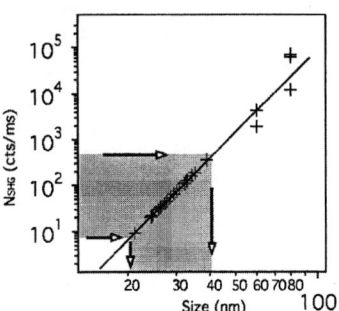

Figure 4. Size estimation of smallest KTP nanocrystals. Black crosses: signal levels of nanocrystals from solutions (b) and (c), line slope equal to 6, red crosses: signal level of smallest KTP nanocrystals in solution (d), which correspond to a size of (30±10) nm

The solution labeled (d) in figure 1 contains the smallest KTP nanoparticles. However, their sizes are not satisfactorily estimated by DLS due to a very weak light scattering signal. For size estimation, we measured the emitted SHG signal in the same conditions as for solutions labeled (b) and (c) for which the average sizes are known. Since the SHG process is coherent, the intensity emitted by a nanocrystal is proportional to the nanocrystal mean size to the power of six. Using this dependency, we can deduce the mean size of particles in solution (d), which is (30±10) nm. (Figure 4)

3 CONCLUSION

In summary, we have isolated photostable KTP nanocrystals. A full investigation of a single KTP nanocrystal is performed leading to its size, its orientation and its SHG efficiency. Nanoparticle sizes as small as 30 nm are obtained. We plan to use these nanocrystals as vectorial probes of optical near-fields, in association with an AFM tip.

We thank F. Treussart for helpful discussions. We are grateful to J. Lautru and A. Brosseau for samples preparations and AFM measurements.

REFERENCES

[1] J. C. Johnson *et al*, *Nano Lett.* **2**, 279 (2002).
[2] J. P. Long *et al*, *Nano Lett.* **7**, 831 (2007).
[3] F. Treussart *et al*, *ChemPhysChem.* **4**, 757 (2003).
[4] M. Böhmer *et al*, *J. Opt. Soc. Am. B* **20**, 554 (2003).
[5] S. Brasselet *et al*, *Phys. Rev. Lett.* **92**, 207401 (2004).
[6] L. Le Xuan *et al*, *Appl. Phys. Lett.* **89**, 121118 (2006).

TuA4
09:30 – 09:45

Iridescent Photonic Nano-Silica for Chemical and Biological Sensing

J. Y. Chyan[1*], and J. Andrew Yeh[1,2]
1 Institute of MicroElectroMechanical Systems
2 Institute of Electronics Engineering
National Tsing Hua University, Hsinchu, Taiwan
*Phone: 886-3-5715131 ext. 33730, E-mail: jychyan@gmail.com

Abstract

Iridescent photonic nano-silica is proposed for chemical and biological sensing. The iridescent photonic nano-silica consists of periodically low and high refractive index nano-silica layers. Its optical characteristics can be controlled by electrochemical etching and thermal oxidation. The validity and the sensitivity of the proposed iridescent photonic nano-silica are demonstrated for organic solution and proteins at 20pg/ml concentrations.

Keywords: photonic, nano-silica, chemical, biological, sensing, electrochemical

1 INTRODUCTION

In nature, structural color can be seen in many creatures such as butterflies and birds. It results from that reflected light experiences Bragg diffraction when it goes through discrete multilayers which consist of organic, inorganic, or composite materials. The discrete multilayers can be considered as layers of alternatively high and low refractive index that lead to optical interference. In order to fabricate artificial discrete multilayers, porous silicon is an attractive candidate. Because the thickness, pore size, porosity of a given layer is controlled by the current density, duration of etch cycle, and etchant solution composition. Porous silicon with a varying porosity gradient provides sharp features in optical reflectivity spectrum. The optical reflectivity spectrum of porous silicon changes with exposing to vapors or liquids. The substitution of air in the pores by the vapors or the liquids causes a repeatable and reversible change in the reflectivity spectrum, owing to an increase in the average refractive index of the layers. The change in reflectivity spectrum makes porous silicon to provide a very sensitive transduction for sensing of vapors [1], proteins [2], DNA, and other molecules that can enter the pores [3]. However, porous silicon is limited by its chemical and mechanical stability. Moreover, porous silicon absorbs visible light energy (1.2eV-1.38eV) significantly, so the reflective visible light decays dramatically when it goes through porous silicon layers. In this paper, iridescent photonic nano-silica (IPNS) is proposed to overcome the drawbacks of porous silicon. Furthermore, the optical properties of the IPNS can be characterized in both reflectance and transmittance which is impossible for porous silicon at visible wavelengths.

2 FABRICATIONS

Porous silicon is first electrochemically etched into a 6" single crystalline silicon substrate (*p*-type, (100), 15mΩ). The etching solution consists of a 7:3 by volume mixture of absolute ethanol and aqueous 49% HF. Etching is carried out in a Teflon cell with a cycling pump. A current density varying between 20 and 65mA/cm^2 is applied for 40 cycles with a periodicity of 30s and 8s, respectively. The freshly etched porous silicon is then placed in an oven in air at 900°C for 3 hours. This step converts the porous silicon into the IPNS completely. For convenience, the IPNS is removed from the silicon substrate and tailored to be centimeter-sized freestanding chips by megasonic treatment (Fig. 1).

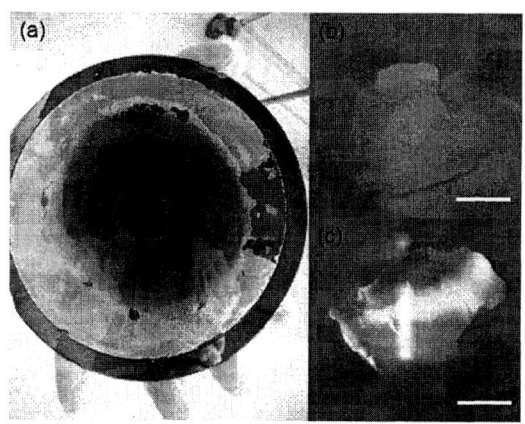

Figure 1: (a) Freestanding IPNS films on the side of 6" silicon substrate. (b) and (c) The same IPNS chip is viewed by different angle. The color change performs the important characteristic of photonic crystals. Scale bar: 1cm

In order to immobilize (3-aminopropyl)trimethoxysilane (APTMS) monolayer on the IPNS chip, the IPNS chip was immersed into an APTMS (1%) ethanol solution for 30min, and then rinsed by pure ethanol. The APTMS-immobilized IPNS chip was further placed in a 120 °C oven for 30min to complete Si–O bond formation. For a-myc(9E10) protein immobilization, the APTMS-immobilized IPNS chip was placed in a PBS solution (pH=7.4) containing 20pg/ml a-myc(9E10), incubated for 12 hours at 4°C.

1-4244-0641-2/07/$20.00 ©2007 IEEE   41

3 RESULTS AND DISSCUSSION

The morphological characterizations of the IPNS chip were performed on a Carl Zeizz Ultra 55 field emission scanning electron microscope (FESEM) at a 5-kV accelerating voltage (see Fig. 2).The IPNS chip consists of low porosity (high refractive index) and high porosity (low refractive index) layers. The thickness of the low porosity layers and the high porosity layers is 307nm and 448nm, respectively. The total thickness of the IPNS chip is 28μm.

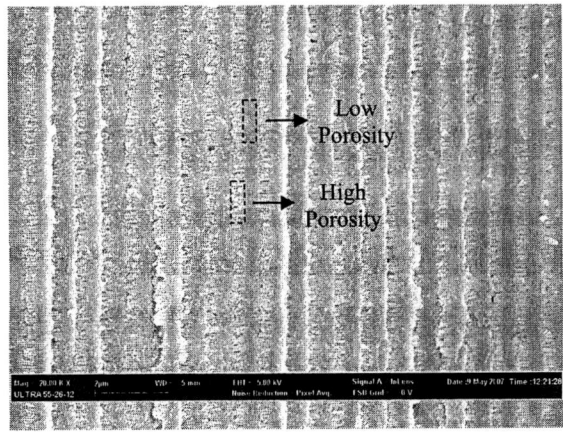

Figure 2: This figure shows the structure of the IPNS chip. It consists of low porosity and high porosity layers. The physical thickness of the low porosity layers and the high porosity layers is 307nm and 448nm, respectively.

The center of the wavelength at which strong reflection occurs in the IPNS film is given by the Bragg condition:

$$\lambda_B = \frac{2}{m}[d_L(n_L{}^2 - \sin^2\theta)^{\frac{1}{2}} + d_H(n_H{}^2 - \sin^2\theta)^{\frac{1}{2}}]$$

where n_L and n_H are low and high refractive indices, d_L and d_H are the correspondent layers thickness. θ is the incident angle, and m is an integer (the order of Bragg condition, m=2 and 3 in this study). The optical characteristics is the presence of a high reflectivity stop band with a peak of transmittance at λ_B. When molecules immobilized on the surface of the IPNS chips, both low and high refractive indices increase, resulting in a shift toward longer wavelengths of its peak position. Figure 3 and 4 show the measured transmission spectra of the IPNS chips by an inverted optical microscope (Olympus IX71) coupled through an optical fiber (Ocean Optics QP1000-2-UV/VIS, N.A. = 0.22) to a UV-VIS CCD spectrometer (BWTek, BRC111A). All spectra were obtained using normal incident and unpolarized light. Figure 3 shows the comparison between the IPNS chips immersed in the air or in the acetone. Figure 4 shows the APTMS-immobilized IPNS

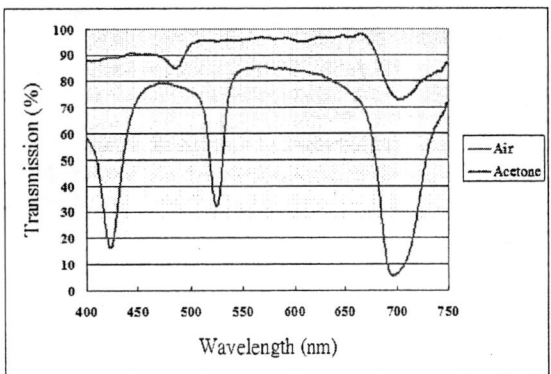

Figure 3: The transmission spectra comparison of the IPNS chip immersed in air or in acetone.

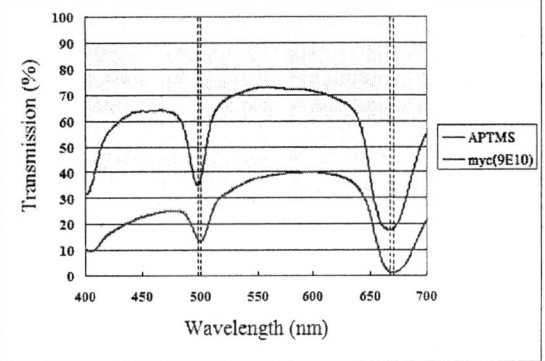

Figure 4: The transmission spectra comparison of the APTMS-immobilized IPNS chip with and without a-myc(9E10) molecule immobilization.

chip conjugated with a-myc(9E10) at 20pg/ml concentration. The peaks shift from 667nm to 670nm for the second order Bragg condition and 497nm to 500nm for the third order Bragg condition.

4 ACKNOWLEDGEMENTS

The authors would like to thank Sino-American Silicon Products Inc. for its grateful financial support and Prof. T. L. Shen for his helpful consultation in biology.

5 REFERENCES

[1] P. A Snow, E. K Squire. P. S. J Russell, and L. T. Canham, "Vapor sensing using the optical properties of porous silicon Bragg mirrors" *J. Appl. Phys.* 86, pp.1781-1784, 1999

[2] B. E Collins, K. P. Dancil, G Abbi, and M. Sailor, "Determining Protein Size Using an Electrochemically Machined Pore Gradient in Silicon"*J. Adv. Funct. Mat.* 12, pp.187-191, 2002

[3] L. D. Stefano, . Rendina, L. Moretti, S. Tundo, and A. M. Rossi, "Smart optical sensors for chemical substances based on porous silicon technology," *App. Optics*, 43, 1, pp. 167-172, 2004

TuA5
09:45 – 10:00

A Novel Single-Cell Surgery Tool Using Photothermal Effects of Metal Nanoparticles

Ting-Hsiang Wu[1], Pao-Yi Tseng[2], Sheraz Kalim[3], Michael Teitell[3] and Pei-Yu Chiou[2]

[1]Department of Electrical Engineering, University of California, Los Angeles (UCLA)

[2]Department of Mechanical and Aerospace Engineering, UCLA

48-121 ENG. IV, 420 Westwood Plaza, Los Angeles, CA 90095-1597, USA

[3]Departments of Pathology and Pediatrics, UCLA

675 Charles E. Young Dr. South, MRL 4-762, Los Angeles, CA 90095-1732, USA

Tel +1-310-825-9091, Fax +1-310-267-0382, E-mail tsw2008@ucla.edu

Abstract

We demonstrate a novel single-cell surgery tool that integrates photothermal effects of gold nanoparticles with microcapillary techniques. A transient hole opening of the cell membrane at the tip of the micropipette was accomplished using laser-induced localized heating of nanoparticles. A control experiment using a conventional glass pipette of the same size without gold nanoparticles is also performed to exclude the effect of direct laser heating. This device has the potential to enable minimum cell damage during operation and provides a direct and convenient access to the cell interior without exerting large mechanical stress to fragile cells.

Keywords: laser, photothermal, gold nanoparticle, cell surgery

INTRODUCTION

Laser-induced photothermal effects of metal nanoparticles have been shown to kill cancer cells in recent studies [1,2]. These methods involved fixing antibodies to gold nanoparticles that target the surface of cancer cells. The gold nanoparticles acted as light absorbing agents and created localized heating upon laser excitation. When excited with a pulsed laser above a certain threshold, transient vapor bubbles from superheating formed around the nanoparticles, which in turn generated extensive damage or holes in the cell membrane, killing the cells.

Here we propose a novel cell surgery device that integrates nanoparticle photothermal effects with microcapillary techniques. Proof-of-concept experiment results are presented here. The conventional microcapillary technique is a versatile tool for performing single cell recording and manipulations. However, it introduces enormous stress to the cell as the microcapillary punctures through the cell membrane. As a result, this procedure often results in cell death, particularly on small or mechanically fragile cells. Current cell surgery methods using laser ablation [3] eliminate the mechanical stress but they require tightly focused light and precise positioning of the injection micropipette at the laser focal spot. The cell surgery device we describe here utilizes photothermal effects of nanoparticles on the tip of a microcapillary pipette. Laser-induced heating of the nanoparticles creates transient holes in the cell membrane as the pipette encounters the cell.

Since the cell damage only occurs at the membrane area in contact with nanoparticles, this device can operate with non- or lightly-focused laser. This way unwanted stress is minimized and possible chemical effects due to strong laser intensity are avoided to ensure the biology of the manipulated cells under study is unaffected.

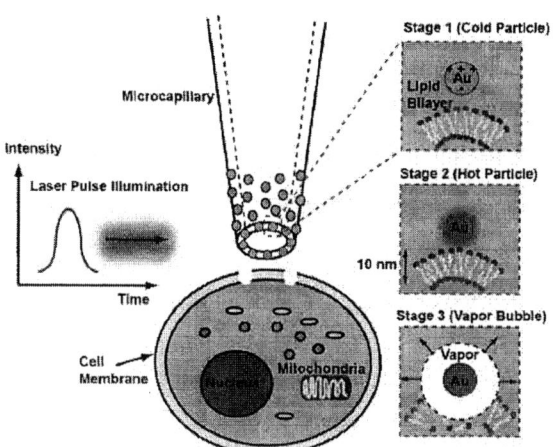

Fig. 1. Schematic of the principle of the photothermal cell surgery tool.

PRINCIPLE AND DEVICE STRUCTURE

Fig. 1 shows a schematic of the cell surgery tool. Gold

1-4244-0641-2/07/$20.00 ©2007 IEEE

nanoparticles are coated onto the tip of a micropipette. Noble metal nanoparticles strongly absorb electromagnetic waves with a frequency close to its surface plasmon frequency, usually in the visible and NIR range for gold nanoparticles [4]. Upon laser pulse excitation, the nanoparticles rapidly heat up due to the absorbed energy, causing superheating and explosive nanoscale bubbles around these particles. This direct heating or cavitation effect from the collapsing vapor bubbles lead to increase in cell membrane permeability or "holes punching" in the membrane. The damage volume can be controlled by laser pulse fluence and the size of nanoparticles. It has been shown that the heated volume extends tens of nanometers from the surface of a 30 nm gold nanosphere by pulsing a 20 nanosecond laser at 0.5 J/cm^2 pulse energy, and the nanoparticle cools down to equilibrium temperature within few nanoseconds after laser pulsing [1,5]. As a result, the rest of the membrane or cell does not have sufficient time to respond and remains mechanically undisturbed. This way a micropipette can penetrate the cell membrane with ease and cell damage is minimized.

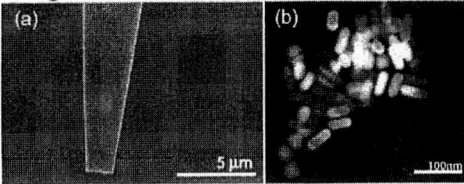

Fig. 2. (a) SEM image of glass micropipette coated with carbon (b) TEM image of synthesized gold nanoparticles.

EXPERIMENT AND RESULTS

In experiments described here, a pulsed laser with wavelength of 532 nm and pulse duration of 5 nanoseconds was used. The laser delivered a fluence of 0.883 J/cm^2 onto a non-focused spot of 11.8 mm^2. Gold nanoparticles were synthesized directly on glass microcapillaries (Fig. 2) using the method described in [6]. Nalm-6 cells (human B cell precursor leukemia) cultured in RPMI were used. Highly localized openings of the cell membrane were generated by the photothermal effect of nanoparticles on glass micropipettes, with the dimension of a typical opening close to the micropipette tip size, around 2 μm (Fig. 3(c)). The cell remained viable after the procedure. A control experiment using a glass micropipette of the same size without gold nanoparticles was also performed. The cell membrane restored its shape instantaneously and showed no sign of hole opening after laser pulsing. We also investigated the laser-induced photothermal effect of an electrically and thermally conductive amorphous carbon coating. A thin film of amorphous carbon was sputtered onto a glass micropipette. Experimental results showed explosive effects extending over a large volume that lysed and killed the cell

instantly under the same laser fluence (Fig. 3(b)). Similar cell lysing and killing effect was observed when the contact area between the gold-nanoparticles-coated micropipette and the cell is too large.

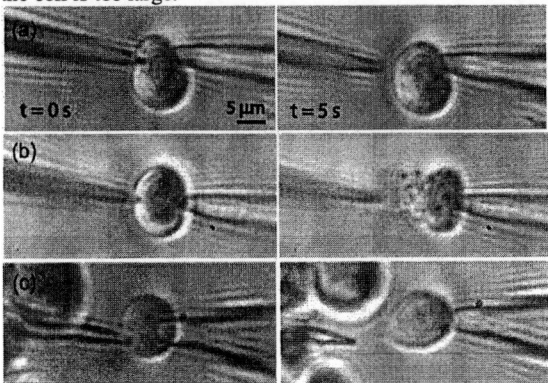

Fig. 3. Onset of laser pulsing (t = 0 s) and after laser pulsing (t = 5s) for (a) Glass micropipette (b) carbon-coated micropipette (c) gold-nanoparticles-coated micropipette.

CONCLUSION

A novel cell surgery tool utilizing photothermal effects of metal nanoparticles is proposed and proof-of-concept experiment is demonstrated. Localized hole opening on the cell membrane is accomplished using a gold-nanoparticles-coated micropipette.

REFERENCES

[1] C.M. Pitsillides, E.K. Joe, X. Wei, R.R. Anderson and C.P. Lin, "Selective cell targeting with light-absorbing microparticles and nanoparticles," Biophys. J., 84, pp.4023–4032, 2003.

[2] D.O. Lapotko et al., "Selective Laser Nano-Thermolysis of Human Leukemia Cells With Microbubbles Generated Around Clusters Of Gold Nanoparticles," Laser Surg. Med., 38, pp.631-642, 2006.

[3] A. Vogel et al., "Mechanisms of femtosecond laser nanosurgery of cells and tissues," Appl. Phys. B-Lasers O., 81 (8), pp.1015-1047, 2005.

[4] G.V. Hartland, "Coherent excitation of vibrational modes in metallic nanoparticles," Annu. Rev. Phys. Chem., 57, pp.403-430, 2006.

[5] V. Kotaidis et al., "Excitation of nanoscale vapor bubbles at the surface of gold nanoparticles in water," J. Chem. Phys., 124 (18), Art. No. 184702, 2006.

[6] Xiaoda Xu et al., "In situ precipitation of gold nanoparticles onto glass for potential architectural applications," Chem. Mater., 16 (11), pp. 2259 -2266, 2004.

TuB1 (Invited)
10:30 – 11:00

Optically controlled, holographic micro-hand

Miles Padgett, Graham Gibson, Jonathan Leach, David Carberry* and Mervyn Miles*
E-mail: m.padgett@physics.gla.ac.uk
Department of Physics and Astronomy, University of Glasgow, Glasgow. UK
*Department of Physics, University of Bristol, Bristol. UK

Abstract

A video image of the operators fingers defines the position of optically trapped micron-sized beads, any movement of the fingers causing a corresponding movement of the beads. The resulting micro-hand is used to position both inert and bio-material with nano-metre precision.

Keywords: Optical tweezers, holography, optical manipulation

1 INTRODUCTION

Optical tweezers rely on the field gradient produced by a tightly focused laser beam to create a force which acts on dielectric particles, drawing them into the position of maximum beam intensity [1]. A laser of a few 10's milliWatts and a x100 microscope objective lens creates pico-newton forces to attract and can confine transparent particles several microns in diameter. Usually, the particles are suspended in a fluid providing both partial buoyancy and a damping force to make the optical trap stable. Optical tweezers can be bought commercially and have found many applications [2], e.g. tweezers have been used in measuring the compliance of bacteria [3] and the forces exerted by motor proteins [4].

Recently optical tweezers have been re-reborn by the availability of spatial light modulators (SLMs) [5]. In such systems an SLM is configured as a computer controlled diffractive optical element allowing a single laser beam to be split forming many independent optical traps. Updating the SLM allows the individual optical traps to be moved independent of each other [6] – hence these systems are termed "holographic optical tweezers" [7]. Unlike other approaches for creating multiple tweezers that use scanning mirrors or acousto-optic modulators to laterally scan the laser between traps [8], an SLM can also display holograms acting as additional lenses creating axial displacements. Arbitrary displacements can be set independent of each other to create complicated 3D structures of micrometer-sized inert spheres or living cells [9] using a variety of hologram design algorithms [10].

Here we report on the application of a revolutionary interface using the operator's fingers to simultaneously determine the position and motion of several optical traps. We use the beads captured in optical traps as the fingertips of a manipulating micro-hand [11].

2 THE SYSTEM CONFIGURATION

Our interface relies on a single fire-wire interface camera above the working area imaging the positions of white map pins attached to the fingers of the operators glove. The height of the beads are inferred from their apparent size in the image. These positions are scaled and fed to the hologram calculation algorithm to produce optical traps at the corresponding positions within the sample cell. For stability, the interface limits the maximum translation velocity of the traps (< 5μm/sec) so that too rapid a movement of the fingers locks the trap positions.

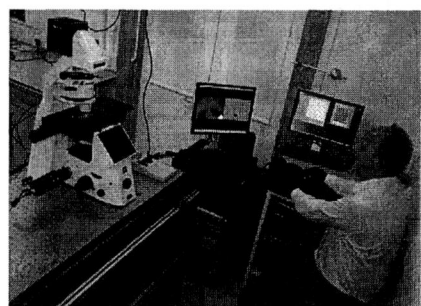

Figure 1.   The micro-hand workstation

The system is based around an inverted microscope with a 1.3NA, x100 objective. The trapping beam is provided by a titanium–sapphire laser (Coherent), pumped by a solid-state laser, emitting up to 4 watts at 835nm. This wavelength is selected to minimise the damage to biological

1-4244-0641-2/07/$20.00 ©2007 IEEE   45

material. The beam is expanded to fill an optically addressed SLM (Hamamatsu), imaged onto the back aperture of the objective lens. A twin processor, dual-core PC is dedicated to the calculation of the holograms, enabling an update rate of order 10Hz, sufficient for real time interactive control. As the tweezers system is based around a conventional microscope, the integration of various accessories such as illumination and filters to enable fluorescence imaging is possible.

3 APPLICATIONS OF THE MICRO-HAND

A traditional limitation of optical tweezers is their reliance on trapping transparent objects. Consequently, the trapping of opaque particles has remained problematic. Circumventing that fundamental problem, we use the digits of the microhand to accurately manipulate micron-sized metallic objects, expanding the range of particles that can be manipulated, see figure 2.

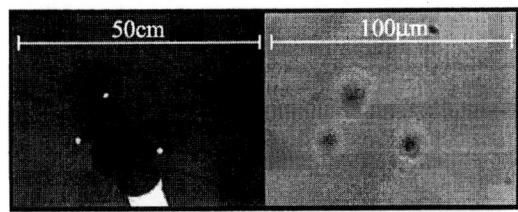
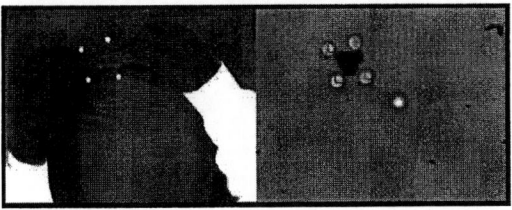

Figure 2. Split screen image showing the mapping of the operators fingers to the position of the trapped beads, in the lower example used to hold a metallic particle

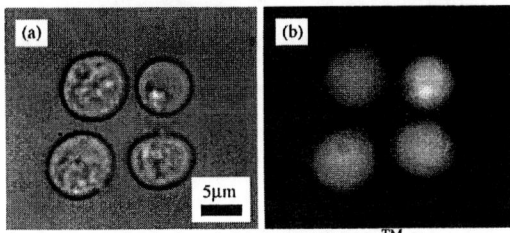

Figure 3. (a) Optical and (b) LIVE/DEAD^{TM} fluorescence images of mouse stem cells after trapping and manipulation with the micro-hand system.

Another limitation of optical tweezers is that the

comparatively high power densities can damage fragile bio-material. Operating the micro-hand system at 830nm is key to minimizing the laser damage. We are currently applying the system to the manipulation of embryonic stem cells, assembling specific geometries to investigate the role that configuration may play in determining the differentiation into cell type [12].

REFERENCES

[1] A. Ashkin, J. M. Dziedzic, J. E. Bjorkman, and S. Chu, "Observation of a single-beam gradient force optical trap for dielectric particles," Opt. Lett. **11**, 288–290 (1986).

[2] J. E. Molloy and M. J. Padgett, "Lights, action: optical tweezers," Contemporary Physics **43**, 241–258 (2002).

[3] S. M. Block, D. F. Blair, and H. C. Berg, "Compliance Of Bacterial Flagella Measured With Optical Tweezers," Nature **338**, 514–518 (1989).

[4] J. T. Finer, R. M. Simmons, and J. A. Spudich, "Single Myosin Molecule Mechanics - Piconewton Forces and Nanometer Steps," Nature **368**, 113–119 (1994).

[5] D. G. Grier, "A revolution in optical manipulation," Nature **424**, 810–816 (2003).

[6] J. Liesener, M. Reicherter, T. Haist, and H. J. Tiziani, "Multi-functional optical tweezers using computer-generated holograms," Opt. Commun. **185**, 77–82 (2000).

[7] J. E. Curtis, B. A. Koss, and D. G. Grier, "Dynamic holographic optical tweezers," Opt. Commun. **207**, 169–175 (2002).

[8] K. Visscher, G. J. Brakenhoff, and J. J. Krol, "Micromanipulation by Multiple Optical Traps Created by a Single Fast Scanning Trap Integrated with the Bilateral Confocal Scanning Laser Microscope," Cytometry **14**, 105–114 (1993).

[9] P. Jordan, J. Leach, M. Padgett, P. Blackburn, N. Isaacs, M. Goksor, D. Hanstorp, A. Wright, J. Girkin, and J. Cooper, "Creating permanent 3D arrangements of isolated cells using holographic optical tweezers," Lab On A Chip **5**, 1224–1228 (2005).

[10] G. Sinclair, P. Jordan, J. Courtial, M. Padgett, J. Cooper, and Z. J. Laczik, "Assembly of 3-dimensional structures using programmable holographic optical tweezers," Opt. Express **12**, 5475–5480 (2004).

[11] G. Whyte, G. Gibson, J. Leach, M. Padgett, D. Robert and M. Miles, "An optical trapped microhand for manipulating micron-sized objects," Opt. Express **14**, 12497-12502 (2006).

[12] J. Leach, D. Howard, G. Gibson, D. Gothard, J. Cooper, K. Shakesheff, M. Padgett and L. Buttery, "Manipulation of live mouse embryonic stem cells using holographic optical tweezers" submitted to Opt. Express (2007)

TuB2
11:00 – 11:15

A Thermo-Pneumatically Actuated Tip-Tilt-Piston Mirror

A. Werber, H. Zappe

Laboratory for Micro-optics, Department of Microsystems Engineering (IMTEK), University of Freiburg,
Georges-Koehler-Allee 102, 79110 Freiburg, Germany
Tel.: +49 761 203 7518, Fax.: +49 761 203 7562, E-mail: werber@imtek.uni-freiburg.de

Abstract

We present a novel, high stroke tip-tilt-piston micro-mirror actuated by thermo-pneumatic forces. A circular silicon mirror plate is mounted on top of three balloons, made of a 50 µm thick elastomer membrane. The balloons are inflated under thermo-pneumatic pressure, thus driving the mirror plate into tip-tilt and piston motion. Very large stroke values of 385 µm as well as tip-tilt angles of 5° were measured. The triangular arrangement of the balloon actuator yields a high flexibility in mirror motion, including tip-tilt in any direction, and the system is completely integrated without the need for any external pressure drivers.

Keywords: Micro-mirror, PDMS, thermo-pneumatics

1 INTRODUCTION

A novel approach for design and fabrication of an integrated tip-tilt-piston mirror (TTPM), based on transparent and highly elastic polydimethylsiloxane (PDMS) membranes and thermo-pneumatic actuation is presented. In previous work, pressure actuated devices, such as tunable membrane lenses [1] as well as thermo-pneumatically actuated mirrors [3] have been shown. Most notably, PDMS membrane based micro-lenses [4], [5] are of considerable scientific interest. This new approach in thermo-pneumatic actuation gives rise to very large deflections in combination with high flexibility in the direction of movement.

2 DESIGN

This new mirror design, as illustrated in Fig. 1, combines tip-tilt motion in any direction with large piston motion. The mirror plate, which is flexibly mounted on the PDMS membrane, is deflected from the bottom by three elastomer balloons, arranged in a triangle, and actuated thermo-pneumatically. Thereby, three-dimensional mirror motion is achieved resulting from the thermal expansion of air enclosed in a sealed cavity, thus generating a surplus of volume that distends the balloon membrane.

The heat is generated by a platinum micro-hotplate which consists of a resistive heater and a thermistor ring structure to allow on-chip temperature measurement.

The dimensions of this arrangement are as follows: the diameter of the mirror plate is 5 mm, with a thickness of 150 µm where the balloon membrane has a diameter of 2 mm in its non-deflected state.

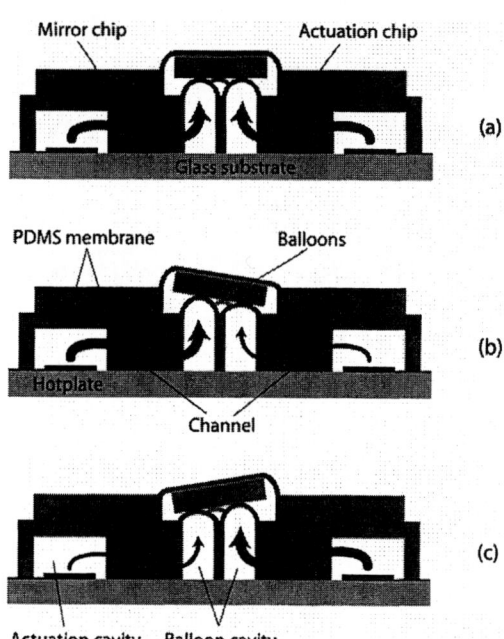

Figure 1. 2D-sketch of TTPM showing three different states of motion: (a) piston motion, (b) negative tip-tilt angle, (c) positive tip-tilt angle. The arrows show the flow of heat-generated surplus of gas volume, where the arrow thickness illustrates the amount of flow.

The TTPM, depicted in Fig. 2, consists of a stack of different circular chip elements: a glass substrate (1 inch in

1-4244-0641-2/07/$20.00 ©2007 IEEE   47

diameter) at the bottom carrying the platinum micro-hotplate, an actuation chip (silicon) that includes the balloons, the micro channels, and the actuation/balloon cavities, and finally the mirror chip (silicon) on top serving as a suspension for the mirror plate. All three chips are fabricated separately and bonded together at the end of processing.

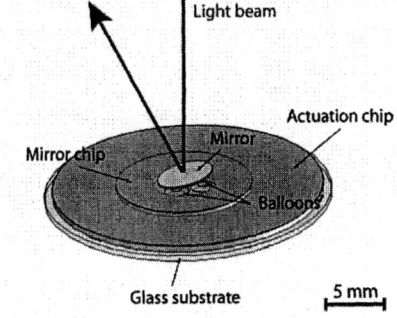

Figure 2. 3D-schematic of a tip-tilt-piston mirror (TTPM). The mirror is pictured in a tilt position actuated by three balloons, arranged in a triangle and actuated thermo-pneumatically.

4 RESULTS

The TTPM was characterized using a white light interferometer to measure the balloon deflection on the actuation chip. The stroke measurement of the mirror plate was accomplished by a laser displacement meter and the tilt angle was measured with an optical laser pointer setup.
Fig. 3 shows the deflection vs. voltage characteristic of a single balloon on the actuator chip without the mirror plate on top. A maximum peak height of 234 μm was achieved.

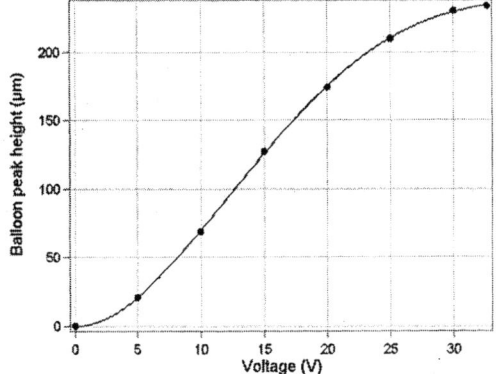

Figure 3. Balloon peak height versus voltage.

The mirror tilt with respect to the actuation voltage was measured with a single balloon actuator, causing a tilt to the mirror plate. A tilt angle of 5° was reached. Fig. 4 shows the tilt angle versus voltage dependence.

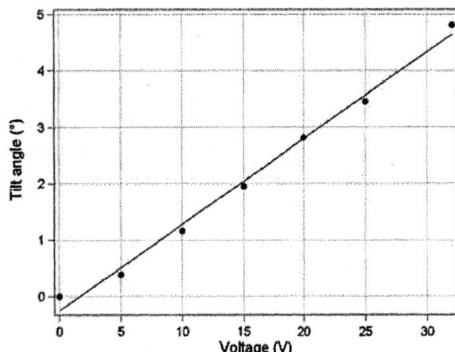

Figure 4. Tilt angle versus voltage.

Aside from the tilt angle, mirror stroke is the most important parameter of the TTPM device. The stroke was determined by actuating all three balloons simultaneously with the identical voltage. A maximum stroke value of 385 μm was measured. Fig. 5 shows an eight point measurement of the stroke versus voltage dependence.

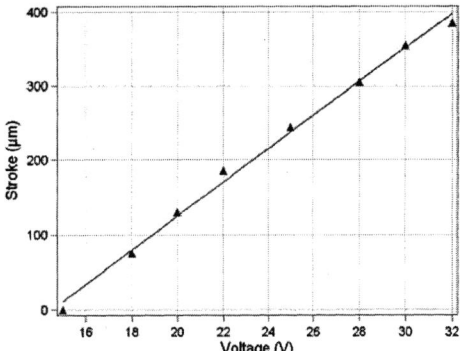

Figure 5. Stroke versus voltage.

Of interest in a thermo-pneumatic system is the temperature characteristic as well as the power consumption of the system. At a voltage of 32 V, the maximum temperature was 158°C and a power consumption of 1500 mW. Moreover, a time constant of $\tau = 1.1$ s for a 10 mHz square waveform actuation was measured.

REFERENCES

[1] A.Werber, H. Zappe, "Tunable micro-fluidic micro-lenses", *Appl. Optics*, 44, pp. 3238-3245 (2005).
[2] A.Werber, H. Zappe, "Thermo-pneumatically actuated, membrane-based, micro-mirror devices", *J. Micromech. Microeng.*, 16, pp. 2524-2531 (2006).
[3] D-Y. Zhang, N. Justis, V. Lien, Y. Berdichevsky, Y-H. Lo, "High-performance fluidic adaptive lenses", *Appl. Optics*, 43, pp. 783-787 (2004).
[4] M. Agarwal, R. Gunasekaran, P. Coane, K. Varahramyan, "Polymer-based variable focal length microlens system" *J. Micromech. Microeng.*, 14, pp. 1665-1673 (2004).

TuB3
11:15 – 11:30

Bi-directionally Driven Metal Cantilevers Developed for Optical Actuation

H. Kwon[1,2], M. Nakada[1,2], Y. Hirabayashi[3], A. Higo[4], M. Ataka[2], H. Fujita[2], H. Toshiyoshi[1,2]

[1]Kanagawa Academy of Science and Technology, Kawasaki, Japan, [2]IIS, the Univ. of Tokyo, Tokyo, Japan,
[3]KITC, Kanagawa, Japan, [4]RCAST, the Univ. of Tokyo, Tokyo, Japan
Tel: +81-3-5452-6277, Fax: +81-3-5452-6250, e-mail: tp_ho@newkast.or.jp

SUMMARY

In this paper we report a silicon solar-cell array to generate voltage (about 4.8V by 16 series PIN diode) to drive cantilever actuators made of stress-controlled combination of chromium-gold-chromium layers for electro-thermal pull-up and electrostatic pull-down motion. The silicon PIN layer was developed by the epitaxial growth on an SOI (silicon-on-insulator) wafer. The metal cantilever actuator was developed by using the reverse sputtering process to minimize the performance variation associated with the patterning error. The bi-directional actuator was developed for potential use in AFM-based data storage.

TARGET APPLICATION

MEMS (Micro electro mechanical systems)-based data storage systems utilizing the atomic force microscopy (AFM) have been developed for a next-generation storage technology [1-4]. Motion controls of MEM devices utilizing light addressing would be a solution to avoid the problem of increasing number of intra-chip electrical interconnection [5]. As a new trial, we propose a solar cell array integrated with micro mechanism as shown in **Figure 1**. Another issue in the AFM type MEM device is the height uniformity of the arrayed tips. In this paper, we propose a simple fabrication process of solar cell arrays and layered-metal cantilevers for up-and-down bi-directional driving.

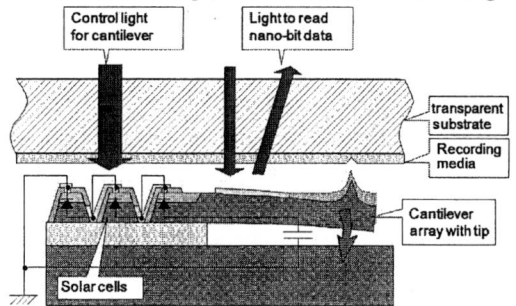

Figure 1: Schematic view of micro opto-electrical mechanism of which motion is controlled by light signal.

SOLARCELL ARRAY

Arrayed solar cells were fabricated and utilized for voltage source to drive micro cantilevers as shown in **Figure 2**. The output voltage from a unit solar cell, series of four cells, and sixteen cells showed voltages of 0.5 V, 1.8 V, and 4.8 V, respectively, under about 630-nm wavelength He-Ne and semiconductor laser beam of about 1.5-mm diameter at 2-5 mW. The output voltage from arrayed cells was found to be

slightly lower than theoretical estimation because of the array size larger than the laser beam spot. Uniformly distributed laser power would improve the output voltage.

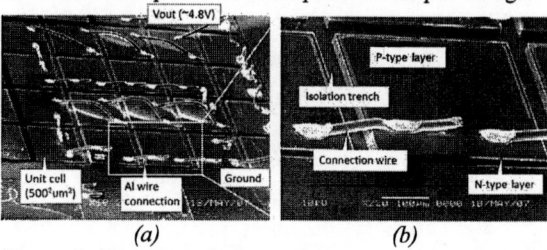

Figure 2: Scanning electron microscopic view of serially connected solar cell arrays: (a) overall view, and (b) close-up view of a unit cell.

BIDIRECTIONAL CANTILEVER

Cantilever was designed to deflect upwards by the electro-thermal Joule heat and also downwards by the electrostatic force; this could be done by changing the interconnection of driving sources as shown in **Figure 3**. Actuator's bi-directional motion was made possible by carefully designing the Cr-Au-Cr layer thicknesses.

The cantilevers were fabricated on silicon wafers. A spin-on-glass was coated for electrical isolation between the silicon substrate and the cantilever electrodes [6]. Multiple layers of chromium, gold, and another chromium layer were deposited by DC sputtering at 500 Watts each. For film stress control, the first chromium was deposited at 2 mTorr for 0.5 minutes, gold at 2 mTorr for 5 minutes, and the second chromium was at 20 mTorr for 3 minutes.

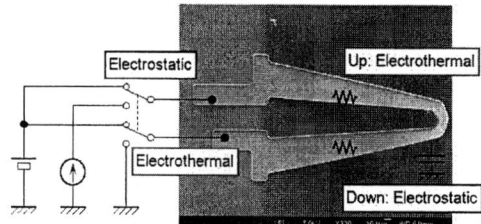

Figure 3: Scanning electron microscopic images of metal-layered cantilevers after process improvement using reverse sputter patterning of Cr, Au, and Cr metal layers. Upward and downward operation was done by electro-thermal and electro-static drive, respectively.

The cantilever was patterned by the series of etching for chromium, gold, and chromium. At the final step, the

1-4244-0641-2/07/$20.00 ©2007 IEEE   49

sacrificial layer was selectively removed for releasing process. The released metal cantilevers were designed to curl up by the intrinsic stress of the layers, as shown in **Figure 3 and 4**. Uniform elevation height is indispensable for the data storage application [3]. However, the first prototype by wet-etching the metal layers resulted in excess side etch of chromium, which made the pattern width difference as shown in the inset of **Figure 4(a)**. The width difference was found to cause the variation of the cantilever elevation. The mean value, maximum deviation, and standard deviation in between cantilevers on 3 inch-size wafer were measured to be 27.0 um, 18.7 um (69.4 % of the mean value), and 4.59 um (17 % of the mean value), respectively.

To overcome this variation problem, we recently utilized the reverse sputtering to pattern the Cr-Au-Cr layers by using an aluminum etching mask. The etch rate of chromium and gold were found to be 30 Angstrom/s and 130 Angstrom/s at RF 400 W power, respectively. The inset picture of **Figure 4(b)** shows the result of well defined cantilevers by reversed sputtering. The curvatures of cantilevers were also made larger. The improved mean value, maximum deviation, and standard deviation in between cantilevers on 3 inch-size wafer were 52.5 um, 4.5 um (8.6 % of the mean value), and 1.31 um (2.5 % of the mean value), respectively. Average values of maximum deviation and standard deviation in one array, so to speak in-chip uniformity, showed 1.33 um (2.5 % of the mean value), and 0.59 um (1.1 % of the mean value), respectively. From the deviation analysis of the cantilever elevation, cantilevers fabricated by the dry etching process had deviations of nearly quarter to that by wet etching process.

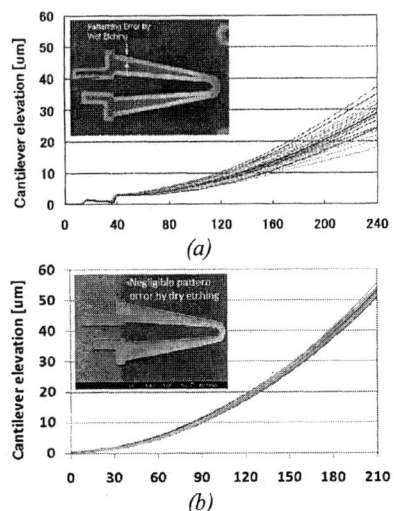

Figure 4: The profiles of cantilever arrays distributed on a wafer which was patterned by (a) wet chemical etching and (b) reversed sputtering, respectively.

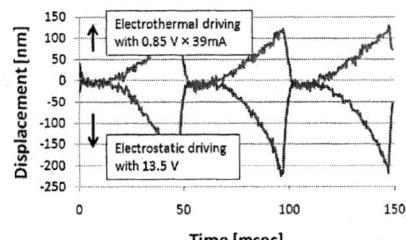

Figure 5: Experimental result of electrostatic and electro-thermal driving of the fabricated Cr/Au/Cr cantilever. The elevation of cantilever was measured at the distance of about 50 um (about one fifth of the length of cantilever) from the anchor because of the measurement limit (numerical aperture) of the laser interferometer.

The elevation of the fabricated metal cantilevers was further increased by the Joule heat using the combination of thermal expansion coefficients and layer thicknesses. On the contrary, the cantilever was electrostatically driven in the downward direction. The cantilever elevation is shown in **Figure 5** as a function of applied voltage and current. Due to the limitation of optical measurement system, we only observed the motion at a spot near the anchor of the metal cantilever to be 130 nm and 220 nm by electrothermal and electrostatic driving, respectively. The motion was found to be large enough to make compensation of the tip height variation.

CONCLUSIONS

In conclusion, solar cells generated voltage of about 4.8V for 16 cell-arrayed to generate voltage source for micro cantilever. The chromium-gold-chromium cantilever with the appropriate thickness combination was found to be suitable for the bidirectional up & down mechanism for the AFM type data storage. The dry etching showed superior patterning of metal-layered cantilevers to wet etching for improved curvature uniformity. For a further study, we are investigating the depositing conditions to reduce the cantilever elevation and the deviation.

ACKNOWLEDGMENTS

This work was supported in part by the Murata Research Foundation. The photomasks used in this work were devel oped using the electron-beam facility at the VLSI Design and Education Center (VDEC) of the University of Tokyo.

REFERENCES

[1] W. P. King, JMEMS, 11(6), pp. 765- 774, Dec. 2002.
[2] H. Nam, MEMS 2005, pp. 247-250.
[3] A. Chand, JMEMS, 9(1), pp. 112-116, Mar. 2000.
[4] C. Hsieh, Trans.on Magn., 41(2), pp.989-991,Feb.2005.
[5] Y. Yamauchi, MOEMS 2004, pp. 164.
[6] H. Kwon, NEMS 2007, pp. 488-492.

TuB4
11:30 – 11:45

Reconfigurable Nanophotonic Systems By Tunable Alignment Between Nanomagnet Arrays

Anthony J. Nichol, William J. Arora, and George Barbastathis
Massachusetts Institute of Technology
77 Massachusetts Ave, Cambridge, MA 02139 USA
Tel +001-617-452-2836, e-mail: anichol@mit.edu

Abstract

We demonstrate self-alignment and tunable displacements of released nanopatterned membranes by use of arrays of nanomagnets. The nanomagnet arrays attract and align when brought into close proximity resulting in sub-200nm accurate self-alignment between membrane segments of >50μm lateral size. The alignment is made reconfigurable by patterning the nanomagnets so that there are multiple stable alignment states. An external field is used to transfer between alignment states by applying a magnetic torque on the arrays and by shifting the nanomagnet polarity. Realignment via folding and unfolding of membranes was demonstrated using 75nm thick cobalt nanomagnets patterned on 1um thick silicon nitride membranes. Accurately aligned reconfigurable 2D or 3D nanophotonic systems, such as active photonic crystals and sheared photonic crystal waveguides, may be fabricated with this method.

Keywords: Nanomagnets, membrane folding, photonic crystals, reconfigurable nanophotonics,

1 INTRODUCTION

We recently showed that nanomagnet arrays patterned on thin silicon nitride membranes can be used for actuation and self-alignment of membranes in stacked or folded configurations [1]. The membranes can additionally be patterned with nanophotonic features and aligned to the necessary tolerances for two dimensional (2D) and three-dimensional (3D) photonic devices. In particular, devices requiring an aligned 3D matrix of features such as 3D photonic crystals (PC) with complex networks of accurately placed defects can be fabricated layer-by-layer with this process.

Many nanophotonic systems require reconfigurability. Zhou et al designed tunable defects in 2D photonic crystals based on the active insertion and removal of material at a point or line in the PC [2]. We are developing a tunable sheared PC wave-guide that requires precise relative movement between two planar halves of a photonic crystal [3]. Many optical MEMS architectures require precise control of the tilt angle and position of micro-fabricated mirrors and gratings. For all of these applications, piezoelectric or electrostatic actuation are obvious choices. However, both can require difficult fabrication and a feedback loop to get the necessary alignment.

We propose a reconfigurable assembly process using the attraction between arrays of nanomagnets. The magnets are patterned such that the membranes can self-align at multiple minimum energy states. The arrays are shifted between alignment states using an external magnetic field. This method may find a unique application in nanophotonics since it is most useful for discrete steps of tens of nanometers at slow actuation frequencies without the need for feedback.

2 NANOMAGNET OPERATION

Magnetic forces on magnets within an external field and attractive forces between magnets become increasingly dominant over gravitational and elastic mechanical forces as all dimensions are scaled down [4]. Nanomagnets are easily introduced into devices with thin-film evaporation and patterning. Furthermore, the difficulty in creating strong permanent magnets with dimensions in microns is overcome by scaling down to the nanoscale since shape anisotropy helps hold the magnetic remanence [5]. These properties make nanomagnets particularly suitable for manipulating thin film membranes (having thickness on the same order as the size of the nanomagnets).

2.1 Nanomagnet alignment

Membranes can be aligned with nanomagnets by their North-South pole interactions, as shown in Figure 1. In Figure 1a, the membranes are magnetized in-plane allowing the magnetic properties to be manipulated by changing the lithographic pattern geometry and size. Magnetic materials that typically do not hold a strong magnetic remanence can be made stronger by shaping slender magnets. In Figure 1b, the nanomagnets are magnetized out of plane, but the possibilities for manipulating the geometry are limited. We previously used the in-plane magnetization method shown in Figure 1a to self-align silicon nitride membranes with cobalt nanomagnets. The force between individual magnets is

1-4244-0641-2/07/$20.00 ©2007 IEEE   51

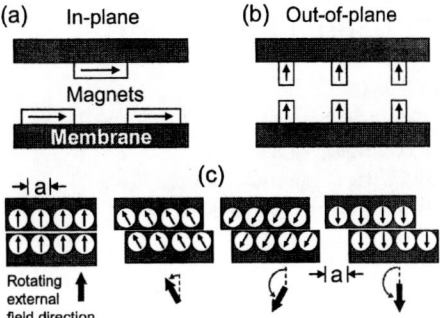

Figure 1: The nanomagnet membrane alignment scheme with arrows indicating magnetic polarization direction. The nanomagnets are polarized (a) in-plane, and (b) out of plane. (c) Top view of small-step shifting by rotating the external magnetic field.

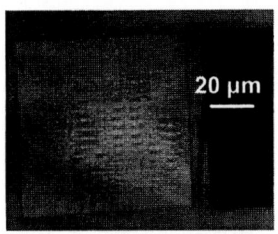

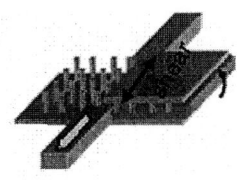

Figure 2: Experimental result showing two silicon nitride membranes aligned to within 200nm after folding.

Figure 3: Proposed implementation of the tunable shear photonic crystal waveguide structure.

approximated as $F = \mu_0 M^2 A^2 / (4\pi d^2)$, where M is the magnetic remanence, A is the magnet cross-sectional area and d is the distance between magnets.

2.2 Tuning by small step shifting

The nanomagnet pattern geometry dictates the final alignment position by local energy minimum. If multiple energy minima are designed into the array, the configuration becomes multi-stable. This is achieved, for example, by placing the magnets in a lattice of period a. The equilibria are then spaced by the same distance a. We showed this experimentally by switching the array alignment by one full period when folding and unfolding two membranes using an external field [1], as shown in Figure 2. The same effect can be achieved by rotating the in-plane magnetization of circular nanomagnets , as shown in Figure 1c. Every rotation of the external field by 180° results in a step of one period, a. This allows discrete, controllable steps dictated strictly by the nanomagnet size and geometry.

3. APPLICATION TO PHOTONIC CRYSTALS

Since the aligned nanomagnet arrays can be separated after assembly, reconfigurable photonic systems based on planar features (e.g. mirrors, zone plates and gratings) can be folded and unfolded (completely or partially) repeatably with precise self-alignment. Photonic crystal structures are well suited to nanomagnet-based reconfiguration because small modifications to the lattice can result in large changes to the band structure. For example, the spacing between photonic crystal layers can be tuned to move the band gap edges, or the size of a waveguide defect can be changed, thereby controlling mode bandwidth and dispersion.

Here, we investigate in more detail the design of a shear discontinuity photonic crystal waveguide [3], shown in Figure 3. Each sheared half-lattice is located on a different segment of the folded membrane. Shear is introduced by moving the segments relative to one another. The waveguide is tunable in the following sense: when the shear equals half the lattice constant, the supported mode is almost dispersion-free and exhibits no mode gap. As the shear decreases gradually to zero, a mode gap is introduced and the group velocity near the mode edge decreases. Since propagation in the mode gap becomes evanescent, this structure can also support a CROW (couple resonator optical waveguide) configuration where the group velocity is even smaller (the minimum is determined by the cavity spacing at zero shear) but can be tuned upwards by increasing the shear [6]. This structure does not enjoy the dispersion management benefits of dynamic tuning [7], but it is more readily implementable.

4 CONCLUSION

We have presented experimental evidence of mechanical alignment and reconfiguration between the folded segments of a membrane that is pre-patterned with nanomagnets. This approach enables novel reconfigurable photonic architecture, including tuning the group velocity by relative shear between two photonic crystal half-lattices.

REFERENCES

[1] Nichol, A et al. *Microelectronic Engineering*. Vol 84, issue5-8. May 2007. p 1168-71
[2] Zhou, W. et. al. *Solid-State Electronics*. Vol 50, issue 6. June 2006, p 908-13
[3] Tian, K. et. al. *Opt. Expr.* **14**(22):10887-10897, 2006.
[4] Gibbs M.R.J. *J. Magnetism and Magnetic Materials* Vol. 290-291 Pt 1. April 2005, p 1298-1303.
[5] Huang, Y.S. et al. *Journal of Physics: Condensed Matter*, v 17, n 25, 29 June 2005, p 3931-41
[6] Tian, K. et. al. "Tunable group velocity in a coupled resonator optical waveguide (CROW) formed by shear discontinuities in a photonic crystal," submitted to *Opt. Expr.*.
[7] Yanik, M. et. al. ""Phys. Rev. Lett., 92, 8 (2004).

TuB5
11:45 – 12:00

High-Accuracy Digital-to-Analog Actuators Using Parallel Spring Array

Won Han and Young-Ho Cho
Digital Nanolocomotion Center
Korea Advanced Institute of Science and Technology (KAIST)
373-1 Guseong-dong, Yuseong-gu, Daejeon 305-701, Republic of Korea
Tel +82-42-869-8691, Fax +82-42-869-8690, E-mail: nanosys@kaist.ac.kr

Abstract

We present a high-accuracy micromechanical digital-to-analog (DA) actuator using parallel interconnection of identical springs. The parallel DA actuator achieves a high-accuracy cascaded actuation because it is independent of spring variation caused by fabrication error. We design the parallel DA actuator using 4-bit digital actuators, whose four different unit displacement of x_1=7.0μm, x_2=5.8μm, x_3=4.8μm, and x_4=4.0μm are designed to make the output displacement linear for the output range of 0~5.4μm. In experimental study, the output range and nonlinearity are measured as 0~4.7μm and 8.65%, respectively. The parallel DA actuator generates the output displacement with the accuracy of 0.29±0.20μm and the precision of 20.4±5.8nm. We experimentally verify the positioning capability of the present parallel DA actuator is suitable for high-accuracy optical manipulation.

Keywords: Digital-to-analog actuators, parallel spring array, high-accuracy actuation

1 INTRODUCTION

Fine positioning functions for lens and mirror manipulation [1] require microactuators with submicron accuracy and precision. Recently, the micromechanical digital-to-analog (DA) actuators [2,3] performing the mechanical modulation of digital strokes have been suggested for high-precision actuation insensitive to electrical noise. However, their positioning accuracy (0.65±0.15μm, [3]) still depends on spring variation caused by fabrication error, because their output displacements are the function of the stiffness ratio of interconnection springs. This paper presents the DA actuator (Fig.1) using parallel interconnection of identical springs, thus achieving the positioning capability independent of spring variation for high-accuracy optical manipulation [1] and telecommunication applications [4].

2 DESIGN AND ANALYSIS

Figure 2 illustrates the parallel DA actuator composed of 4-bit digital actuators with identical springs (k). From force equilibrium conditions, the output displacement, y, is given by

$$y = \frac{1}{4}\left(b_1 x_1 + b_2 x_2 + b_3 x_3 + b_4 x_4\right) \qquad (1)$$

where $x_i(i=1,2,3,4)$ are the digital unit displacement and $b_i(i=1,2,3,4)$ are the digital digits with the value of 1 or 0 depending on the ON or OFF state of the digital actuator, i, respectively. From Eq.(1), it is noted that the output displacement, y, is independent of the stiffness, k, of the springs, thus reducing the output displacement uncertainty caused by spring variation. Each digital unit displacement (x_1,x_2,x_3,x_4) is designed to make output displacement as possible as linear. The digital unit displacement 1, x_1, is designed as 7.0μm, which is the maximum displacement possible to design for the output range of 0~5.4μm. The digital unit displacement 4, x_4, is determined to 4.0μm by the minimum etching window for deep RIE (Reactive Ion Etching) process of 20μm-thick silicon. The other digital unit displacements, x_2 and x_3, are designed as 5.8μm and 4.8μm, respectively, for the minimum nonlinearity of 9.03% in the output range of 0~5.4μm. Other geometric dimensions are reported in [4].

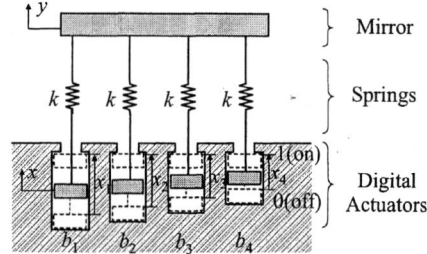

Figure 1. Simplified model of the parallel DA actuator using parallel spring array.

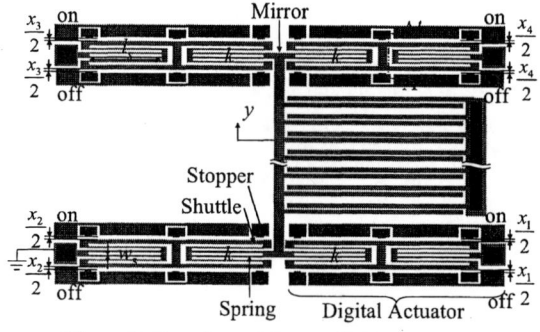

Figure 2. Top view of the parallel DA actuator

1-4244-0641-2/07/$20.00 ©2007 IEEE

Table 1. Estimated and measured output characteristics

Characteristics	Estimated values	Measured values
Output range	0~5.0μm	0~4.7μm
Nonlinearity	8.84%	8.65%
Accuracy	0μm	0.29±0.20μm**
Precision*	14.1nm	20.4±5.8nm**

* The measured precision indicates the doubled standard deviation (2σ) of the output displacement obtained from three repeated measurements, and the estimated precision is calculated by the side wall roughness of the stoppers in the order of ±10nm.

**The values denote the mean and standard deviation for 16 different digital modes.

3 EXPERIMENTAL RESULTS

The parallel DA actuator is fabricated by the single-mask fabrication process [4] using SOI wafers. Figure 3 shows a SEM photograph for the fabricated device. For characterization of output displacement, the parallel DA actuator is driven by 100Hz, 25V pulse signals. The output displacement is measured by the laser interferometer, LDV (Laser Doppler Vibrometer). The output results are obtained from three repeated measurements at each mode. Figure 4, 5, and 6 show the output displacement, accuracy, and precision, respectively, for varying digital actuation modes. The parallel DA actuator shows that the measured output range and nonlinearity (Fig.4) are 0~4.7μm and 8.65%, respectively. Table 1 shows the measured accuracy of 0.29±0.20μm reduced by 55.4% of the accuracy (0.65±0.15μm) reported in the previous DA actuators [3]. The precision is measured as 20.4±5.8nm.

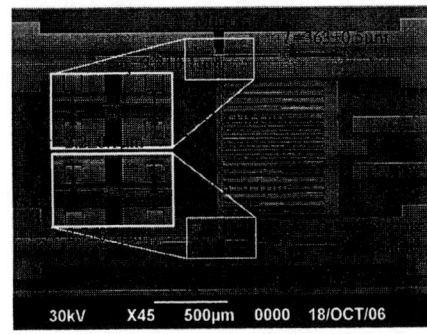

Figure 3. SEM photograph of the fabricated devices.

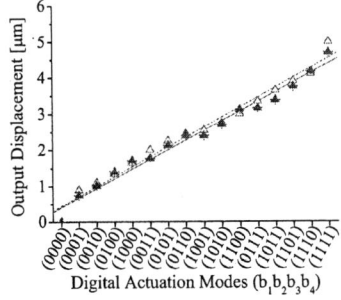

Figure 4. Estimated (empty symbols) and measured (filled symbols) output displacement for 4-bit digital modes.

4 CONCLUSIONS

We developed the micromechanical digital-to-analog (DA) actuator using parallel interconnection of identical springs for high-accuracy cascaded actuation independent of spring variation. From the experimental results, the parallel DA actuator showed the output linearity of 8.65% for the output range of 0~4.7μm. For entire digital actuation modes, the accuracy and precision was measured as 0.29±0.20μm and 20.4±5.8nm, respectively. Therefore, we experimentally verified the high-accuracy and high-precision positioning function of the parallel DA actuator for optical manipulation and telecommunication applications.

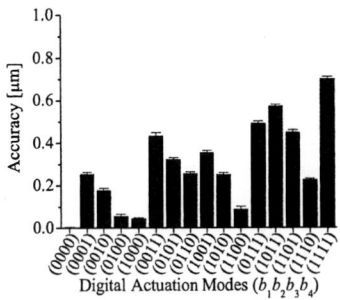

Figure 5. Measured accuracy for 4-bit digital modes.

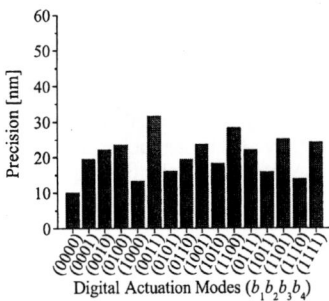

Figure 6. Measured precision for 4-bit digital modes.

ACKNOWLEDGMENT

This work has been supported by the National Creative Research Initiative Program of the Ministry of Science and Technology (MOST) and the Korea Science and Engineering Foundation (KOSEF) under the project title of "Realization of Bio-Inspired Digital Nanoactuators."

REFERENCES

[1] M. C. Wu, *Proceedings of the IEEE*, vol.85, pp.1833-1856, 1997.

[2] H. Toshiyoshi, D. Kobayashi, M. Mita, G. Hashiguchi, H. Fujita, J. Endo, and Y. Wada, *J. Microelectromech. Syst.*, vol.9, pp.218-225, 2000.

[3] R. Yeh, R. A. Conant, and K. S. Pister, in *Proc. 10th Int. Conf. Solid-State Sensors and Actuators*, 1999, pp.998-1001.

[4] W. Han and Y.-H. Cho in *Proc. Int. Conf. IEEE MEMS*, 2007, pp.815-818.

TuP1
14:30 – 17:30

Novel Large Area Applications Using Optical MEMS

V. Viereck[1], Q. Li[1], J. Ackermann[1], A. Schwank[2], S. Araujo[2], A. Jäkel[1], S. Werner[1],
N. Dharmarasu[1], J. Schmid[2] and H. Hillmer[1]
[1] Institute of Nanostructure Technologies and Analytics
[2] Institute of Electrical Engineering – Efficient Energy Conversion
University of Kassel
Heinrich-Plett-Str. 40, 34132 Kassel, Germany
Tel +49 561 804 4536, Fax +49 561 804 4488, E-mail: volker.viereck@ina.uni-kassel.de

Abstract

We focus on the potential of Optical MEMS for large area applications and discuss why it is very beneficial to use micro system technologies, especially Optical MEMS, on large scale. We present micromirror applications for daylight guiding into buildings and for light concentration in solar technologies as well as the corresponding technological approaches.

Keywords: Micromirrors, daylight guiding, large areas, low cost approach

1 INTRODUCTION

In microelectronics there is a general tendency towards continuously reduced structures. On the other hand there is another trend towards comparatively huge structures on larger areas, such as display technologies and photovoltaics. The two latter applications are not mainly based on conventional semiconductor materials, but also on low-cost thin- film technologies and materials like sputtered metals / oxides and vapor deposited amorphous materials.

A trend towards smaller structures is also visible in MEMS. Applications with non-semiconductor materials are rare. But applications for MEMS and especially Optical MEMS on large areas are very rare. Examples are microshutters for display applications [1].

We identify a huge potential to use micro systems not only when it is useful or necessary for size reasons, but to use them because of considerably enhanced stability, increased resonance frequencies, increased lifetimes and increased efficiency of actuating forces in direct comparison to those forces causing material fatigue [2].

Micromirrors for scanning applications [3] for example use their benefiting properties of very high resonance frequencies with very high mechanical stability at the same time, compared with macroscopic solutions. Related applications are micromirror arrays for beamer applications [4].

2 LARGE AREA APPLICATIONS

In this paper we introduce these advantages of Optical MEMS into applications where micromirrors are planned to be placed on areas of several square meters [5]. Our favorite application is the guiding of daylight with large area micromirror arrays, especially for

- Sunlight concentration onto photovoltaic cells and
- the guiding of daylight into work- and living rooms.

We focus on the second application and deal with the advantages which large area micromirror arrays could have in comparison to conventional solutions. These advantages involve longer lifetime, reduced power consumption and considerably reduced costs since we propagate low-cost materials for large scale Optical MEMS.

These micromirror arrays can be embedded between the two panes of a standard double glazing window. The micromirrors have to be actuated from an out-of-plane position nearly 90° to an in-plane position. The mirrors have to be distributed side by side over the whole window. Due to their size the pane only reveals a homogenous grey tone which is tunable but shows no structure details (Fig 1). Such a setup is comparable to standard blinds with reflective surfaces, but has a lot of advantages:

1-4244-0641-2/07/$20.00 ©2007 IEEE

- protected between the two panes against wind, weather and defilement
- no material fatigue, no maintenance
- not visible from outside the building
- nearly free and "uncut" outlook through a window in "mirrors-open-position" because the eye cannot seize the single mirror elements
- different mirror positions in different window areas can easily be implemented, high flexibility

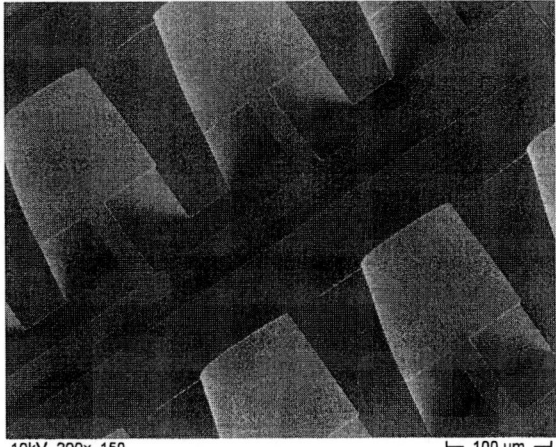

10kV, 200x, 150   ⊢ 100 μm ⊣

Fig. 2: SEM-micrograph of low-cost micromirrors

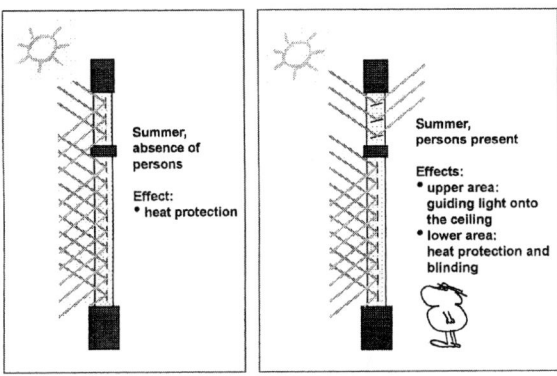

Fig 1: Schematic cross section of a window in summer. Millions of micromirrors are indicated by a small number of elements to show the functionality. The system considers daytime- and seasons-variable sun positions and is based on reflection.

3 TECHNOLOGICAL IMPLEMENTATION

We use a low cost approach for electrostatically actuated micromirrors which meet the requirements for the application mentioned above.

We process the micromirrors on a glass substrate, many of these substrates shall be interconnected in a modular way. There is no need to address each single mirror individually, but only groups of several square centimeters or more. Therefore, a very simple grid is necessary.

We only use sputter deposition of metal-layers and transparent conductive oxides (TCOs) as well as PECV-Deposition of dielectrics, processes which are very common in processing commercial photovoltaic-cells. For processing the mirrors shown in Fig 2 we need five deposition steps and three photolithography steps.

We are able to actuate groups of micromirrors from a nearly 90° position to the in-plane position with an actuation voltage of 80 volts. The leakage current is about $8 \mu A / cm^2$.

4 REFERENCES

[1] M. Pizzi, V. Koniachkine, M. Nieri, S. Sinesi and P. Perlo
Electrostatically driven film light modulators for display applications,
Microsystem Technologies, Vol 10, 17-21 (2003).

[2] H. Hillmer, J. Daleiden, S. Irmer, F. Römer, C. Prott, A. Tarraf, M. Strassner, E. Ataro and T. Scholz
Potential of micromachined photonics: miniaturization, scaling and applications in continuously tunable vertical air-cavity filters,
(invited) SPIE Proc. series Vol. 4947, 196-211 (2003).

[3] M. Tani, M.Akamatsu, Y. Yasuda, H. Fujita and H. Toshiyoshi
A laser display using PZT-actuated 2D Optical Scanner,
Technical Digest of Optical MEMS 2005, A2, P. 9, ISBN 0-7803-9278-7

[4] Hornbeck, L.J.
Digital Light Processing and MEMS: an overview
Advanced Applications of Lasers in Materials Processing, 1996/Broadband Optical Networks/Smart Pixels/Optical MEMs and Their Applications, 1996. IEEE/LEOS 1996 Summer Topical Meetings:
Volume , Issue , 5-9 Aug 1996 Page(s):7 - 8

[5] H.Hillmer, J.Schmid
European Grand Prix of Innovation Awards 2006

TuP2
14:30 – 17:30

Characterization of an Improved, Real-Time MEMS-Based Phase-Shifting Interferometer

R Kant   D Garmire H C oo and R S Muller

Department of Electrical Engineering Stanford University Stanford CA USA

7X Allen CIS Building Extension 4   Via Palou Mall Stanford CA USA 94 5

Fax   95 847 6678 E mail rik9@stanford edu

Berkeley Sensor   Actuator Center University of California Berkeley CA USA

Abstract

We describe and present detailed performance c aracterizations for an en anced version of our MEMS Based P ase S ifting Interferometer MBPSI) t at ac ieves   times denser motion reconstruction t an our original system We measure t e noise level to be ≤±6 nm λ   for a 66 nm laser) and t e frequency resolution to be ≤   Hz for   Hz motion captured at   Hz We ave successfully tracked a piezo based actuator driven wit an arbitrary waveform composed of transients ≤   Hz

Keywords: optical MEMS, Phase-shifting, interferometer, transient motion

1 INTRODUCTION

We   ave greatly increased t e motion reconstruction resolution of our previously reported MEMS Based P ase S ifting Interferometer MBPSI)] from   Hz to   Hz T is improvement enables t e reconstruction of real time motion wit   times more samples resulting in more precise measurements   T e improvement is ac ieved t roug a combination of ig er frame rate of t e CMOS imager and newly developed post processing routines

Our MBPSI uses a resonating MEMS mirror as t e p ase s ifting element in conjunction wit a strobing laser as s own in Figure   ] Sync ronizing t e strobes wit t e resonating element eliminates t e settling time delay typically found in conventional p ase s ifting interferometers allowing our MBPSI to ac ieve camera frame rate limited captures similar to digital olograp ic microscopy   ] In addition to employing faster imager rates increase from 9 to   Hz) we utilize a simple yet ig ly effective new post processing algorit m described in Figure   w ic quadruples t e time resolution of measurements

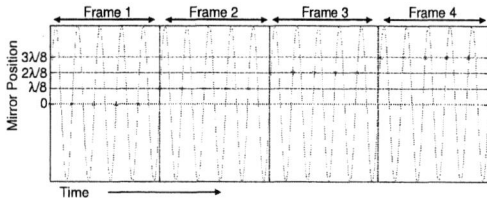

Figure 1. T e laser is strobed in sync denoted by ') wit t e resonating MEMS mirror to generate t e four distinct p ase s ifts required for surface reconstruction wit in four camera frames

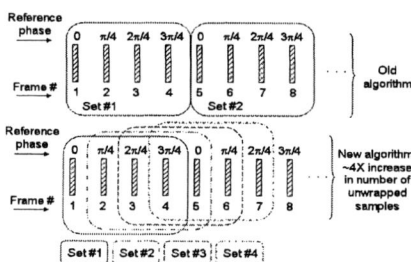

Figure 2. Our new post processing routine divides images into sets for reconstruction by using   images from t e previous set plus t e next image yielding a 4X increase in t e number of reconstructed points

We ave also c aracterized t e capabilities of t e system by using a piezo bending actuator as t e test sample T e noise level is ≤±6 nm λ   for a 66 nm laser) and t e frequency resolution of t e measurement is better t an   Hz for tracking   Hz motions captured at   Hz Finally we ave successfully applied our system to reconstruct arbitrary non periodic motions of a PZT actuator

2 EXPERIMENT SETUP

We use a modified Twyman Green configuration interferometer as s own in Figure   T e test sample is a mirror mounted on a piezo driven bending actuator

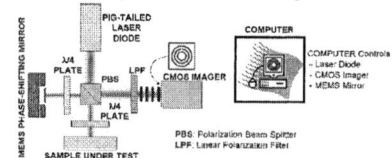

Figure 3. MBPSI experimental setup in Twyman Green configuration

1-4244-0641-2/07/$20.00 ©2007 IEEE

T e image area was set to 6 by 64 pixels corresponding to a 6 μm by 64 μm mirror surface T e camera gain and brig tness were constant for all experiments T e laser pulse widt was set to   μs T e laser strobes per frame were 8 and 5 for frame rates of   and   Hz respectively Every experiment was conducted wit in a capture window of 4 seconds Data were processed using t e 4 frame p ase unwrapping algorit m T e position at eac point in time was calculated as t e mean over t e entire imaged area T e motion plots were generated by subtracting t e initial position from eac of t e subsequent positions

3 RESULTS

3.1 Noise level

T e noise level was determined by continuously measuring a static surface over 4 seconds and analyzing t e measured position variations T e istograms of position variations s ow Gaussian distribution wit standard deviation σ) of 47nm   5 and 9 nm for   and   Hz frame rates respectively For Gaussian distribution 95   of t e values fall wit in ± σ of t e mean value yielding a noise limit of ≤±6 nm for t e system ±λ   for 66 nm laser) at

   Hz Hig er frame rates s ow ig er noise levels because t e number of laser strobes decreases resulting in a lower signal to noise ratio of t e captured images

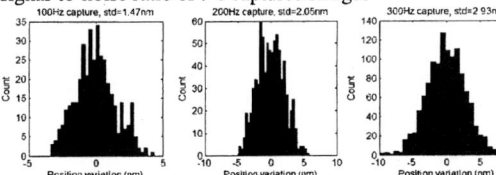

Figure 4 Measured position variation of a static surface over time at different frame rates yields a base noise level of 6nm   σ) for   Hz

3.2 Frequency-resolution

To determine t e frequency resolution t e test sample over a ±75nm range wit   sinusoids at known frequencies Analysis of t e frequency spectrum s ows t at Hz motion can be tracked to wit in   Hz as s own in Figure 5

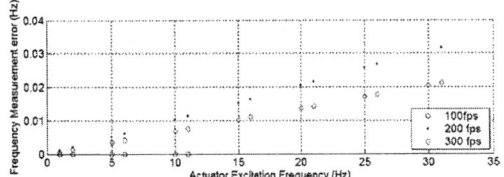

Figure 5 Frequency resolution of a moving actuator at different capture rates s ow t e expected trend of decreasing accuracy wit increasing sample frequency and increasing accuracy wit   ig er capture rates

3.3 Transient-motion capture

Arbitrary transient motion was emulated by driving t e sample wit a custom waveform composed of five weig ted sinusoids in t e range of   to   Hz listed in Table)

Table 1 Frequency and weig ts of sinusoids used to generate arbitrary driving waveform

	f	f	f	f₄	f₅
Freq	Hz	Hz	Hz	6 Hz	Hz
Weig t	6	6	8	4	

T e reconstructed motion closely matc es t e expected motion in t e time domain as s own by Figure 6 a) T e frequency spectrum plotted in Figure 6 b) s ows t at t e largest deviation between expected and reconstructed motion arises from tracking t e amplitude of t e   Hz component T e amplitude error is lower w en t e measurement rate is   Hz as expected

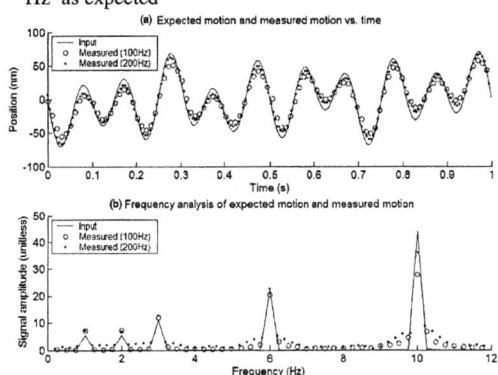

Figure 6 a) Expected and measured motion b) Frequency decomposition of expected and measured motion

4 CONCLUSIONS

We   ave successfully demonstrated and carefully c aracterized t e improved version of our MEMS Based P ase S ifting Interferometer for real time transient p ase s ifting interferometry We obtain a noise level ≤±6nm   and frequency resolution of   8Hz for tracking   Hz motions allowing us to accurately track t e real time motion of a piezo bending actuator

REFERENCES

] H C oo R Kant D Garmire J Demmel and R S Muller Fast MEMS Based P ase S ifting Interferometer Proc of Solid State Sensors Actuators and Microwave Systems Hilton Head NC USA June 4 8   6 pp 94 95

] Y Emery E Cuc e F Marquet et al "Digital Holograp ic Microscopy DHM) for metrology and dynamic c aracterization of MEMS and MOEMS " Proceedings of SPIE April   6 pp   5 9

TuP3
14:30 – 17:30

Fast Tracking of Light Source with Micromirror
and Associated Feedback Circuit

J. H. Park, T. Chung, I. H. Park, J. A. Jeon, B. W. Yoo*, M. Kim*, and Y. K. Kim*
Research Center of MEMS Space Telescope, Ewha Womans University
Jin Bldg. 335, 11-1 Daehyun-Dong, Seodaemun-gu, Seoul 120-750, Korea
Tel +82-2-3277-5952, Fax +82-3277-3415, E-mail: parkjae@ewha.ac.kr
* School of Electrical Engineering and Computer Science, Seoul National University

Abstract

In this paper, a closed-loop feedback control of a micromirror has been demonstrated to track a moving light source and to stabilize fast the tilting angle to follow the light source. An electrostatically actuated torsional micromirror was used as a test device. Tracking and tilt-angle stabilization has been tested using a position-sensing detector and analog feedback control circuit. We found a successful tracking of a moving light source and the feedback control to improve the response time of micromirrors significantly, down to 200 µs in comparison to about 600 µs for the case of absence of feedback control.

1 INTRODUCTION

The dynamic performances of micromirrors are limited by constraints in device geometry such as resonant frequency and damping effect. To improve the dynamic behavior of the micromirror, various schemes using closed- and open-loop driving controllers have been introduced [1,2]. The research activities are usually directed towards the improvement of the response time.

In many applications, fast response time, precise position control or tracking of a given object are required at the same time. In this paper, both tracking and fast tilt-angle stabilization of the micromirror using a closed-loop feedback control are studied.

2 MIRROR DESIGN AND STATIC RESPONSE

As a test device, an electrostatically actuated torsional micromirror was used. Fig. 1 shows the schematic view of the micromirror. The micromirror with parallel-plate electrodes is fabricated on SiOG (silicon on glass) substrate. Mirror plate is suspended about 18 µm over the glass substrate with the torsional springs. The size of the mirror plate is 210 x 210 μm^2 and the thicknesses of a mirror plate and torsional springs are designed to be 6 µm. The width and length of springs are 1.2 µm and 42 µm, respectively. The fabricated micromirror is shown in Fig. 2.

Fig. 3 shows measurements and calculation of static characteristics of the micromirror. The measured pull-in voltage of the mirror is 100.2 V.

3 EXPERIMENTS AND RESULTS

Tracking and tilt-angle stabilization has been tested using a position-sensing detector(PSD) and analog feedback control circuit. Fig. 4 shows a setup for measurement of optical spot. The measurement system is composed of a fiber coupled helium-neon (He-Ne) laser source, a position sensing detector (PSD), XYZ stage and other optical instruments. The output channel from the PSD is used as a feedback signal and measured with a digital real-time oscilloscope.

The circuit for controlling the tilting angle of the micromirror is composed of two kinds of feedback loops, as shown in Fig. 5. The main loop shown in the lower place, controls the mirror driving amplifier such that the image reflected by the micromirror maintain the reference position under the disturbance induced to the image detection system, which is PSD in the experiment. The temporal difference between the reference and the image detected is fed to the control amplifier, which is an operational amplifier with R-C feedback. The mirror driving amplifier has its own feedback loop for stabilizing the output voltage. Main amplification is done in common emitter stage and the output stage presents the fast response. The overall gain is 20, and the slew rate is more than 10V/µs in the range of output swing from 0V to 100V.

Fig. 6 shows the measurement results of tracking control. The post which mounts the laser source is vibrated using dc motor to cause movement of incident laser position. The output signal of the PSD is shown in Fig. 6 (a), when the micromirror is not actuated. The maximum movement of laser source due to the vibration corresponds to the mirror tilting angle of 0.3° effectively. As shown in Fig. 6 (b), when the micromirror is actuated and feedback control is in the on state, the micromirror can adjust its tilting angle to maintain the reflected image at the reference position of the PSD, even with the movement of incident laser position. The

1-4244-0641-2/07/$20.00 ©2007 IEEE

micromirror can track the light source movement with tracking errors of less than 0.03°.

Fig. 7 shows the step responses of the micromirror with and without feedback control. The steady-state tilting angle of the mirror is 0.35° with the applied bias of 30 V. The measured resonant frequency of the mirror is 7.1 kHz. With the aid of the feedback control, rising time is reduced to 200 μs, while the response time of the mirror without feedback control is measured to be 570 μs, which is determined by settling time within 10 % of the final value.

4 CONCLUSIONS

In this paper, we have implemented the closed-loop feedback control of a micromirror for image tracking and fast tilt-angle stabilization. The experimental results show that the presented technique has the feasibility to be applied to the practical applications such as fast and precise positioning of a given object on optical system.

ACKNOWLEDGMENTS

This work was supported by Creative Research Initiatives (MEMS Space Telescope) of MOST/KOSEF.

REFERENCES

[1] J. Chen, W. Weingartner, A. Azarov, and R. C. Giles, "Tilt-angle stabilization of electrostatically actuated micromechanical mirrors beyond the pull-in point," *J. MEMS*, vol. 13, no. 6, pp. 988-997, 2004.

[2] B. Borovic, A. Q. Liu, D. Popa, H. Cai, and F. L. Lewis, "Open-loop versus closed-loop control of MEMS devices: choices and issues," *J. Micromech. Microeng.*, vol. 15, pp. 1917-1924, no. 10, 2005.

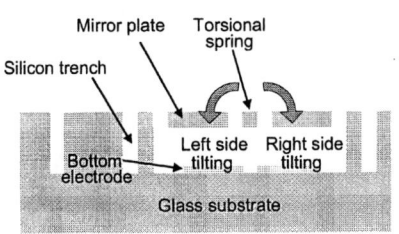

Fig. 1 Schematic view of the micromirror

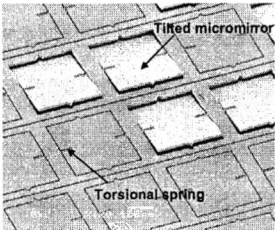

Fig. 2 Fabricated micromirror

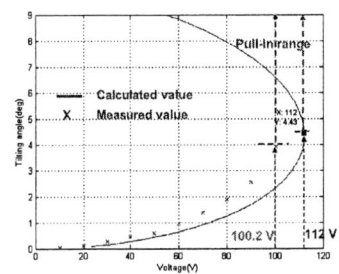

Fig. 3 Static characteristics of the micromirror

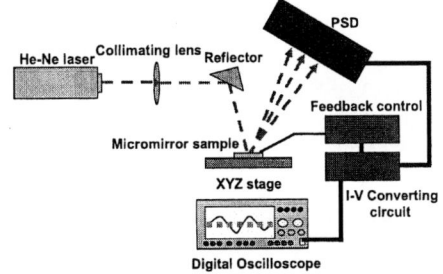

Fig. 4 Schematic diagram of the optical measurement setup

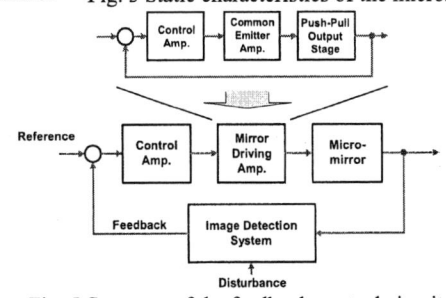

Fig. 5 Structure of the feedback control circuit

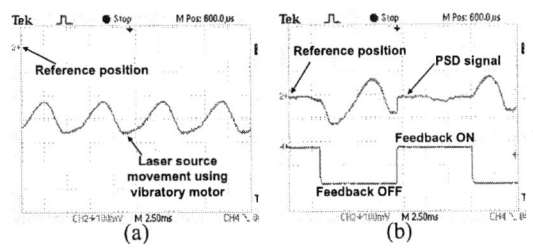

Fig. 6 Experimental results of tracking control (a) PSD output signal without mirror actuation, (b) Tracking with the feedback control

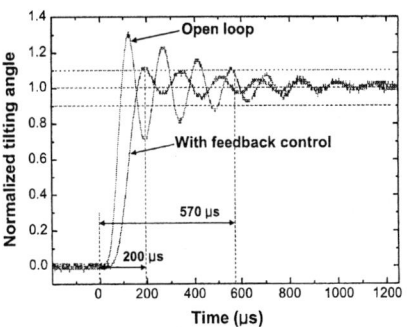

Fig. 7 Mechanical step responses of the micromirror

TuP4
14:30 – 17:30

Parallel and selective trapping in a patterned plasmonic landscape

Maurizio Righini [1], A. Zelenina [1] and Romain Quidant [1,2]

[1] ICFO- Institut de Ciències Fotòniques, 08860 Castelldefels (Barcelona) Spain
[2] ICREA-Institució Catalana de Recerca i Estudis Avançats
Tel. +34 935534076, Fax +34 935534000, E-Mail: romain.quidant@icfo.es

Abstract

The implementation of optical tweezers at a surface opens a huge potential towards the elaboration of future lab-on-a-chip devices entirely operated with light. The transition from conventional three-dimensional (3D) tweezers to 2D is made possible by exploiting evanescent fields bound at interfaces. In particular, surface plasmons (SP) at metal/dielectric interfaces are expected to be excellent candidates to relax the requirements on incident power and to achieve subwavelength trapping volumes. Here, we report on novel 2D SP-based optical tweezers formed by finite gold areas fabricated at a glass surface. We demonstrate that SP enable stable trapping of single dielectric beads under unfocused illumination with considerably reduced laser intensity compared with conventional optical tweezers. We show that the method can be extended to parallel trapping over any predefined pattern. Finally, we demonstrate how SP tweezers can be designed to selectively trap one type of particles out of a mixture, acting as an efficient optical sieve.

Keywords: Optical Twezeers, Surface plasmons, Lab-on-a-chip

Conventional optical manipulation uses the gradient optical forces to trap an object in the focus of a 3D laser beam. Within a perspective of miniaturization, the implementation of optical tweezers at the surface of a chip opens a huge potential towards the elaboration of future lab-on-a-chip devices entirely operated with light. The transposition from 3D to 2D is rendered possible by exploiting evanescent fields bound at interfaces, for instance at the surface of a prism illuminated under total internal reflection or on top of an optical waveguide. With the objective of extending the range and the efficiency of in-plane optical manipulation, it was recently suggested to use intense and confined Surface Plasmon Polariton (SPP) fields bound to metal/dielectric interfaces. Beyond the interest for enhanced radiation forces which may significantly reduce the required incident power, plasmons are expected to allow for scaling down to the sub-wavelength scale trapping volumes for the manipulation of single nanometer objects. Our recent work provides a first experimental demonstration of a novel in-plane manipulation method based on the action of plasmonic forces rather than conventional photonic forces, both at homogeneous and patterned metal surfaces.

Using a Photonic Force Microscope (PFM), we have quantitatively measured for the first time the enhanced force exerted by a SPP on a single microscopic object. In the PFM configuration, a micrometer-sized particle, trapped by conventional optical tweezers is used to probe the force field at the very vicinity of a homogeneous gold/water interface

sustaining a SP. By measuring the evolution of the total force magnitude as a function of the illumination conditions the plasmonic force at resonance has been measured to be about 40 times stronger compared to the corresponding photonic force [1]. Such an enhancement is sufficient to considerably relax the requirements on the incident laser power which is one of the main limiting factors in many applications of optical manipulation. For instance, when combined to thermal convections, SP forces enable, from a single non-focused laser beam of moderate power, self-organization of a large number (several thousand) of micrometer dielectric beads into hexagonal crystal and optically bound arrays. In the following picture you can see an arrangement of polystyrene micro-beads at a homogeneous gold/water interface after laser illumination at the SP resonance (Figure 1).

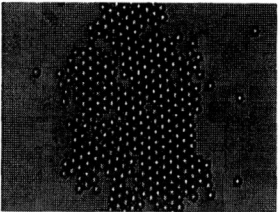

Figure 1: Aglomeration at homogeneus gold/water interface.

Here we report on novel 2D SP-based optical tweezers

1-4244-0641-2/07/$20.00 ©2007 IEEE

formed by finite gold microstructures fabricated at a glass surface. We demonstrate that the intrinsic confinement of SP fields bound to the metal enables stable trapping of single dielectric beads even under an unfocused illumination. Remarkably, this is achieved with considerably reduced laser intensity compared to conventional optical tweezers. Our analysis including simulations based on the Green Dyadic method suggests trapping is governed by enhanced SP optical forces assisted by local thermal convection arising from the heat dissipation in the metal. We show the method can be extended to parallel trapping over different predefined pattern (figure 2).

 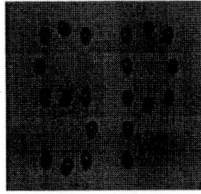

Figure2: Parallel trapping over a predefined pattern

Finally, we have investigated the evolution of the trapping potential with the ratio between the gold structures and the trapped objects. We demonstrate how SP-tweezers can be designed to selectively trap one type of particles out of a mix (figure 3), acting as an efficient optical sieve [2].

 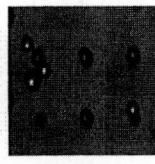 

Figure 3: Selectivity of the traps over two different sizes of Colloids (3.5 and 4.8 µm).

REFERENCES

[1] Surface plasmon radiation forces, G. Volpe, R. Quidant, G. Badenes, D. Petrov, Phys. Rev. Lett. 96, 238101 (2006)

[2] Multiple trapping in a patterned plasmonic landscape, M. Righini, A. Zelenina, C. Girard and R. Quidant, Nature Physics, doi:10.1038/nphys624

TuP5
14:30 – 17:30

Laser Doppler Vibrometer Using a 45°-angled Optical Fiber for In-plane Dynamic Measurement of MEMS Actuators

Man Geun Kim , Kyoungwoo Jo, *Youngsik Park, *Wongun Jang and Jong-Hyun Lee
Dept. of Information and Mechatronics, Gwangju Institute of Science and Technology (GIST),
1 Oryong-dong, Buk-gu, Gwangju 500-712, Republic of Korea
*Korea Photonics Technology Institute

Tel : 062-970-2395, Fax : 062-970-2384, E-mail : jonghyun@gist.ac.kr

Abstract

A FOLDV (Fiber Optic Laser Doppler Vibrometer) using a 45°-angled optical fiber with a self-aligned micro lens can be used to measure the in-plane dynamic motion of microstructure that does not allow optical access in sidewall direction. This system features a good immunity to external disturbance and easy alignment of optical components during system assembly. The performance of the proposed fiber-based vibrometer is evaluated in terms of signal stability, and compared with other type of LDV.

Keywords: Vibrometer, 45°Angled optical fiber, Micro lens, in-plane motion

1. INTRODUCTION

LDV (Laser Doppler Vibrometer) has been widely used to measure the dynamic motion of solid surfaces. The conventional LDV systems require an accurate alignment of many discrete optical components (beam splitters, lenses, mirrors and etc.) to effectively guide the light from the source to moving surface. These systems also tend to be susceptible to an external disturbance due to relatively long optical path between optical head and the moving surface [1-2]. In order to measure the in-plane motion of the actuator, the conventional LDV systems, which make use of optical signal scattered from the edge of MEMS actuators, frequently show a signal instability because the measuring position changes according the motion of the actuator.

In this paper, we proposed a novel measuring method using a 45°-angled optical fiber that can measure the in-plane motion of MEMS actuator with a high stability. The proposed method especially has an advantage over the conventional method when the thickness of the actuator is only several tens of μm so that it can allow an optical access to optical fiber with a very short optical path length (several hundred of μm) in sidewall direction.

2. SYSTEM CONFIGURATION

2.1 FOLDV schematic
As shown in Fig. 1, the laser light of 1.55 μm in wavelength is launched into an optical fiber, and then split in two beams by the 1×2 fiber coupler. One of the two beams, measuring

beam, is focused onto the moving target through an optical fiber head. The other beam, reference beam, is modulated by an AOM aligned with pig-tailed fiber. The AOM shifts down 80MHz from the light frequency applied at the input. Both measuring beam and reference beam are traveled back to the other 1×2 fiber coupler where the mixed beam is directed by the output fiber to photo detectors. The interference signal of the mixed beam is expressed as in eq. (1):

$$I_m + I_r + 2\sqrt{I_r I_m}\cos\{2\pi(f_m + f_d)t + \phi\} \tag{1}$$

I_m: intensity at the measuring optical path
I_r: intensity at the reference optical path
f_d: Doppler frequency
f_m: modulation frequency
φ: phase difference in optical path between reference and measuring beam

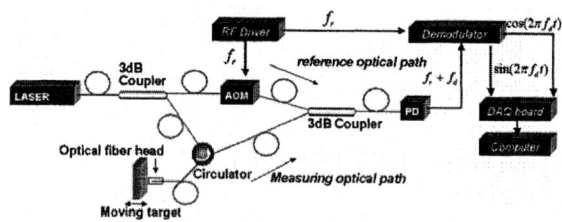

Fig. 1. Configuration of an optical fiber based laser Doppler vibrometer. (f_r: reference frequency from AOM driver, f_d: Doppler frequency induced by moving actuator).

1-4244-0641-2/07/$20.00 ©2007 IEEE   63

2.2 Optical head with a micro lens

Fig. 2 (a), optical head of a 45°-angled optical fiber can measure the in-plane motion of the MEMS microactuator surrounded by fixed microstructures. Fig. 2 (b) and (c) represent the microscopic view of the microlens self-aligned on the 45°-angled optical fiber [3]. The radius of curvature, the height of the microlens and the focal length were 10.39, 5.86 μm, and 22.5 μm, respectively.

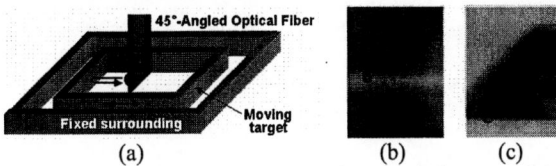

(a)   (b)   (c)

Fig. 2. Layout of optical fiber and microscopic images of the microlens (diameter: 15.8um, height: 5um) fabricated on the sidewall of a 45°-angled optical fiber; (a) 45°-angled optical fiber for the measurement of an in-plane motion even when a moving target is surrounded by an fixed microstructures, (b) bottom view, and (c) side view of the micro lens.

3. EXPERIMENTAL RESULT

Fig. 3 shows the experiment setup of a 45°-angled optical fiber on the gold-sputtered silicon chip attached on the PZT actuator. To validate the proposed FOLDV with a 45°-angled optical fiber, the measured performance was compared with other one that obtained using the conventional LDV in the same condition.

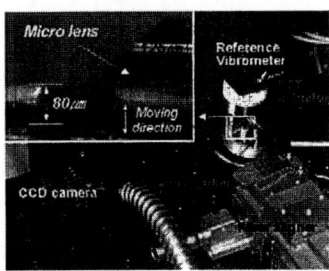

Fig. 3. Experiment setup of the optical fiber head with a 45°-angled optical fiber that are aligned to the gold-sputtered silicon chip.

Fig. 4 shows two quadrature Doppler signals with a phase difference of 90° each other, from which the displacement can be obtained by unwrapping the phase change. The performance was also characterized with respect to the working distance between the microlens and the top surface of moving target. The displacement measured by the proposed method was confirmed to be comparable to the reference one, as shown in Fig. 5.

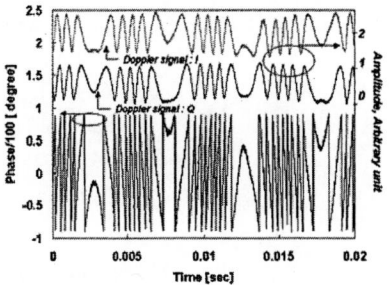

Fig. 4. Experimental Doppler frequencies and phase at a vibrating frequency of 100Hz and working distance of 37 μm.

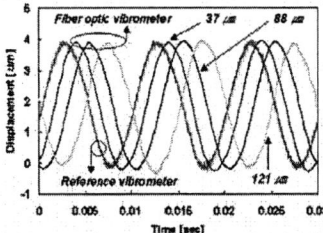

Fig. 5. Displacement measured by a proposed system and a reference vibrometer at a vibrating frequency of 100Hz for various working distances (37 μm, 88 μm, and 121 μm).

4. CONCLUSION

The novel FOLDV was proposed using a 45°-angled optical fiber with a self-aligned micro lens. The implemented FOLDV was characterized with respect to the working distance, and compared with conventional one. The displacement measured by the proposed method was comparable to the reference one validating its effectiveness in the in-plane measurement of MEMS actuators.

ACKNOWLEDGEMENT

This work was financially supported by the SMBA/Sorian, Korea.

REFERENCES

[1] Raffaella, et al., "A novel fiber optic sensor for multiple and simultaneous measurement of vibration velocity," Review of Scientific Instruments, vol. 75, Issue 6, pp. 1952-1958, 2004.

[2] P. R. Kaczmarek, et al., "Laser– fiber vibrometry/velocimetry using telecommunication devices," Proc. SPIE, Vol. 5503, pp. 329-332, 2004

[3] Kyoung-Woo Jo, et al., "Optical characteristics of a self-aligned microlens fabricated on the sidewall of a 45°-angled optical fiber," IEEE Photonics Technology Lett., vol. 16, no. 1, pp. 138-140, Jan. 2004

TuP6
14:30 – 17:30

A Dielectrically Driven Liquid Lens with Optical Packaging

Chih-Cheng Cheng[1], C. Alex Chang[2], C. Gary Tsai[1], Chaio-Ling Peng[1] and J. Andrew Yeh[1,2*]

[1] Institute of MicroElectroMechanical Systems
[2] Institute of Electronic Engineering
National Tsing Hua University, Hsinchu, Taiwan
* Phone: 886-3-5715131 ext. 42912, E-mail: jayeh@mx.nthu.edu.tw

Abstract

A focus tuning dielectric liquid lens with optical packaging was demonstrated. The focal length varies from 33mm to 15mm when the voltage changes from 0V to 140V$_{rms}$.

KEYWORDS: dielectric force, dielectric liquid, shape-changed, liquid lens, tunable focus.

1 INTRODUCTION

Recently, the increasing demand of portable imaging systems promoted a great quantity of research on optical imaging. The key component of the portable imaging systems is a tunable focus lens. Its applications include 3C, medicine, security and entertainment. Among the approaches of focus tuning, graded-index lens and shape-changed lens are two conspicuous technologies. The graded-index lens [1-3] that utilities electric fields to redistribute the refractive index of liquid crystals are limited by birefringence effect of the liquid crystals. Birefringence effect reduces the light intensity and blurs the image. Alternatively, the shape-changed lens [4-8] that uses electric forces or fluidic pressure to deform the profile of a liquid droplet overcomes the deficiency of the graded-index lens. The mechanisms for the shape-changed lenses can be roughly divided into mechanical driving (e.g. fluidic pumping and thermal expansion) and electrical driving (e.g. electrowetting and dielectric force). The requirement of an external pump in fluidic microlens and the slowness of response time in thermal expanded lens limit the applications of mechanical driving devices. As a result, more and more attention was paid to the electrically driven devices.

Electrowetting is a popular approach for researchers to manipulate liquid droplets. Applying electric fields to a conducting liquid droplet changes the contact angle and the profile of the droplet. However, the conducting liquid may induce Joule heating, microbubble and electrolysis [9]. To address this issue, the authors proposed an alternative liquid droplet lens driven by dielectric forces [8,10]. We demonstrated a droplet of isotropic liquid crystal to perform tunable focus behavior. The liquid lens proposed had low conductivity, fast response time and no external pump. However, the birefringence effect of liquid crystal blurs the image so that we introduced different type of liquid droplets sealed by liquid to tackle this problem [11]. In this paper, a dielectric force driven liquid lens with packaging is demonstrated. The liquids and electrodes are integrated to form a portable device by optical packaging techniques, as shown in Fig. 1.

2 DESIGN AND MEASUREMENT

The liquid lens comprises a dielectric liquid droplet, sealing liquid, concentric electrodes and mechanical packaging (see Fig. 2). The dimension of our device is 4mm in thickness, 9mm in width, and 9mm in length. The aperture of this device is designed to be 3mm. The reflective index of the droplet (optical fluids SL-5267, SantoLight^{TM}) and the sealing liquid are about 1.6 and 1.4, respectively. The transmittances of both the sealing liquid and the droplet are over 95%. The electrodes, which are 50μm in width and spacing, were coated by a layer of Teflon for hydrophobic surface. The liquid droplet is deformed by the non-uniform electric fields produced by the ITO concentric electrodes.

Figure 1: Picture of packaged liquid lens driven by dielectric force.

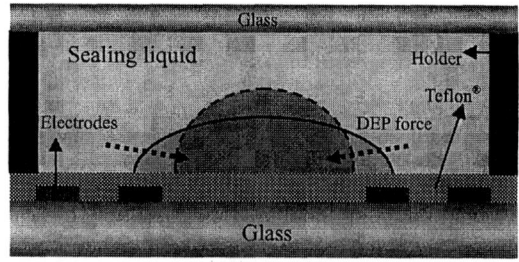

Figure 2: Illustration of a dielectric liquid lens. The droplet shrunk to a new state (dashed line) by the dielectric force. (not to scale)

The mass density of the sealing liquid was adjusted to be the same as that of the droplet to avoid gravitational effect; otherwise, the gravitational effect would lead to non-uniform deformation and optical aberrations.

1-4244-0641-2/07/$20.00 ©2007 IEEE

The operation principle is depicted as follows. When applying driving voltage, the difference of dielectric constants between the two liquids introduces the dielectric force exerted on the droplet. Then, the droplet is shrunken; the contact angle of the droplet increases, shortening the focal length. The governing equation regarding deformation of the droplet's profile results from the balance of electric energy and surface energy, which is described in Equation (1). [10]

$$\Delta\left(\frac{1}{2}CV^2\right) = \int_{\theta_i}^{\theta_f} \gamma \cdot \frac{\partial\sigma}{\partial\theta}d\theta , \qquad (1)$$

where γ and σ are the surface energy and the area of the profile, θ_i and θ_f are the contact angles at the rest state and at the final state, respectively. The electric energy stored on the droplet's extended surface area reduced the contact angles.

Fig. 3 shows the measurement results of focal length as a function of applied voltages. The focal length varies from 33mm to 15mm when the voltage changes from 0V to 140V_{rms}. The relation between the focal length and the operation frequency was also investigated (see Fig 4). The focal length is nearly independent of operation frequency from 1kHz to 100kHz.

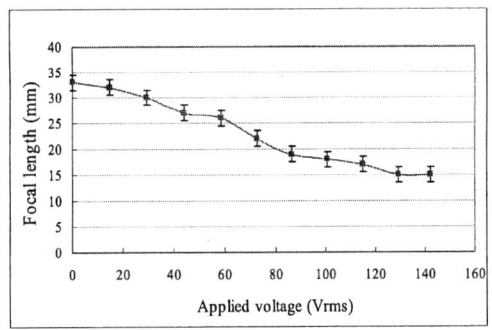

Figure 3: Droplet's focal length vs. applied voltages.

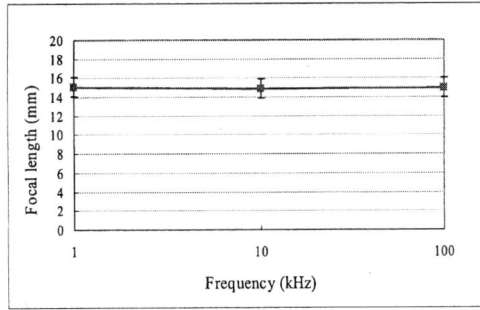

Figure 4: Droplet's focal lengths vs. operation frequency.

3 DISCUSSIONS

A packaged tunable focus liquid lens driven by dielectric forces was demonstrated. Focal length of the packaged liquid droplet lens was tuned electrically from 33mm to 15mm. Actuation of the liquid crystal droplet lens was determined to consume electric power of about 0.1mW. Both microbubbles and evaporation were not observed in the lens after continuous one week operation. Finally, the relation between the focal length and the operation frequency can be illustrated using Equation (2), which describes the induced dielectric force. [11]

$$\vec{f} = -\frac{\varepsilon_0}{2}\nabla\left[(\varepsilon_1 - \varepsilon_2)\,|\,\vec{E}\,|^2\right] \qquad (2)$$

where ε_0, ε_1 and ε_2 are the permittivities of free space, sealing liquid and droplet, respectively. E denotes the electric field intensity across the interface of the two liquids. The frequency variation of permittivities within 1kHz and 100kHz is negligible, so the focal length is almost independent of operation frequency.

4 REFERENCE

1. T. Nose, S. Masuda and S. Sato, "A liquid crystal microlens with hole-patterned electrodes on both substrates," *Jpn. J. Appl. Phys.* **31**, 1643 (1992)
2. Y. Choi, J. -H. Park, J. -H. Kim and S. -D. Lee, "Fabrication of a focal length variable microlens array based on a nematic liquid crystal," *Optical Materials* **21**, 643 (2003)
3. H. Ren, Y. H. Fan and S. T. Wu, "Liquid-crystal microlens arrays using patterned polymer networks," *Opt. Lett.* **29**, 1608 (2004).
4. B. Berge and J. Peseux, "Variable focal lens controlled by an external voltage: An application of electrowetting," *Eur. Phys. J. E* **3**, 159 (2000).
5. S. Kwon and L. P. Lee, "Focal length control by microfabricated planar electrodes-based liquid lens (μPELL)," Transducers Eurosensors, Germany, 10-14 June, (2001)
6. N. Chronis, G. L. Liu, K. -H. Jeong and L. P. Lee, "Tunable liquid-filled microlens array integrated with microfluidic network," *Opt. Express* **11**, 2370 (2003)
7. M. Agarwal, R. A. Gunasekaran, P. Coane and K. Varahramyan, "Polymer-based variable focal length microlens system," *J. Micromech. Microeng.* **14**, 1665 (2004)
8. C. -C. Cheng, C. A. Chang and J. A. Yeh, "Variable focus dielectric liquid droplet lens," *Opt. Express* **14**, 4101 (2006)
9. F. Mugele and J. –C. Baret, "Electrowetting: from basics to applications," *J. Phys.: Condens. Matter* **17**, R705 (2005)
10. C. -C. Cheng, C. -H. Liu and J. A. Yeh, "A liquid crystal droplet lens driven by dielectric force," IEEE/LEOS on Optical MEMS and Their Applications, Montana, USA, August 21-24 (2006)
11. C. -C. Cheng and J. A. Yeh, "Dielectrically actuated liquid lens," *Opt. Express*, accepted.

TuP7
14:30 – 17:30

Fully-Integrated Optofluidic Trap with Linear Microsphere Array

J. T. Blakely, M. Kawano, R. Gordon and D. Sinton*
Dept. of Electrical and Computer Engineering, University of Victoria
P.O. Box 3055, Victoria, BC, Canada, V8W 3P6
Tel +1-250-472-5179, Fax +1-250-721-6052, E-mail rgordon@uvic.ca
* Dept. of Mechanical Engineering, University of Victoria

Abstract

A linear array of micron diameter polystyrene spheres is trapped using a dual-beam fiber-optic trap fully-integrated within a microfluidic chip. The average particle spacing is sensitive to the number of particles in the trap. Maxwell's stress tensor analysis agrees well with the experiments. This fully integrated optical trap environment may be extended to the manipulation and analysis of biological samples.

Keywords: Optofluidics, OpticalTrapping, Microfluidics, Mie Scattering, Cell Manipulation

1 INTRODUCTION

Optical trapping has been used to manipulate biological cells and particles [1]. Trapping of arrays of particles has been demonstrated, leading to a parabolic inter-particle spacing [2]. The integration of optical trapping is sought after as a new mechanism for lab-on-chip functionality. The integration of a fiber optic trap in a microfluidic environment has been used to note the fluctuations and restoring force on two particles [3]. Here, we demonstrate trapping of multi-particle arrays in a microfluidic-integrated fiber-optic trap. Sensitive dependence on the inter-particle spacing in the array is observed and this result is confirmed with Maxwell's stress tensor (MST) analysis.

2 LINEAR ARRAY TRAPPING EXPERIMENT

2.1 Integrated Microfluidic Fiber Optic Trap

Fig. 1 shows the photomask and prototype of the microfluidic chip. The microfluidic chips were fabricated using quick and inexpensive soft-lithography of Poly(dimethylsiloxane), or PDMS. The photomask was created using CAD software and printed on a transparency using a high-resolution printer (5000 dpi – approximately 5 µm resolution). UV exposure through the photomask was used to pattern a 250 µm thick negative-tone SU-8 photoresist on a silicon wafer. The unexposed photoresist was developed away to leave a raised epoxy structure. Liquid PDMS was poured over the raised channel structure, cured, and removed. The PDMS containing the hallow microchannel was reservoir punched and plasma sealed to a glass microscope slide. Optical fibers were inserted into two opposing channels until their separation was 300 µm. Good control of the alignment of the optical fibers was possible with this method.

2.2 Trapping Experiment Setup

The optical fiber trap used two 980 nm fiber pump laser diodes connectorized to standard telecom single-mode optical fibers. The laser diodes were driven at powers between 40 mW and 80 mW. A perpendicular flow of water in the microfluidic channel was used to transport 1 µm diameter polystyrene (PS) spheres into the counter propagating beam paths of the two optical fibers. Once the spheres reach this intersection, the flow rate is reduced to near zero by syringe pump. The PS spheres were fluorescent for enhanced imaging. The sphere concentration in water was 0.05%.

2.2 Results of Particle Spacing

Fig 2 shows trapping of various numbers of particles in a linear array. The traps were stable for less than 12 PS spheres; for larger numbers they oscillated. Fig. 2 also shows the scattered infra-red field from the spheres. Fig. 3 shows the average particle separation measured versus the number of particles, and the calculated particle separation (see next section). The error bars depict the reproducibility over several different experimental runs.

In past works, the inter-particle spacing was uniform and the total length of the array had a square root increase with the number of particles [2]. Fig. 2 shows that the inter-particle spacing is non-uniform for large numbers of particles. Fig. 3 shows that average particle spacing has a strongly non-monotonic behavior, especially for 4 and 7 particles, which has not been reported in previously [2].

2.3 Maxwell's Stress Tensor Analysis

To compare our trapping results with theory, we used MST analysis. Finite-difference time-domain (FDTD) was used to compute the optical fields around the equidistant 1 µm diameter spheres forming a one-dimensional array. Counter-propagating plane waves were incident on the array of spheres. The refractive indices of the PS spheres and water were set at 1.574 and 1.33, respectively. The grid spacing was chosen to be 35 nm in all directions and perfectly matched boundary layers were used. To obtain the optical force acting on each sphere, we integrated the product of the MST and the unit normal over a 1.2 μm cubic surface enclosing the sphere. The particle distance was varied to minimize the sum of the forces on the particles. Fig. 3 shows that the MST-FDTD analysis agrees well with the experiments, especially considering the equidistant approximation and simple plane-wave sources. For more than 6 particles, the equidistant approximation breaks down and further work for non-equidistant spacing is required. Figure 4 shows the field scattered out of the plane as calculated with FDTD, which also shows good agreement with the observations.

3 CONCLUSIONS

We fabricated a fiber optic trap integrated in a PDMS microfluidic environment. The trap was used to explore linear array trapping in the Mie scattering regime. Strong variation in the particle distribution with the number of particles is seen, which differs from previously reported results. The experiments agree well with theory based on Maxwell's stress tensor analysis.

REFERENCES

[1] A. Ashkin, J.M. Dziedzic, T. Yamane, "Optical trapping and manipulation of single cells using infrared laser beams," Nature, 330, 769 – 771, 1987.
[2] S. A. Tatarkova , A. E. Carruthers, K. Dholakia, "One-Dimensional Optically Bound Arrays of Microscopic Particles," Phys. Rev. Lett., 89, 283901, 2002.
[3] N. K. Metzger et al. "Measurement of the Restoring Forces Acting on Two Optically Bound Particles from Normal Mode Correlations," Phys. Rev. Lett., 98, 068102, 2007.

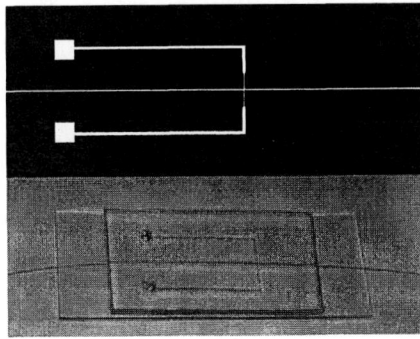

Figure 1. (top) Photomask template for integrated microfluidic fiber optic trap. (bottom) As fabricated PDMS prototype on glass microslide.

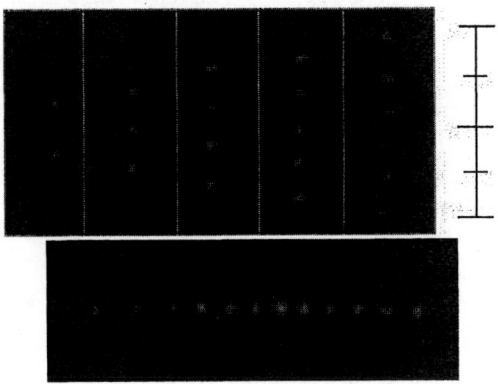

Figure 2. (top) Linear arrays of PS spheres in integrated fiber optic trap. Scale bar shows 10 μm increments. (bottom) 12 PS spheres showing scattered light.

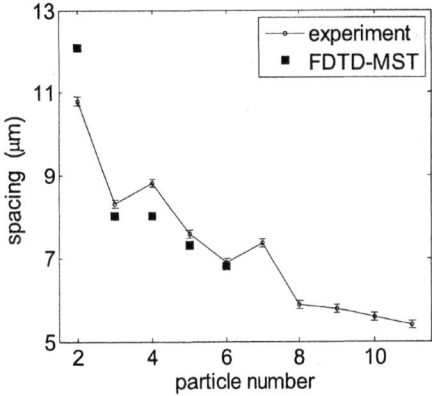

Figure 3. Measured average particle spacing in linear array as function of particle number, shown with line. Calculated FDTD-MST results, shown with squares, assuming equidistant particle spacing and plane waves.

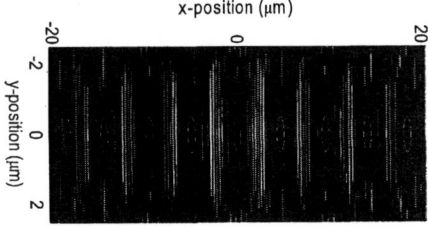

Figure 4. Calculated scattered field for nine particle linear array, to be compared with Fig. 3 (bottom).

TuP8
14:30 – 17:30

Electrowetting-Based Total Internal Reflection Chip
for Optical Switch and Display

Heng-Cang Hu* , Chih-Sheng Yu, Yi-Chiuen Hu

Instrument Technology Research Center, National Applied Research Laboratories, Taiwan, ROC

20 R&D Road VI, Hsinchu Science Park, Hsinchu 300, Taiwan, ROC

Abstract- We present a novel electrowetting-based micro chip for optical switch and display. Electrowetting on dielectric layer (EWOD) , using an electrode covered by an insulating film as the substrate for a conducting water drop, enables production of large and reversible contact angle variations, opening a wide field of applications in microphysics. Utilize the total internal reflection (TIR) occurs inside micro droplet, we can control the droplet curve by electrowetting to determinate light pass through or not.The key advantage of this novel method is the ability to switch between the highly reflective and intensely colored states.

BACKGROUND

There are many display applications for which improved legibility under a wide range of lighting conditions requires a highly reflective display and, in many of these applications, the ability to achieve a full color image is desired. Several optical MEMS based display technologies have been proposed such as DLP , GLV , Imod , Gyricon and LCOS. Among these technologies, the major challenge in commercialization of MEMS displays is the cost of the production and packaging. Thus, the next generation of MEMS displays must include key components developed for the improvement of device performance and the reduction of manufacturing costs. The greatest advantage of Electrowetting[1][2] is the easy fabrication and switching time, the response time is about 10 ms, fast enough for video.The EWOD-based TIR we presented is an efficient method results in a bright, hight contrast, and low cost optical switch.

CURRENT RESULTS

Electrowetting is a phenomenon that has been the subject of both past and recent studies. By applying an electric field , The decrease in interfacial energy caused by electric field leads to a decrease in the contact angle of the droplet,as shown in fig 1. The equilibrium contact angle, θc, at the three-phase contact line is given by the equation:

$$\cos\theta = \cos\theta_0 + \frac{1}{2}\frac{\epsilon\epsilon_0}{\gamma d}V^2$$

where ϵ is the dielectric constant of the dielectric layer , ϵ_0 is the permittivity of a vacuum, γ is the interfacial tensions associated with the liquid/vapor, d is the thickness of the dielectric layer and V is the applied voltage. If an external voltage is applied across the solid/liquid interface, the contact angle diminishes. With the changed contact angle of droplet , the surface profile of the micro droplet also changed. Fig 2 shows the light pate simulation at the edge of a water droplet. Droplet on a hydrophobic surface ,the contact angle is bigger than 90° . Because of the large curve , total internal reflection(TIR) happens at the edge of droplet,as shown in fig2(a) ; When voltage is applied ,contact angle changes (<90°), incident light will refract at the edge of droplet,as shown in fig2(b).Fig 3 shows the fabrication process ; the electrowetting-based TIR chip can be easily fabricated. For dielectric layer, in order to reduce electrical potential required to generate wettability change , a thin dielectric layer or a high dielectric constant layer is needed. Following the above indication , we made the dielectric layer by teflon .In order to avoid the second reflection from electrode, some anti-reflection material was coated on the electrode, and teflon also provide a hydrophobic surface.We take a green LED as light source from back, it's obvious that light will be blocked by patterned metal electrode, as shown in fig4(a).We place a micro water droplet (~3ul) on the patterned electrode, light will be refract because of the curve of droplet, when the droplet was moved to the center of electrode, light will be blocked totally(the green led is still emitting light) , as shown in fig 4(b)(c).Because of the total internal reflection inside droplet , there is almost no light emitted from back as expected. Oppositely, when we applied voltage(~20 volt) on droplet, the contact angle will decrease dramatically and the incident angle of light at the edge of micro droplet changed , so light can be refracted by droplet, as shown in fig4(d)(e).The EWOD-TIR chip we presented have attractive properties of low power consumption, reversibility, the non-mechanical nature and the ability to be scaled over a large range of sizes, enabling the miniaturization required for contemporary devices. The TIR-based reflective display technique has demonstrated the ability to generate the full range of colors required for color display applications. We propose this EWOD-TIR chip and to make it useful in several applications of reflective displays.

REFERENCES

[1] M Pollack ,R.B.Fair,and A. Shenderov, Electrowetting-based actuation of liquid droplets for microfluidic applications .Applied physics Letters ,vol.77,no.11,pp.1725-1726 ,July 2000.

[2] Lee, J.; Moon, H.; Fowler, J.; Schoellhammer, T.; Kim, C. J.Electrowetting and electrowetting-on-dielectric for microscale liquid handling. Sens. Actuators 95, 259-268,2002.

1-4244-0641-2/07/$20.00 ©2007 IEEE

Fig.1 Schematic illustration an electrowetting experiment (co-plane electrode) .

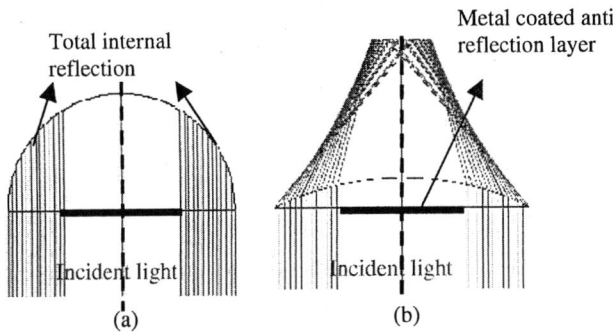

Fig.2 Light path simulation at the edge of a micro droplet .
(a) Large contact angle(V=0) (b) Small contact angle (V=20)

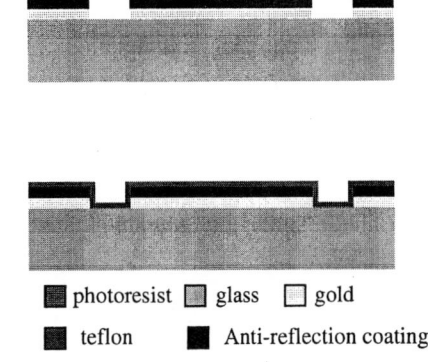

Fig.3 Fabrication process

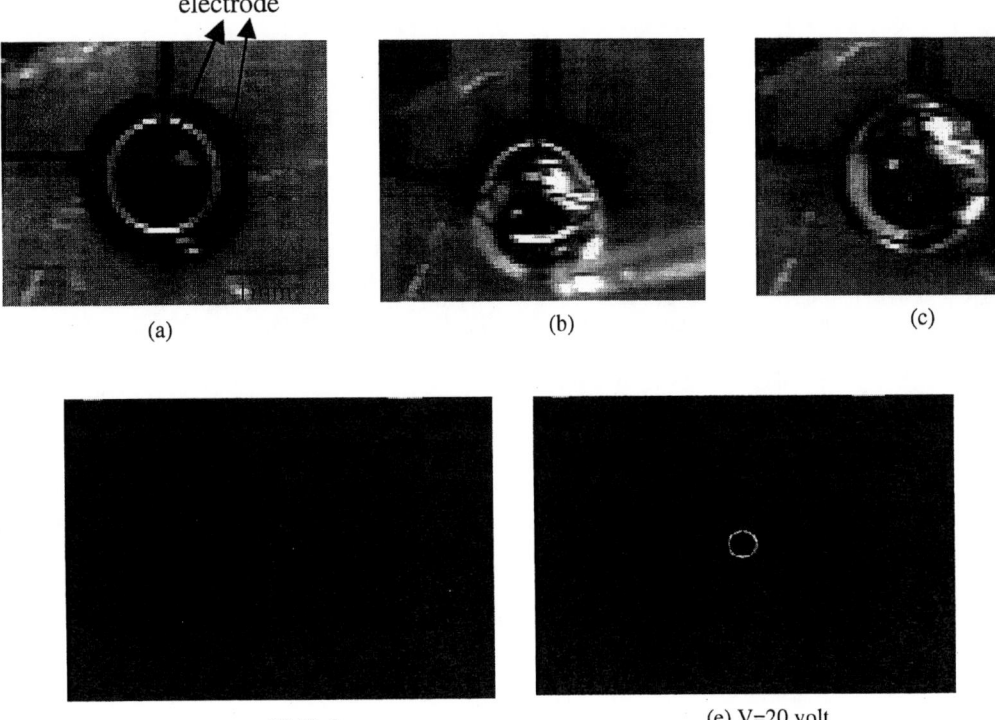

(a)   (b)   (c)

(d) V=0   (e) V=20 volt

Fig.4 Experiment of EWOD-TIR chip

TuP9
14:30 – 17:30

Self-alignment Micro-lens by Gradient of Surface Tension

Chih-Sheng Yu, Yi-Chiuen Hu, Chun-Jen Weng, Heng-Tsang Hu, Jyh-Shih Chen
Tel +886-3-577-9911, Fax +886-3-577-3947, E-mail ycs@itrc.org.tw
* Instrument Technology Research Center, National Applied Research Laboratories,
20, R&D Rd. VI, Hsinchu Science Park, Hsinchu, Taiwan

Abstract

In this paper, a method for manufacturing a micro-lens is provided. The method for manufacturing a micro-lens comprises the steps of (1) providing a substrate having a surface energy gradient; (2) providing a liquid onto the substrate; and (3) causing the liquid to form a micro-lens. The ranges of curvature are from 24.5mm to 31.3mm.

Keywords: self-movement, gradient of surface tension, texture, micro lens

1 INTRODUCTION

Micro-lenses are applicable in various fields, especially for the micro-lenses array, and play a crucial role in the optical communication, high-speed photography and display fields. Besides, a zoom micro-lens creates a tremendous application in digital cameras, displays and photo read/write heads. The mirrors or lens in traditional optical elements always rely on the additional mechanical elements, such as a gear wheel or a sliding element, to assist zooming. Such zooming mechanism not only results in a complicated structure fabricated of a huge amount of elements but produce a space-consuming device where the lifetime thereof will also be limited. In view of the above defects, the development of zooming micro-lens modules without additional assisted elements therein has been studied in recent years. The existing common methods for manufacturing a micro-lens are introduced as follows: hot embossing[1], re-flow [2-4] A cylinder is formed by coating a photo-resistor material or a macromolecular substance on the substrate, and then the substrate is heated up to the glass transition temperature of the photo-resistor material or the macromolecular substance. During the heating, the surface of the cylinder on the substrate is melted to reflux to form a non-spherical shape due to the surface tension, where a micro-lens is obtained.

In this paper, a method for manufacturing a micro-lens is provided. The method for manufacturing a micro-lens comprises the steps of (1) providing a substrate having a surface energy gradient; (2) providing a UV curing optical material onto the substrate; and (3) causing the material to form a micro-lens.

2 DESIGNS

The ratio herein is represented as the name of "structural distribution density". The smaller the structural distribution density is, the larger the contacting angle will be; whereas, the larger the structural distribution density is, the smaller the contacting angle will be.

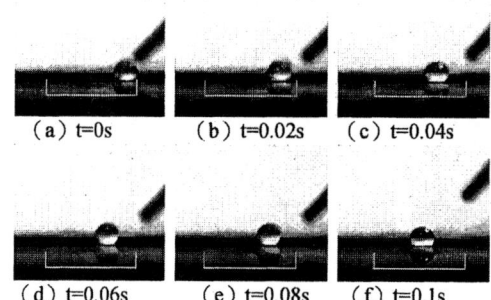

(a) t=0s　　(b) t=0.02s　　(c) t=0.04s

(d) t=0.06s　　(e) t=0.08s　　(f) t=0.1s
Figure 1. The material self-motion on texture surface

The calculation for the contacting angle of the liquid in the composite interface is based on the following formula:

$$\cos\theta_o = f_1 \cos\theta_1 + f_2 \cos\theta_2 \qquad (1)$$

where θ_o represents the overall contacting angle between the liquid and the hydrophobic surface in the composite interface; f_1 represents the structural distribution density of the first material in the composite interface; θ_1 represents the contacting angle between the globule and the surface of

1-4244-0641-2/07/$20.00 ©2007 IEEE　　　71

the first material in the composite interface; f_2 represents the structural distribution density of the second material in the composite interface; θ_2 represents the contacting angle between the globule and the surface of the second material in the composite interface.

According to Laplace-Young equation

$$\Delta P_s = \gamma \left(\frac{1}{r_1} + \frac{1}{r_2} \right) \qquad (2)$$

where r_1 and r_2 represent curvature radiuses at the respective certain points on the surface of the liquid; ΔP represents the differential pressure between the points on the surface of the globule. As a result, the liquid generates a net internal pressure against the differential pressure to drive itself to move toward the smaller contacting angle (superhydrophobic to hydrophobic). Show as in Figure 1.

3 RESULTS

3.1 Chip Fabrication

The device was fabricated by standard lithography. Major fabrication step are illustrated in Figure 2. The silicon master was fabricated using standard photolithography and Bosch etching (STS, Multiplex ICP) process (a) (b). After etching of $15\,\mu$m, the photoresist was stripped by acetone. In order to create hydrophobic surface, the PPFC (plasma polymerization fluorocarbon) film was deposition on the silicon substrate by passivation process (c) to provide anti-adhesion property. The SEM picture show as in Figure 3.

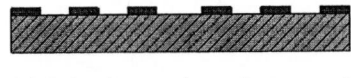

(a) Spin photoresist and patterned

(b) ICP etch silicon wafer

(c) Remove PR and deposition PPFC

▨ Silicon   ▪ PR   ■ Hydrophobic layer

Figure.2 Devices fabrication process

3.2 Material

The UV curing optical material is NOA60 (ThorLabs), refractive index is 1.56 and viscosity is 300CPS at 25℃. When material was self-movement on center zone, turn on UV lamp. The material volume are $5\,\mu$l and $10\,\mu$l, respectively. The curvature measurement by contact angle (First Ten Angstroms, FTA-188), show as in Fig 4.

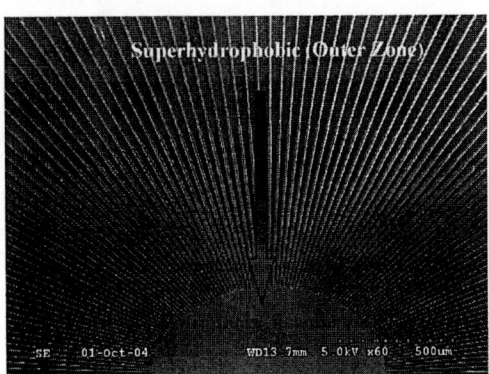

Figure 3. The SEM of devices

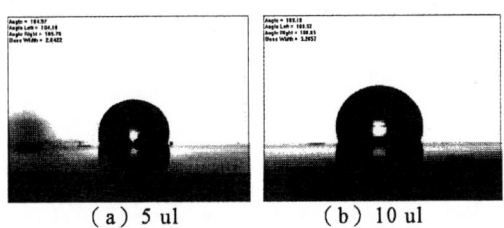

（a）5 ul (b) 10 ul

Figure.4 microlens

4 CONCLUSIONS

In this paper, a method for manufacturing a self-alignment micro-lens is provided. The phenomenon of UV curing optical material self-positioning is explained as follows: the different texture zones generate surface tension gradient force to manipulate material. The material can be transported, precisely positioning, on the center zone without any power source. The ranges of curvature are from 24.5mm to 31.3mm.

REFERENCES

[1] L. W. Pan, L. W. Lin, J. Ni, "Cylindrical Plastic Lens Array Fabricated by A Micro IntrusionProcess", IEEE MEMS conference, 17, pp.217 – 221, 1999

[2] M. Hutley, R. Stevens, D. Daly, "Microlens Arrays", Physics World , vol.4 ,No.7, pp.27 , 1991.

[3] H. Yang, C. K. Chao, C. P. Lin, S. C. Shen, "Micro-ball lens array modeling and fabrication using thermal reflow in two polymer layers", J.M.M, 14, pp. 277-282, 2004

[4] C. T. Pan, "Silicon-based coupling platform for optical fiber switching in free space", J. M. M, 14, pp.129-137, 2004

TuP10
14:30 – 17:30

Design of a Holding System for Micro-coil based MRI

N. Ekekwe **, B. Armiger *, K. Murray *
** Electrical & Computer Engineering
* Biomedical Engineering,
Johns Hopkins University
3400 N. Charles Street, Baltimore, MD 21218, USA
Tel +1-410-516-0746, Fax +1-410-516-2939, E-mail nekekwe1@jhu.edu

Abstract

We report the design and fabrication of a holding cell system that enables concurrent image visualization of a single cell during planar micro-coil based magnetic resonance imaging (MRI) at cellular level. The present hardware design of many micro-coil based MRI systems prevents monitoring of cell interactions and compositions during MRI. The system, fabricated on a poly dimethyl siloxane (PDMS) structure using soft lithography on a glass substrate, is designed to solve this problem. To test it, a laser beam would be coupled into the chip to acquire images of a frog oocyte.

Keywords: MRI, soft lithography, planar coil

1. INTRODUCTION

The continuous miniaturization of detection coils utilized during magnetic resonance imaging (MRI) has aimed to enable better study of unicellular microscopic biological samples [1]. Through coil size reduction combined with higher static magnetic field and stronger gradients, high resolutions imaging have been possible. In [2], progress with major developments in magnetic resonance microscopy is reviewed. In general, though solenoid microcoils have dominated practical microcoil-based MRI, planar microcoils fabricated by photolithographic techniques have also emerged [1, 3, 4].

These advances in micro-coils have sustained magnetic resonance imaging (MR) to be conducted on single molecules. While this imaging technique allows better understanding of molecular composition and interactions, visual imaging during MRI is not possible due to the design of the imaging hardware. Largely, the solution to this problem would not come from redesigning the hardware and as a result, a holding cell must be designed to allow visual imaging to take place concurrently during MRI.

From many considerations, the easiest method to optically view a cell while conducting MRI is to design a holding cell that has an optical input coupled with an optical output. The input side will guide a laser to the cell's location, and the output will gather the laser signal after it passes through the cell. The signal will then be carried to a photodiode or a charge-coupled device. The blocking of the light by the cell will produce images which on analyses would provide information about the cell.

2. DESIGN DESCRIPTION

The design of the holding system must consider the electromagnetic properties of the coil used for the signal detection [1]. This is important for the signal-to-noise ratio (SNR) of the system. Equation 1 [5] shows that the unitary magnetic field ($B_{1u}(r)$) generated by a planar coil increases with the reduction of coil diameter (volume, dV_s), indicating that smaller coils are more efficient in detecting microscopic samples.

$$B_{1u}(r) = \frac{\xi_0(r)}{M_{xy}(r).\omega_0.dV_s}$$ (1)

where ω_0 is the resonance frequency, $\xi_0(r)$ the signal contribution of a small sample volume dV_s, at position **r** and

$$M_{xy}(r) = M_0(r)\sin(\gamma B_{1u}(r)(I/2)\tau)$$ (2)

with $M_0(r)$ the magnetization at equilibrium, γ the gyromagnetic constant, I the excitation current amplitude and τ the pulse duration. (2) is valid when the same coil in (1) is used for the sample excitation. The implication of this analysis is that smaller system would produce better detection results and the holding system must be designed to support small coils.

Specifically, the two main functions of the holding device are to 1) stabilize a biological cell so that MRI can be conducted and 2) to guide fiber optic cables into place to allow visual imaging of the cell to be simultaneously taken. These guidelines yielded a final mask design as shown in Figure 1. The central channel will be used for the target cell and buffer, with any extraneous fluid being removed by using a five micron drainage channel. The two arching portions will

1-4244-0641-2/07/$20.00 ©2007 IEEE

be used to feed fiber optic cable into position so that a visual image of the cell can be captured. The dimensions of the chip are designed to allow the device to be used in a cellular MRI machine at Johns Hopkins University, Baltimore.

Figure 1: Holding cell mask design. The curving channels are designed so that fiber optic cables can be fed into position to view a cell held in the central channel. The two smaller channels in the central region are drainage channels that have been added so that a pressure differential can be generated across the holding cell. This will allow a specimen to be placed at the bottom of the cell by generating a flow.

3. FABRICATION TECHNIQUE

The device was fabricated from PDMS by using soft lithography [6] techniques in the Johns Hopkins University clean room facility. PDMS is the material of choice due to the fact that it will not interfere with the MRI and it will not damage the biological sample. The fabrication steps are shown Figure 2 with final system in Figure 3.

4. EXPERIMENTAL SETUP

Figure 4 shows the setup to test the fabricated system. It involves coupling laser light through a chopper via a fiber optic cable into the holding cell system. The light will pass through the cell (under MRI) and collected from the other end of the system. This light is sent through a photodiode and subsequently to equipment (e.g. oscilloscope) where analyses and visualization would take place.

5. DISCUSSION

The major challenge in this design is the difficulty of feeding the fiber optic cable through the channel into the PDMS. Using a stiffer PDMS material with a big channel would solve this problem. Besides, any other resilient, biologically friendly polymer would also serve as an acceptable substitute. A stiffer polymer will provide resistance to the cable and force it to follow the channel.

6. CONCLUSION AND FUTURE PLAN

We have reported the design, development and fabrication techniques of a holding cell system for use in microcoil based MRI for unicellular cells. The system is designed to enable concurrent imaging and visualization of cells during MRI. Through two channels that enable coupling of light into the cell, images of cells can be examined. The system would be verified by putting a frog oocyte or any cell inside the holding cell, image it in MRI chamber with concurrent visual imaging

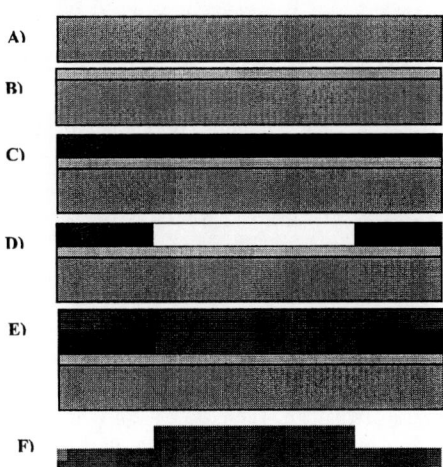

Figure 2: The step for fabricating the PDMS version of the holding cell. Green represents the glass, gray symbolizes the aluminum adhesion layer, blue is the SU-8 template, and red denotes the PDMS. The process begins with a clean glass slide (A) that is coated with aluminum (B). SU-8 is processed onto the aluminum layer (C) and is developed with the desired features (D). PDMS is then applied to the completed template (E) and is removed; thereby leaving the features imprinted in the PDMS layer (F).

Figure 3: Fabricated system on PDMS

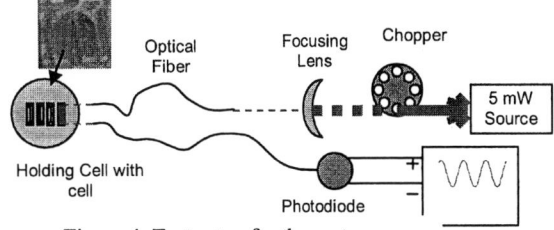

Figure 4: Test setup for the system

REFERENCES

[1] C. Massin, S. Eroglu, F. Vincent, **B. Gimi**, P.-A. Besse, R.L. Magin and R.S. Popovic, " Planar microcoil-based Magnetic Resonance Imaging of cells", IEEE Intl Conf. on Solid-State Sensors, Actuators, and Microsystems, Boston, MA, June 2000

[2] P.G. Glover, P. Mansfield, "Limits to magnetic resonance microscopy", Reports on progress in physics, vol. 65, 2002

[3] J. B. Aguayo, S.J. Blackband, J. Schoeniger, M.A. Mattingly and M. Hintermann, "Nuclear magnetic resonance imaging of a single cell", Nature 322, 190 - 191 (10 July 1986)

[4] B. Gimi, S. C. Grant, Richard L. Magin, Gary Friedman, "Investigation of NMR Signal-to-Noise for RF Scroll Microcoils." IEEE IntL. IEEE-EMBS Special Topics Conf. on Microtechnologies in Medicine & Biology, October 2000

[5] D. I., Hoult, R. E. Richards, "The signal-to-noise-ratio of the nuclear magnetic resonance experiment". Journal of Magnetic Resonance, vol.24, 1976, pp 71–85

[6] Y. Xia and G.M. Whitesides, "Soft Lithography", In Angew. Chem. Int. Ed. Engl. 37, 551-575, 1998

TuP11
14:30 – 17:30

Glass Reflowed Microlens Array and its Optical Characteristics

Sung-Kil Lee, Man Geun Kim, Kyung-Woo Jo and Jong-Hyun Lee
Department of Mechatronics, Gwangju Institute of Science & Technology (GIST),
1 Oryong-dong, Buk-gu, Gwangju 500-712, Republic of Korea
Tel +82-62-970-2395, Fax +82-62-270-2384, E-mail jonghyun@gist.ac.kr

Abstract

We have demonstrated glass microlens array fabricated by thermal reflow process to take advantage of durable lens material and simple process. The microlens was designed using ray tracing, and was realized by controlling the diameter and height of cylindrical glass pattern. The glass pattern (Diameter x Height = 250 x 28um) was formed by etching a glass plate with HF solution for 4 min (E/R: ~7um). Then the glass pattern was reflowed at 850℃ that is above the glass transition temperature (T_g = 820℃), in the furnace to form a spherical shape. The profile of microlens was measured using a confocal microscopy and the diameter of the spherical shape is larger than 80 percent of the mask pattern size. The fabricated microlens was evaluated in terms of focal length (~600um) and beam waist (~12um) using an IR source (λ=1550nm).

Keywords: Microlens Array, Glass Reflow, Focal Length, Wet Etching, Transition Temperature

1. INTRODUCTION

Until now, microlens array has been studied actively due to the increasing demand in optical communication, display, storage and biochemical systems. Many methods to fabricate the microlens were introduced using surface tension [1,2], grayscale lithography [3], isotropic propagation, pattern transfer and solid deformation [4]. Typical lens material in these methods is based on polymer such as photo-resist (PR), polycarbonate (PC), cyclic olefin copolymer (COC) and so on. These materials have some drawbacks such as swelling, scratching and chemical instability.

Thus, glass has been widely used as an optical material to improve robustness. Previously, glass microlens was fabricated using pressure difference between cavity and surrounding, which is similar to complicated hot embossing process [5]. In this paper, we will introduce the new fabrication method to fabricate the glass microlens by thermal reflow process that has not been extensively studied so far. The fabricated microlens array will be evaluated in terms of focal length and beam waist.

2. FABRICATION

The glass material used was Pyrex 7740 due to the similar coefficient of thermal expansion to that of silicon. Microlens array was fabricated by following process steps as shown in figure 1. The first step is anodic bonding process between the glass plate (200um in thickness) and silicon substrate (500um in thickness) for 15 min, followed by glass thinning process for 23 min (E/R: 7um/min) in HF (49%) solution to reduce the glass thickness down to 40um. For the glass patterning, Cr/Au was sputtered on glass substrate and photo-resist was patterned for the etch mask during Cr/Au etching process. Then, the glass was etched again for 4 min so that the etched glass substrate remains at 12um in thickness. Some part of glass plate outside of lens area was intentionally left to increase the lens curvature after reflow process. Finally, glass was reflowed for 20 minute at 850℃ that is higher than glass transition temperature (T_g= 820℃) in furnace.

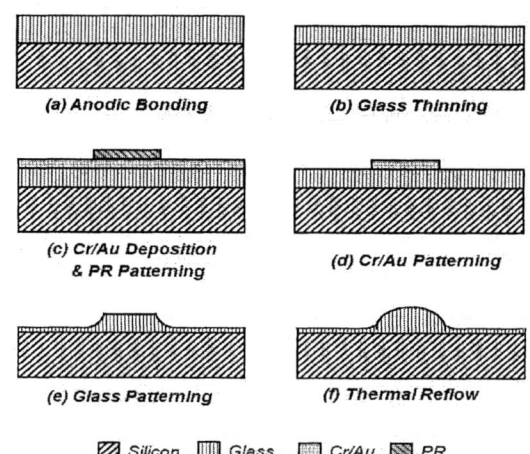

Figure 1. Process flow for the fabrication of a glass microlens by thermal reflow method.

1-4244-0641-2/07/$20.00 ©2007 IEEE

The SEM image of the fabricated microlens array was shown in figure 2. The lens profile was experimentally obtained by confocal microscopy as shown in figure 3, where the lens diameter and roughness were 300 um and 27nm, respectively. The diameter of the spherical shape is larger than 80 percent of the mask pattern size, while the overall profile is rather similar to sinusoidal shape due to the reflow of the glass remained between lenses.

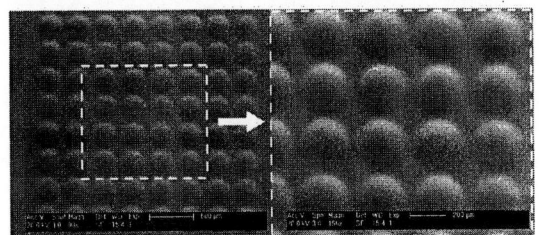

Figure 2. SEM image of microlens array fabricated by thermal reflow process.

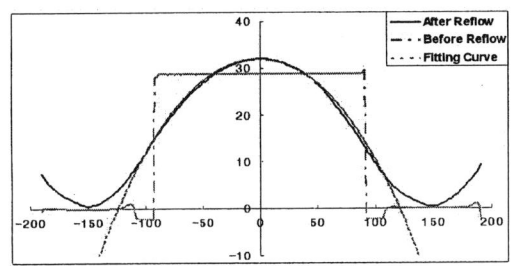

Figure 3. Profiles of microlens(solid line) overlapped with etching profile(dash line) and circular fitting curve(dot line).

3. OPTICAL CHARACTERIZATION

Optical characteristics were measured using IR CCD camera, GRIN lens and optical stage (6-Axis), as shown in figure 4. To measure the beam radius, light intensity of microlens array was scanned along Z-axis and the beam images were captured by IR CCD camera. Finally, the estimated beam profile was fitted by Gaussian function to obtain beam radii along z-axis. The focal length and minimum beam radius were 600 um and 12 um, respectively, as illustrated in figure 5. Experimental result is in good agreement with theoretical values calculated based on Gaussian beam approximation.

4. CONCLUSION

A glass microlens was successfully fabricated using thermal reflow process. The focal length was controlled by glass thickness. This fabricated glass lens features high durability.

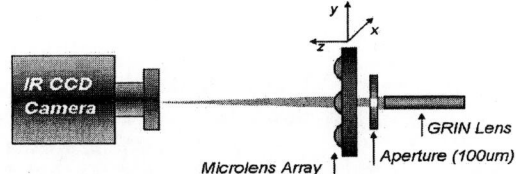

Figure 4. Schematic of experiment setup for measuring the focal length and the beam waist of fabricated microlens.

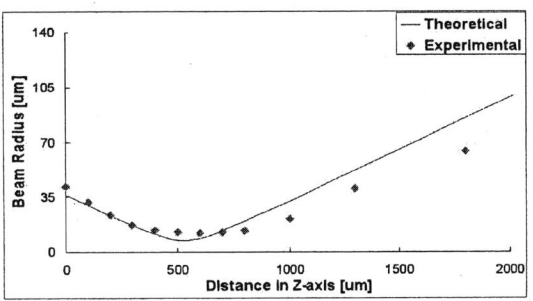

Figure 5. Experimental and theoretical comparison of the beam radii along z-axis for glass reflow microlens.

ACKNOWLEDGMENT

This work was financially supported by the GTI/GIST, Korea.

REFERENCES

[1] Kyoung-Woo Jo, et al., "Optical characteristics of a self-aligned microlens fabricated on the side wall of a 45°-angled optical fiber," IEEE Photonics Technology Lett., vol. 16, no. 1, pp. 138-140, January, 2004

[2] Daniel M. Hartmann, et al., "Characterization of a polymer microlens fabricated by use of the hydrophobic effect," Optics Lett., vol. 25, no. 13, pp. 975-977, July 1, 2000

[3] X.-C. Yuan, et al., "Cost-effective fabrication of microlenses on hybrid sol-gel glass with a high-energy beam-sensitive gray-scale mask," Optics Express, vol. 10, no. 7, pp. 303-308, April 8, 2002

[4] C.-T. Chang, et al., "Fabrication of plastic microlens array using gas assisted micro-hot-embossing with a silicon mold," Infrared Physics & Tech. 48, pp. 163-173, 2006

[5] P. Merz, et al., "A novel micromachining technology for structuring borosilicate glass substrates," Transducers '03, Boston, June 8-12, 2003, pp. 258-261

TuP12
14:30 – 17:30

Photothermally Actuated Microcantilever Beams Using Nanoparticles

C. N. Chen[1], C. M. Hsieh[1], C. G. Tsai[2], J. A. Yeh[1,2] and C. Lee[3]
[1]Institute of Electronics Engineering, National Tsing Hua University,
[2]Institute of Micro-Electro-Mechanical Systems, National Tsing Hua University,
101, Section 2 Kuang Fu Rd., Hsinchu, TAIWAN
Tel 886-3-5715131-33730, Fax 886-3-5745454, E-mail d913771@oz.nthu.edu.tw
[3]Department of Electrical & Computer Engineering, National University of Singapore,
4 Engineering Drive 3, SINGAPORE

Abstract

Photothermally actuated microcantilever beams based on localized surface plasmons (LSPs) of nanoparticles (NPs) were used to modulate absorption efficiency. The bimaterial beams were composed of 200 nm thick poly(vinylpyrrolidone) (PVP) embedded with silver (Ag) NPs and 370 nm thick silicon nitride. NPs were designed to have a strong absorption at about 532 nm and convert the absorbed light energy to heat. The beams had a vertical deflection of 30μm under the excitation of 532nm green laser with an intensity of 500 W/cm². The corresponding mechanical and thermal response times were 0.27 ms and 10 ms, respectively.

Keywords: Nanoparticles, photothermally, microcantilever beam

1 INTRODUCTION

Photothermally actuated microcantilever beams are applied in various applications including self-resonant systems [1], photothermal spectroscopy [2], and high precision motion transducers [3], etc. Among these applications, the light-absorbing materials are adopted in photothermal beams while the peak of the absorption spectrum is fixed and incapable for modulation. Consequently, the used materials become constrains for applications. In this paper, NPs coated on the beams are proposed to induce photothermal actuation. Based on LSPs, the strong light absorption occurs below specific NPs size when incident wavelength excites LSPs of NPs [4]. The strong light absorption resulted in photothermal effect that bends the beams. Surface plasmon (SP) absorption band of NPs can be tunable by choice of particle materials, sizes, size distribution, shapes and embedded medium. The beams coated with NPs have tunable photothermal efficiency with respect to different incident wavelengths.

2 DEVICE FABRICATION AND SYSTEM SETUP

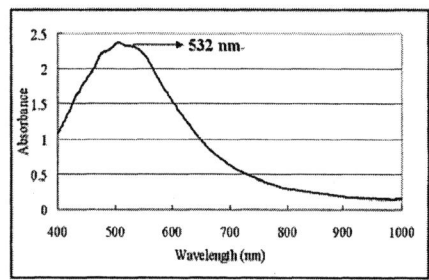

Figure 1: Absorption spectrum of Ag NPs embedded in a PVP. The maximum of absorbance was 0.21 at wavelength of about 532nm.

The Ag nanoparticles embedded in PVP are synthesized by the polyol process [5] and its absorption spectrum is shown in Fig 1. The spectrum indicated that the NPs had a maximum photothermal effect at wavelength of about 532nm. A narrow frequency response was achieved by improvement of size uniformity of NPs. The beams coated with NPs were fabricated as follows: a 370 nm thick silicon nitride layer was deposited on patterned silicon by LPCVD. Then, the 200 nm thick PVP including NPs was spinned onto the silicon nitride to form a bimaterial film. The structures were etched using O₂ plasma and dielectric reaction ion etch (DRIE). Finally, the silicon under beams was removed using potassium hydroxide (KOH) bath. The bimaterial beams made of PVP and silicon nitrite deflected when excited by light source. The beams had a length 350μm, a width 40μm, and a thickness 570nm.

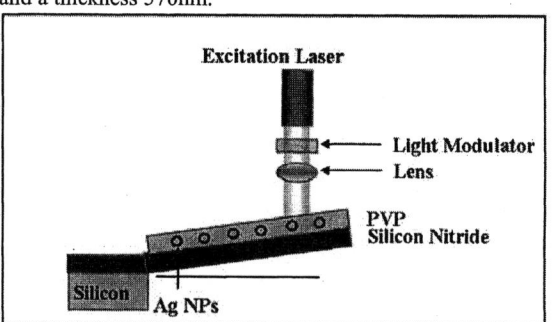

Figure 2: Schematic representation of measurement setup. The laser was focused on the beams that converted the absorbed light energy to heat. The deflections of the beams were captured by a CCD camera.

Schematic representation of measurement setup is shown in Fig. 2. Two kinds of lasers with wavelengths of

1-4244-0641-2/07/$20.00 ©2007 IEEE   77

532nm and 632nm were used for comparison of light absorption efficiency of NPs. The light intensity is defined as the laser power divided by the laser spot area. In the experiment, the laser power was adjusted by the light modulator. The observation of beam deflection was captured by a CCD camera through a microscope.

3 RESULTS

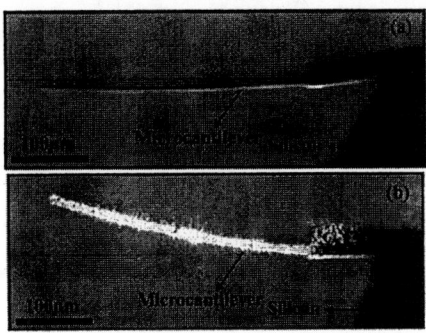

Figure 3: The deflections of beams using photothermal actuation induced by laser with a wavelength of 532nm. (a) the initial state with no excitation (b) the beam deflected after excitation.

Fig. 3 shows the beams response to photothermal effect induced by laser excitation. The maximum deflection was 30μm when illuminated by a green laser with an intensity of 500W/cm². The bending of the beams was due to the thermal gradient in the thermal bimorph where the distinct thermal expansion coefficients of PVP and silicon nitrite result in elongation difference.

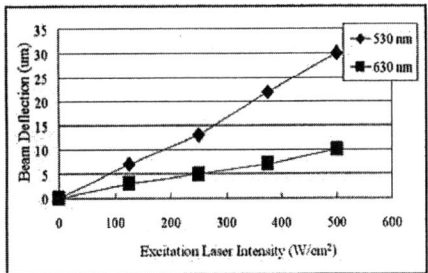

Figure 4: The deflection of beams with respect to different intensities of two kinds of excitation lasers at 532nm and 632nm, respectively. The deflection was proportional to light intensity. The laser spot size of 30μm was focused on the beams by a 5X objective lens.

The bending characteristics of the beams were compared by illuminating two kinds of excitation lasers with wavelengths of 532nm and 632nm, respectively. The light intensity with an increment of 125 W/cm² ranged from 0 to 500W/cm². Fig. 4 shows that the deflection was proportional to laser intensity in both cases. The deflection by green laser was three times larger than that by red light

because light absorption capability had difference of three times.

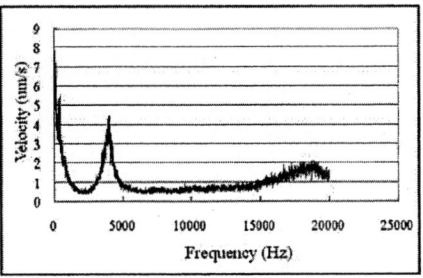

Figure 5: Beams having mechanical resonance frequency at 4kHz measured by laser doppler velocimety.

Mechanical resonance frequency of the beams measured by laser doppler velocimety was 4 kHz, shown in Fig. 5. In thermal response test, the rise time of the beams measured by CCD camera was 13ms. Mechanical resonance frequency and rising time of silicon nitrate were estimated to 3.6kHz and 10ms according to $f = 1/2\pi \sqrt{k_{mechanial}/m_{eff}}$ from simple harmonic motion frequency and $\tau = l^2 \rho c / k_{thermal}$ from the lumped element method. It was found that the mechanical and thermal response characteristics of the beams were predominated by substrate layer of silicon nitride. Studies are sill under investigation for improving the response characteristics of photothermal microcantilever beams by choosing appropriate materials in replace of silicon nitride.

4 CONCLUSION

Due to the photothermal effect enhanced by LSPs of NPs, the beams had tunable efficiency in different incident wavelengths. The maximum deflection of photothermally actuated beams occurred with excitation of LSPs.

REFERENCES

[1] K. Aubin, M. Zalalutdinov, T. Alan, R. B. Reichenbach, R. Rand, A. Zehnder, J. Parpia and H. Craighead, J. Microelectromech. Syst., vol. 13, pp. 1018-1026, 2004.

[2] A. Wig, E.T. Arakawa, A. Passian, T.L. Ferrell and T. Thundat, Sensor. Actuat. B-Chem., vol. 114, pp. 206-211, 2006.

[3] A. Sampathkumar, T. W. Murray and K. L. Ekinci, Appl. Phys. Lett., vol. 88, 223104, 2006.

[4] D. Boyer, P. Tamarat, A. Maali, B. Lounis and M. Orrit, Science, vol. 297, pp. 1160-1163, 2002.

[5] Y. Sun and Y. Xia, Science, vol. 298, 2002

TuP13
14:30 – 17:30

Vertical Comb-Drive MEMS Mirror for Optical Spectrum Sensing

Daisuke Inoue*, Fumikazu Oohira*, Kazuya Yamamoto*, Masahiro Kondo*, Takaki Harada*,
Ichirou Ishimaru*, Gen Hashiguchi*, Maho Hosogi*
Tel +81-87-864-2341, Fax +81-87-864-2341, E-mail: s06g505@stmail.eng.kagawa-u.ac.jp
* Kagawa University, Department of Intelligent Mechanical Systems Engineering,
2217-20, Hayashi-cho, Takamatsu-shi, Kagawa, Japan

Abstract

We have proposed a vertical comb-drive MEMS mirror which can be applied for a spectrometer with phase-shifting method, and fabricated the mirror device using micro fabrication technology. The configuration of the MEMS mirror was designed with three vertical electrostatic comb-drive actuators and hinges at the periphery of the mirror. In the proposed spectrometer, the phase of the light which is reflected at the movable mirror is changed, and obtain the interferogram is obtained by the movable mirror which is driven precisely. We evaluated the moving characteristics of the fabricated MEMS mirror and confirmed that the MEMS mirror could be driven vertically to about 30μ m and also could be tilted. Using this MEMS mirror, we configured the spectrometer with a source of monochromatic light ($\lambda = 405$ nm), and evaluated the spectroscopic characteristic in principle. Therefore, we have confirmed that the fabricated MEMS mirror can be applied to the spectrometer with phase-shifting method.

Keywords: vertical comb-drive actuator, spectrum sensing, phase shift, MEMS, SOI

1 Introduction

The MEMS mirrors fabricated by the micro fabrication technology which are small size and low electric power have been widely applied for many fields such as optical switch, laser scanner, attenuator etc.[1][2]. We aim to apply the MEMS mirror for not only tilting use such as optical switch etc. but also the optical sensing use such as the spectrometer with phase-shifting method [3] by the vertically driven movable mirror. In this study, we designed and fabricated this MEMS mirror for the spectrometer. We evaluated the moving characteristics of the MEMS mirror and confirmed that the MEMS mirror could be driven vertically to about 30μ m. Using this MEMS mirror, we configured the spectrometer with a source of monochromatic light ($\lambda = 405$ nm), and confirmed the spectroscopic characteristic in principle.

2 Principle measurement

Figure 1 shows the principle configuration of the spectrometer with phase-sifting method. A light from an object is changed to a parallel light, and the half of the parallel light is reflected at the area of the movable mirror and the other half of the light is reflected at the area of the fixed mirror. In this case, when the movable mirror is moved, the phase difference is generated at the half area of reflected light. This generates the interference between the different phase lights when the two lights converge at the imaging surface. The interferogram is a summation of the interference intensity of each object light spectrum. Therefore, the relative intensity of each wavelength can be calculated by Fourier-transform analysis of the obtained interferogram. This principle is two-dimensional Fourier spectroscopy with phase-shifting interferometer between the object lights. Hence, by using all object lights converged by an objective lens, the two-dimensional spectroscopy for a very week light can be obtained.

3 Configuration of MEMS mirror

Figure 2 shows the configuration of the MEMS mirror proposed in this study. In order to apply the MEMS mirror to the spectrometer with phase-shifting method, the movable mirror must be driven parallel to the fixed mirror in order to generate the phase-shift. For this purpose, the configuration of the MEMS mirror was designed with three hinges at the periphery of the mirror and three vertical electrostatic comb drive actuators that can generate the big driving force. When the different voltage is applied to each actuator, the movable mirror can be vertically driven but and tilted. This configuration makes it possible to adjust the mirror to the parallel position and to drive parallel to the fixed mirror. The shape of the hinges is curved, and the hinges and combs are designed at three-points with symmetry for each other.

3 Fabrication of MEMS mirror

The vertical electrostatic comb-drive actuators may slide to the horizontal direction, therefore they have a possibility to contact between the fixed teeth and movable teeth in the substrate plane. Therefore, in order to drive the movable mirror stably, the self-aliment technique between fixed teeth and movable teeth has been realized using the delay-masking process in the fabrication process[4][5]. In this study, we devised the process for the MEMS mirror fabrication based on this method. Using the micro machining technologies with DRIE, we fabricated a MEMS mirror on SOI (Silicon on Insulator) wafer: the top Si layer of 25μm thickness, the bottom Si layer of 125μm thickness and the buried oxide layer of 1μm thickness. Figure 3(a),(b) shows the SEM images of the fabricated MEMS mirror. Figure 3(a) shows the overview from the top surface. Figure 3(b) shows the enlarged view of the teeth from the top surface. As shown in figure 3(b), the gap between the movable teeth and the fixed teeth could be fabricated symmetrically by the devised process.

4 Characteristics of fabricated MEMS mirror

We evaluated the characteristics of the fabricated MEMS mirror. By applying the same voltage to three vertical comb-drive actuators, the movable mirror could be vertically driven. The displacement of the movable mirror was measured by the laser displacement sensor with the resolution of 10nm. Figure 4 shows the relation between the applied voltage and the measured and the designed displacement of the movable mirror. The movable mirror was actuated with large displacement in proportional to the squared voltage, and the displacement of 10 μm was achieved at about 85V, and the displacement of about 30 μm at 130V. In addition, when the different voltage is applied to each actuator, we confirmed that the

1-4244-0641-2/07/$20.00 ©2007 IEEE

MEMS mirror could be tilted.

Next, we examined the principle experiment with this MEMS mirror to apply for the spectrometer with phase-shifting method. As a principle experiment, a source of monochromatic light ($\lambda = 405$ nm) was used at the object position shown in figure1. When the movable mirror is actuated, the phase difference is generated between the reflected light of the movable mirror and the fixed mirror, and then, the interferogram obtained in each pixel of a charge coupled device (CCD) camera should show a cosine wave. One wavelength of this cosine wave should be consistent with the wavelength of the source of the monochromatic light. The spectroscopic characteristics of the used monochromatic light can be calculated by the Fourier-transform analysis of this interferogram. In this experiment, the movable mirror was vertically actuated at 40nm pitch to about 7μm. Figure 5 shows the obtained interferogram. Figure 6 shows result of the Fourier-transform analysis of the interferogram. As shown in figure 6,we could obtain the peak of the characteristic as the same value as the light source wavelength of 405nm.

5 Conclusions

In this study, we aim to apply the MEMS mirror to the spectrometer with phase-shifting method. For this purpose, the MEMS mirror with vertical electrostatic comb-drive actuators was designed and fabricated, and then the driving characteristics of this device were evaluated. Using this MEMS mirror, we configured the spectrometer using a source of monochromatic light ($\lambda = 405$ nm), and evaluated the spectroscopic characteristic in principle. As the result, we have confirmed that the fabricated MEMS mirror can be applied to the spectrometer with phase-shifting method.

References

[1] Y.Sakai,T.Yamada,S.Ide,K.Mori,A.Ishizuka,O.Tsuboi,T.Matsuyama,Y. Ishii,andM.Kawai,OpticalMEMS,Waikoloa,Hawaii, Aug. 18-21, 2003.

[2] H.Obi,H.Fujita,H.Toshiyoshi,"Desing for high stable electrostatic torsion mirror with vertical-comb actuators"IEICE Transaction,2004

[3] Yuseke Inoue, Ichirou Ishimaru , Toshiki Yasogawa , Katsumi Ishizaki "Variable phase-contrast fluoresce spectrometry for fluorescent stained cell",Appl. Phys. Lett. 89, 121103 (2006)

[4] KRISHNAMOORTHY U, LEE D and SOLGAARD O: "Self-Aligned Vertical Electrostatic Combdrives for Micromirror Actuation", Microelectromechanical Syst , Vol.12, No.4, pp.458-464 (2003.8)

[5] Makoto Mita, Yoshio Mita, Hiroshi Toshiyoshi and Hiroyuki Fujita : "Delay-Masking Process for Silicon Three-Dimentional Bulk Structures", T.IEE Japan, Vol.119-E, No.5, pp.310-311 (1999)

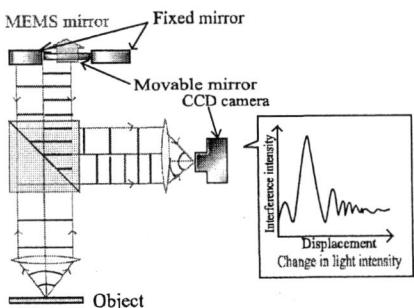

Fig.1 Principle configuration of the spectrometer with phase-sifting method

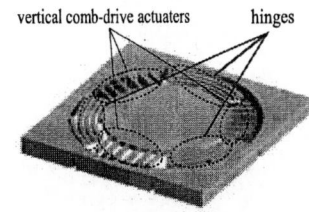

Fig.2   Configuration of the MEMS mirror

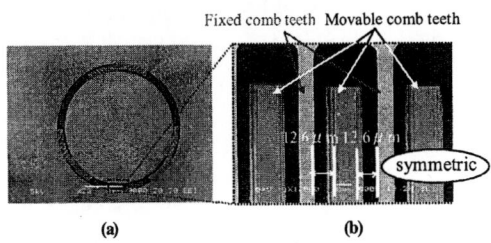

Fig.3 SEM images of the fabricated MEMS mirror
(a) Overview from the top surface, (b) Enlarged view of the teeth from the top surface

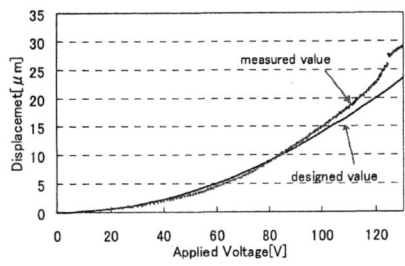

Fig.4 Relation between the applied voltage and the displacement of the movable mirror

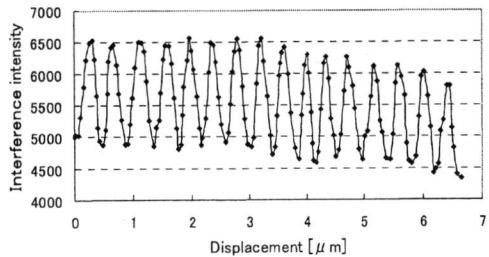

Fig.5 Interferogram

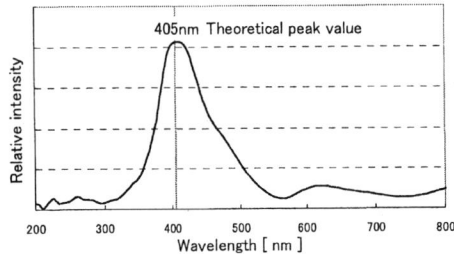

Fig.6 Fourier-transform analysis of the interferogram

TuP14
14:30 – 17:30

Drift-Free Single Crystalline Silicon Micromirror with Floating Field Limiting Shields

Byung-Wook Yoo, Yun-Ho Jang*, Kyoungsik Yu**, Jae-Hyoung Park*** and Yong-Kweon Kim

School of Electrical Engineering and Computer Science, Seoul National University, Seoul, South Korea

Tel +82-2-880-1793, Fax +82-2-873-9953, E-mail despina1@snu.ac.kr

*Image development team, Samsug Electronic Co., Ltd., South Korea

**Korea Electrical Engineering & Science Research Institute, Seoul, South Korea

***Department of Physics, Ewha Womans University, Seoul, South Korea

Abstract

This paper presents the charging mitigation method at single crystalline silicon micromirror by means of improving the geometry around electrodes. We have now determined that tilt angle drift and stiction are due to the charging effect, and that the effect can be enormously reduced by floating field limiting shields around the electrodes and dielectric geometry. While 0.5 degrees drift was measured at the 83 % voltage of the pull-in voltage from the previous micromirror model, the micromirror considering drift-free design showed ±0.005 degrees drift at the same percentage voltage of it. At the 97.7 % of the pull-in voltage, stiction occurred after about 220 seconds from the previous micromirror model. On the other hand, newly designed micromirror did not show any pull-in effect even at above percentage of it.

Keywords: single crystalline silicon, micromirror, floating field limiting shields, drift-free, stiction

1 INTRODUCTION

This paper describes a drift-free micromirror design, which suppresses the electrostatic charging effects using floating field limiting shields. Charging effects adversely affect the reliability and accuracy of electrostatic microactuators, because they can cause the unwanted drift of actuator deflection [1]. Although several approaches, such as ac driving and argon packaging have been proposed [2], it is still of great interest to investigate new approaches that can mitigate the charging effects. In this paper, we present metallic shielding structures on top of exposed dielectric surfaces can reduce the charging effects.

2 MICROMIRROR DESIGN AND FABRICATION

We have fabricated torsional micromirrors with electrostatic parallel-plate actuators on single crystalline silicon and glass materials using a bulk-micromachining process. Our micromachining process is modified from our previous methods [3], and schematically described in Figure 1. The bottom substrate is glass material (Pyrex^{TM} corning 7740), while the micromirror plate ($210 \times 210~\mu m^2$) and torsional spring structures ($1.2 \times 42~\mu m^2$) are defined on the silicon layer. The silicon layer thickness is measured to be 7 μm. Fabrication processes for type A and B are illustrated in

Figure 2. To reduce the charging in the dielectric substrate, we have covered the exposed dielectric surface around the bottom electrodes of the parallel-plate actuators with the field-limiting gold shields as shown in Figure 3 and 4 (micromirror type A). To compare the tilt angle stability of new micromirror design, we also made micromirrors without the shielding structures (micromirror type B).

3 EXPERIMENTAL RESULTS

Metallic shield structures around bottom electrodes are electrically floating, and thus attract the electric field from charges in the dielectric layer. The shields also inhibit direct injection or trapping of charges. To obtain the micromirror tilt angle as a function of time and input voltages, we used an optical spot measurement setup with a duolateral position sensitive detector ($10 \times 10~mm^2$, ±0.3% error) and a data acquisition system. When the input voltages to the micromirrors are maintained at 83 % of the pull-in voltage, the tilt angle variation for the micromirror type A is less than only ±0.005 degrees for the first 5000 seconds after the voltage is applied (Figure 5(a)). The type B, however, shows a tilt angle drift of 0.5 degrees for the same duration. In Figure 5(b), the tilt angle drift gradually stops after around 1750 seconds, because the electrical charge accumulation in the dielectric layer becomes saturated due to the electrical repelling force. When the input voltage to the micromirror type B is maintained at 97.7 % of the pull-in voltage, the pull-in phenomenon occurs after 220 seconds due to the injected and/or trapped charges in the dielectric layer (Figure 6(b)). In comparison, the micromirror type A does not show any pull-in phenomenon even at the 98.7 % of the pull-in voltage, and the tilt angle variation is measured to be less than ± 0.01 degree for 8000 seconds (Figure 6(a)).

4 CONCLUSION

Floating metallic shields reduce the potential gradient out of charges in dielectric so that the tilt angle drift from the charging effect seems to largely disappear in the micromirror type A. It means there are barely induced charges which can influence the micromirror with the floating metal shields. The micromirror, therefore, employing this technique will improve the long-term stability.

ACKNOWLEDGMENTS

This work was supported by Creative Research Initiatives (MEMS Space Telescope) of MOST/KOSEF.

REFERENCES

[1] K. W. Goosen, et al., "Charging effects in electrostatically-actuated membrane devices," *the SPIE Conf. on Miniaturized Systems with Micro-Optics and MEMS 1999*, Santa Clara, CA, vol. 3878, pp. 407-415.

[2] Herbert R. Shea, et al., "Effects of electrical leakage currents on MEMS reliability and performance," *IEEE Trans. on Device and Materials Reliability*, vol. 4, no. 2, pp. 198-207, June 2004.

[3] Y. H. Jang, et al., "Characterization of a single-crystal silicon micromirror array for maskless UV lithography in biochip applications," *J. Micromech. Microeng.*, vol. 16, no. 11, pp. 2360-2368, 2006.

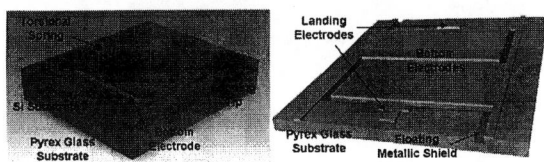

Figure 1. Schematic diagram of a micromirror device and a glass substrate (micromirror type A)

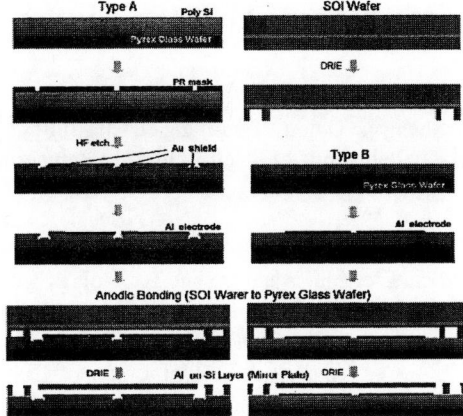

Figure 2. Process flows for the micromirror type A and B

Figure 3. SEM images of a micromirror type A (left) and its bottom electrodes with floating metallic shields (right).

Figure 4. Magnified SEM image of floating metallic shield

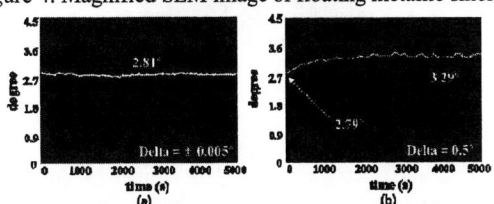

Figure 5. Tilt angle drift at 83 % of pull-in voltage (a) Type A with ± 0.005 degree drift (b) Type B with 0.5 degree drift

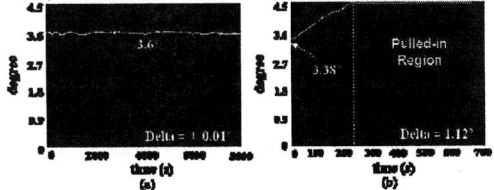

Figure 6. Tilt angle drift nearby the critical pull-in angle (a) Type A with ± 0.01 degree drift (b) Pulled-in type B

TuP15
14:30 – 17:30

A Two-Axis MEMS Scanner Driven by Radial Vertical Combdrive Actuators

Sheng-jie Chiou[1], Tien-liang Hsieh[1], Jui-che Tsai[1], Chia-Wei Sun[2], Dooyoung Hah[3], and Ming C. Wu[4]

[1] Graduate Institute of Electro-Optical Engineering and Department of Electrical Engineering, National Taiwan University

No. 1, Sec.4, Roosevelt Rd., Taipei 10617, Taiwan

Tel +886-2-2363-5251 Ext. 247, Fax +886-2-3366-3686, E-mail: jctsai@cc.ee.ntu.edu.tw

[2] Medical Electronics and Device Technology Center, Industrial Technology Research Institute, Hsinchu, Taiwan

[3] Department of Electrical and Computer Engineering, Louisiana State University, Baton Rouge, LA 70803, USA

[4] Department of Electrical Engineering and Computer Sciences and Berkeley Sensor and Actuator Center (BSAC), University of California, Berkeley, CA 94720-1774, USA

Abstract

We report a two-axis MEMS scanner driven by radial vertical combdrive actuators. The device is fabricated by a five-layer polysilicon surface micromachining process. A cross-bar spring structure consisting of lower and upper torsion springs is designed to achieve two rotational degrees of freedom, enabling the dual-axis rotation. Both the vertical comdrive actuators and the torsion springs are hidden underneath the mirror to achieve a small form factor. Mechanical rotation angles of ±5.4° at 42V and ±2.4° at 63V are obtained for rotation about the lower and upper springs, respectively.

Keywords: Two-axis scanner, radial comb-drive actuators, small form factor, surface micromachining

1 INTRODUCTION

Advances in optical fiber communication, display technologies, and biological imaging and tomography have been the major driving forces for the development of MEMS scanners [1]. Particularly, two-axis scanners are of great interest as they enable the two-dimensional (2-D) steering of optical beams. Among all the actuation mechanisms, electrostatic actuation has been one of the most popular approaches for driving MEMS devices. Parallel-plate electrostatic actuators have simple structures; however, their displacements or rotation ranges are limited by the pull-in effect. Vertical combdrive actuators are more attractive as they are free from the pull-in effect and also offer larger force densities.

Typically, a gimbal structure exists in a dual-axis vertical-combdrive micromirror to achieve two rotational degrees of freedom (DOFs) [2]. It is desirable to eliminate the gimbal so that the device form factor can be reduced. Vertical combdrive actuators in conjunction with the leverage mechanism were used in gimbal-less two-axis scanners [3,4]. The device structures were relatively complicated and required delicate mechanical designs. In this paper, we propose using radial vertical combdrive actuators to eliminate the need of gimbals in two-axis scanners. Similar radial-comb arrangement was demonstrated previously by others; however, their achieved rotation angle was relatively small (2 mrad) [5]. In our

device, a cross-bar spring structure consisting of lower and upper torsion springs is deployed to obtain two rotational DOFs, enabling the dual-axis rotation. Both the actuators and the torsion springs are hidden under the mirror to achieve a small form factor. Mechanical rotation angles of ±5.4° (42V) and ±2.4° (63V) are attained for rotation about the lower and upper springs, respectively.

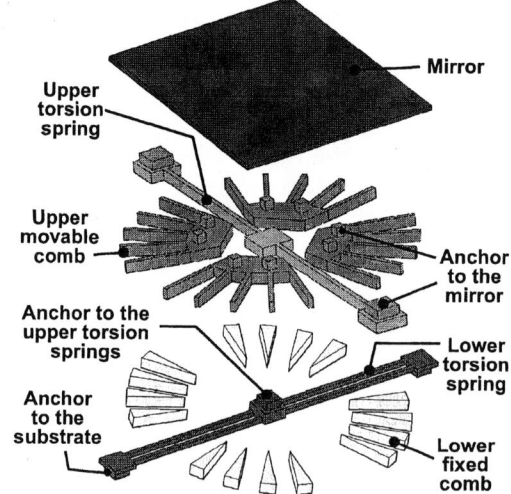

Figure 1. Schematic of the two-axis MEMS scanner with radial vertical combdrive actuators and a cross-bar spring structure.

1-4244-0641-2/07/$20.00 ©2007 IEEE

2 DESIGN AND FABRICATION

Figure 1 is the schematic drawing of the two-axis scanner. It is imaginarily dismantled into three tiers for a clearer illustration. The device is fabricated using SUMMiT-V surface micromachining process offered by Sandia National Laboratory. It has five polysilicon layers, including one nonreleasable ground/shield layer (mmpoly0) and four structural layers (mmpoly1 to mmpoly4). The lower torsion springs and the fixed combs are made of the mmpoly1 (1-μm thick) layer and the laminated mmpoly1+mmpoly2 stack (2.5-μm thick), respectively. Both the upper torsion springs and movable combs are fabricated with the mmpoly3 layer (2.25-μm thick). The top polysilicon layer, mmpoly4 (2.25-μm thick), is used for the mirror. The CMP (chemical mechanical planarization) process before the deposition of the top two polysilicon layers eliminates the topography underneath the mirrors. They also provides a large gap spacing (10.75 μm) between the mirror and substrate.

3 DEVICE CHARACTERIZATION

Figure 2 shows the SEM photos of the fabricated devices. The mirror in Figure 2(a) is intentionally cut into a circular shape to reveal the underlying structures. This particular device has one lower torsion spring on each side, different from the nominal device (Figure 1) which has two on each side. The upper torsion springs are not shown in the photo as they are hidden under the bars extruding from the circular mirror. Figure 2(b) is photo of the nominal device.

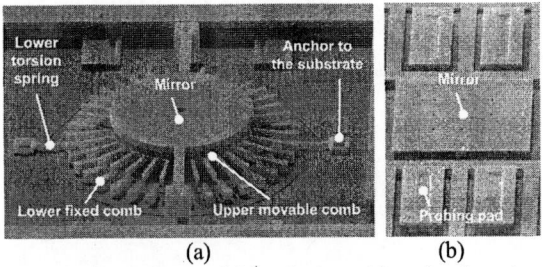

(a)   (b)

Figure 2. SEM photos of (a) a device with a circular mirror and (b) the nominal device.

Figure 3 is the DC characteristic of the nominal device. The mechanical rotation angles are ±5.4° at 42V and ±2.4° at 63V for rotation about the lower and upper springs, respectively. Unlike a normal combdirve actuator, the gaps between fingers change during rotation, leading to lateral instability. The rotation about the upper springs is more susceptible to lateral instability due to the asymmetric spring structure and therefore has a smaller scan range. With a symmetric spring design, the scan angles for the two axes are predicted to balance. The resonant frequencies are first obtained with the modal analysis using ANSYS (Figure 4). They are 18 kHz and 24 kHz for the aforementioned two rotation modes, respectively. Experimental measurements of the resonant frequencies are still in progress.

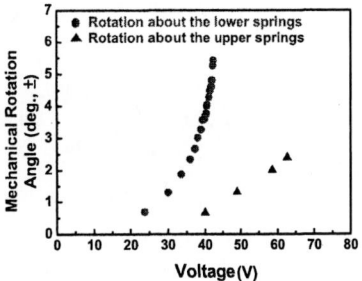

Figure 3. DC characteristic of the nominal device.

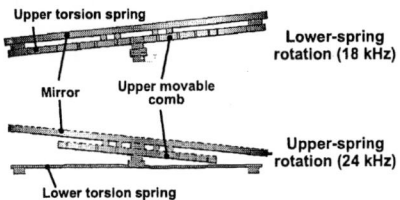

Figure 4. Modal analysis for the resonant frequencies. The fixed combs are not shown.

4 CONCLUSION

We have demonstrated a two-axis micromirror driven by radial vertical combdrive actuators. Mechanical rotation angles of ±5.4° (42V) and ±2.4° (63V) are achieved for rotation about the lower and upper springs of the cross-bar spring structure, respectively.

ACKNOWLEDGEMENT

The authors would like to thank Nan-Fu Chiu and Prof. Chii-Wann Lin of the Institute of Biomedical Engineering and Department of Electrical Engineering, National Taiwan University, for their assistance with the SEM images.

REFERENCES

[1] M. C. Wu et al., "Optical MEMS for lightwave communication," *J. Lightw. Technol.*, vol. 24, pp. 4433-4454.

[2] W. Piyawattanametha et al., "Surface- and bulk-micromachined two-dimensional scanner driven by angular vertical comb actuators," *J. Microelectromech. Syst.*, vol. 14, pp. 1329-1338.

[3] J. C. Tsai et al., "Design, fabrication, and characterization of a high fill-factor, large scan-angle, two-axis scanner array driven by a leverage mechanism," *J. Microelectromech. Syst.*, vol. 15, pp. 1209-1213.

[4] V. Milanović et al., "Gimbal-less monolithic silicon actuators for tip-tilt-piston micromirror applications," *J. Select. Topics Quantum Electron.*, vol. 10, pp. 462-471.

[5] W. Noell et al, "Compact and stress-released piston tip-tilt mirror," *Proc. of IEEE/LEOS Optical MEMS 2006*, Big Sky, Montana, Aug. 2006, pp. 40-41.

TuP16
14:30 – 17:30

Pull-in Analysis of Scanners Actuated by Electrostatic Vertical Combdrives

Daesung Lee and Olav Solgaard

E. L. Ginzton Laboratory, Stanford University, Stanford, CA 94305, USA

Tel: +1-650-723-6104, Fax: +1-650-725-7509, E-mail: daesung28@gmail.com

Abstract

This paper presents an analysis of pull-in due to in-plane twist in MEMS scanners actuated by vertical combdrives with general comb gap arrangements and cross-sections. The analysis is based on a 2-DOF actuator with a single voltage control. A closed-form pull-in deflection angle expression is obtained by accurately modeling the capacitance between the movable and fixed combs. The analysis results are in good agreement with simulations, and allow optimization of scanner designs by combining pull-in deflection angle, capacitance maximum angle, and available torque.

Keywords: Pull-in, vertical combdrive, comb gap, scanner, in-plane twist

INTRODUCTION

The preferred electrostatic actuators for large out-of-plane rotation of MEMS are vertical combdrives due to their high force, high speed, relatively low operating voltage, and relatively large deflection. Staggered vertical combdrives with movable and fixed combs patterned by two separate DRIE steps in two silicon layers have inherent misalignment that is detrimental to the operation of scanners. To solve this problem, several fabrication techniques have been developed enabling self-aligned vertical combdrives including both staggered and angular types [1-3].

In designing scanners, it is not sufficient to consider only capacitance maximum angle and available torque, and pull-in behavior treated as a 1-DOF problem [4] does not give complete design equations. In this paper, we present a 2-DOF pull-in analysis of scanners actuated by vertical combdrives. The analysis is not confined to a particular type of vertical combdrives, but is applicable to general comb cross-sections with general comb gap arrangements. The derived expressions for maximum stable deflection angles are verified by simulations.

METHODOLOGY

The pull-in analysis is based on 2-DOF actuators with a single voltage control utilizing co-energy representation of the total energy [5]. Failure through transversal motion is treated in our earlier work [1]. However, the dominant failure mode is pull-in due to in-plane twist (ϕ), which is analyzed in detail in this paper. The mechanical spring is assumed to be linear and decoupled along the main deflection (θ), and ϕ with spring constants k_θ and k_ϕ. The capacitance between the movable and fixed combs is computed by overlap-area calculations. Definitions of the design parameters are specified in Fig. 1 where $A=0.5(l_1+l_2)$

and $o=l_2-l_1$. The capacitance of the misaligned combs can be expressed as

$$C(\theta,\phi) \cong N\varepsilon_0 \int_{l_1}^{l_2} h(\theta,t)\left(\frac{1}{g_1-t\phi}+\frac{1}{g_2+t\phi}\right)dt \quad (1)$$

where $h(\theta,t)$ is the vertical engagement between the movable and fixed combs at distance t from the axis of rotation of scanners at an actuation angle θ, N is the total number of movable combs, and ε_0 is the permittivity of free space. The transversal shift of the movable combs is $t\phi$. For staggered vertical combdrives, $h(\theta,t)=t\theta+z_0$ where z_0 is the initial engagement between the movable and fixed combs. For angular vertical combdrives, $h(\theta,t)=t\theta-(t-t_0)\theta_0+z_0$ by introducing two additional parameters, θ_0 and t_0 that are, respectively, the initial tilt between the movable and fixed combs, and the distance from the hinge to the axis of rotation.

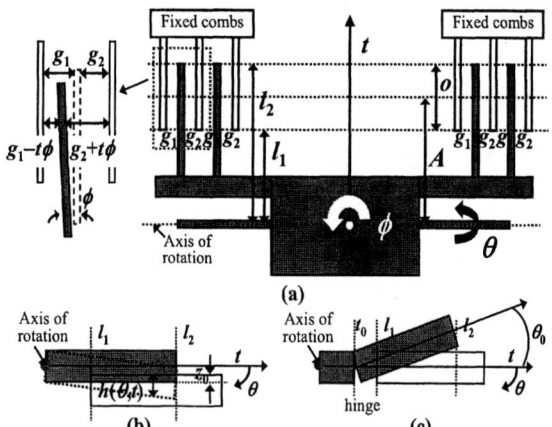

Figure 1. Design parameter definitions of scanner actuated by vertical combdrives for capacitance calculation: (a) Top view, (b) Side view (staggered vertical combdrive), (c) Side view (angular vertical combdrive)

1-4244-0641-2/07/$20.00 ©2007 IEEE

The maximum forward deflection angle, θ_{PI} is determined by solving three simultaneous equations, $\frac{1}{2}\frac{\partial C}{\partial \theta}V^2 = k_\theta \theta$,

$$\frac{1}{2}\frac{\partial C}{\partial \phi}V^2 = k_\phi \phi \;, \quad \text{and} \quad \left(\frac{1}{2}\frac{\partial^2 C}{\partial \theta \partial \phi}V^2\right)^2 = k_\theta\left(k_\phi - \frac{1}{2}\frac{\partial^2 C}{\partial \phi^2}V^2\right)$$

which are two balanced torque equations and a dominant stability equation, respectively. For symmetric capacitance with respect to ϕ, i.e., $g_1=g_2=g$, $\phi=0$ is directly obtained from the second equation. Also the third equation is simplified to $\frac{1}{2}\frac{\partial^2 C}{\partial \phi^2}V^2 = k_\phi$, which can be combined with the first one to solve for θ_{PI}. For $h(\theta,t)=t\theta +z_0$, the equation for θ_{PI} can be expressed as

$$\theta_{PI}\left(\theta_{PI} + z_0 \frac{Y^*(0)}{X^*(0)}\right) = \frac{k_\phi}{k_\theta}\frac{X(0)}{X^*(0)} \qquad (2)$$

where $\quad X(\phi) = \int_{l_1}^{l_2}\left(\frac{t}{g_1 - t\phi} + \frac{t}{g_2 + t\phi}\right)dt \quad$ and

$Y(\phi) = \int_{l_1}^{l_2}\left(\frac{1}{g_1 - t\phi} + \frac{1}{g_2 + t\phi}\right)dt$. Especially for $z_0=0$,

$$\theta_{PI} \cong \sqrt{\frac{g^2}{A^2 + 0.25o^2}}\sqrt{\frac{k_\phi}{2k_\theta}} \cong \frac{g}{A}\sqrt{\frac{k_\phi}{2k_\theta}} \;. \qquad (3)$$

For asymmetric capacitance with respect to ϕ, i.e., $g_1 \neq g_2$, it is difficult to solve for θ_{PI} due to non-zero ϕ. Therefore we approximate the integral in (1) as

$$C(\theta,\phi) \cong C_0 \cdot (\theta + \theta_1)\left(\frac{1}{g_1 - A\phi} + \frac{1}{g_2 + A\phi}\right) \quad \text{which can be}$$

applicable to both staggered and angular vertical combdrives. Figure 2 shows the normalized maximum forward deflection

normalized to $\frac{g}{A}\sqrt{\frac{k_\phi}{2k_\theta}}$ versus the normalized misalignment

for $\theta_1=0$.

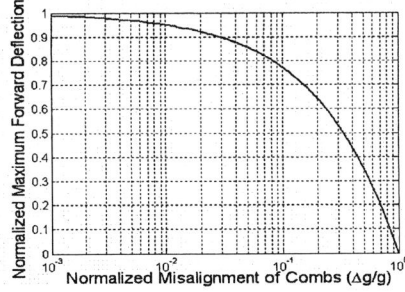

Figure 2. Normalized maximum forward deflection as a function of normalized misalignment of combs

SIMULATIONS AND DISCUSSIONS

To verify the validity of our analysis in (3) and Fig. 2, we performed simulations with Coventorware for perfectly aligned and misaligned combs (25%, 50%) with two values of A, 200μm and 320μm. The dimension of the mirror and the torsional spring is 500×500×30μm and 250×6×30μm, respectively. And g is 4μm and o is 100μm. For perfectly aligned combs, θ_{PI} obtained from simulations is 3.27° for A=200μm and 2.20° for A=320μm. θ_{PI} obtained from (3) is 3.57° for A=200μm and 2.28° for A=320μm, in good agreement with simulations. The capacitance maximum angle, $\theta_{cap\_max} \cong 30\mu m/l_2$ is calculated to be 6.84° for A=200μm and 4.65° for A=320μm, which implies that pull-in does occur before the capacitance maximum is reached.

The condition under which θ_{PI} is equal to θ_{cap_max} comes in the form of the ratio of the thickness of the combs to the comb gap, which can be shown to be smaller than typical DRIE aspect ratio limits ranging from 15 to 40. In this example, the ratio is 3.91 for A=200μm and 3.68 for A=320μm. Another key observation is that θ_{PI} is proportional to g/A whereas the available torque with an approximation of $(2N\varepsilon_0 Ao/g)V^2$ is inversely proportional to g/A.

For 25%, 50% misaligned combs, the normalized maximum forward deflection obtained from simulations is 0.53, 0.34 for A=200μm and 0.54, 0.32 for A=320μm. These are in good agreement with the corresponding analytical values from Fig. 2, 0.59 and 0.35, which implies that the approximation of the capacitance is quite accurate for our analysis.

CONCLUSIONS

We have derived the analytical formula of pull-in deflection angle of scanners actuated by electrostatic vertical combdrives with general comb gap arrangements and cross-sections. Our analysis is shown to be in good agreement with simulations. Pull-in deflection angle should be combined with the capacitance maximum angle and the available torque to correctly optimize scanner designs. Also the combs should be perfectly aligned to maximize pull-in deflection angle independent of the comb cross-sections.

REFERENCES

[1] U. Krishnamoorthy, et. al., *JMEMS*, vol.12, pp. 458-464, Aug. 2003.
[2] D. Lee, et. al., in *Hilton Head* 2004, pp. 352-355.
[3] P. Patterson, et. al., in *Proc. Int. Conf. MEMS*, Las Vegas, NV, Jan. 2002, pp. 544-547.
[4] D. Hah, et. al., *JSTQE*, vol.10, no.3, pp. 505-513, 2004.
[5] D. Elata, et. al., *JMEMS*, vol.12, pp. 681-691, 2003.

TuP17
14:30 – 17:30

Micromirrors for Multiobject Spectroscopy: Large Array Actuation and Cryogenic Compatibility

S. Waldis, P. Ayyalasomayajula, W. Noell, N. F. de Rooij
Institute of Microtechnology, University of Neuchatel
2000 Neuchatel, Switzerland
Email: severin.waldis@unine.ch

F. Zamkotsian
Laboratoire d'Astrophysique de Marseille
13248 Marseille Cedex 4, France
Email: Frederic.Zamkotsian@oamp.fr

Abstract—Micromirror arrays are being developed dedicated for astronomical instrumentation. We report on the strategy to handle a large number of micromirrors needed for covering a 0.5m x 0.5m large active surface and on cryogenic characterization.

I. INTRODUCTION

The observation of the formation of primary galaxies is important to understand our origins. The light coming from these faraway objects is very faint and shifted to the infrared. Multi object spectroscopy (MOS) is the central method for studying many isolated objects simultaneously, using a slit mask in the focal plane of the telescope for blocking spoiling sources and background light. This masks, performing object selection, nowadays are static perforated sheets or complex fiber-optics based systems. In the future, micro-electromechanical systems (MEMS) could provide remote controllable, reconfigurable slit mask, increasing the scientific efficiency of a MOS.

In the framework of the studies on the future European Extremely Large Telescope (E-ELT) we are developing a micromirror array (MMA) based reflective slit-mask. Another MEMS solution, a micro-shutter based slit mask, is being developed for the James Webb Space Telescope by Moseley et al. at NASA [1]. The E-ELT has a projected focal plane diameter in the meter range and will be cooled to cryogenic temperatures for infrared MOS. We estimate that MMA are suited for realizing large slit masks – covering an active surface of 0.5m x 0.5m is aimed.

We realized so far small arrays of 5x5 micromirrors potentially suited for object selection in a MOS. The micromirrors, $100\mu m$ x $200\mu m$ in size, are fabricated using a combination of surface and bulk micromachining. They show an optically flat surface (peak-to-valley deformation of only 7nm) and can be electrostatically tilted by 20° (mechanical tilt-angle) at 90V. The micromirrors are equipped with a stopper system, allowing a precise and uniform tilt-angle over the whole array [2]. We are currently working toward large area coverage and cryogenic operation - two characteristics of our MMA still to be shown for completing the demonstration of the suitability of the MMA for future MOS.

II. LARGE ARRAYS AND THEIR ACTUATION

Covering a, in MEMS terms huge, surface of 0.5m x 0.5m with one large micromirror array is not realistic. A

feasible approach is using smaller arrays as paving stones or tiles to cover an "arbitrary" large surface. This mosaicking approach requires an advanced concept for the mechanical and electrical interfacing of the individual tiles to each other and to the substrate. As for the uniformity of the tilt-angle within one micromirror array, the orientation of the arrays itself must be very uniform over the whole surface. This is a even bigger challenge considering the cryogenic environment. We are currently working on a concept using spring loaded electrical contacts and electrostatic clamping. The advantage of the mosaicking approach is that the individual building blocks can be selected regarding reliability and yield. In a first phase we set the size of one building block to an array of 200x100 micromirrors. Figure 1 shows a subset of a 200x100 array currently being fabricated. One micromirror is $100\mu m$ x $200\mu m$ in size, which yields in an array dimension of 2cm x 2cm.

Actuation of a micromirror of the generation currently being fabricated will require about 100V for a tilt-angle of 20°. Implementing standard CMOS circuitry for generating or switching this voltages on the MEMS chip is not possible due to the cryogenic environment. Therefore the actuation voltages must be generated outside the cold environment (and thus away from the MEMS chip) and wired to the micromirrors. Direct wiring, that is one wire per mirror is not practical due to the huge number of micromirrors: 5000x2500 (12.5 millions) micromirrors of $100\mu m$ x $200\mu m$ in size will be needed to cover an area of 0.5m x 0.5m. A column-line actuation scheme reduces the number of required individual voltages from $n \cdot m$ to $n + m$. Here a voltage V_i is set to line i and V_j to the column j, where $V_i, V_j < V_a$ and V_a being the voltage needed to switch the mirror into the ON state. Hence the all the mirrors remain unactuated except for the mirror in the intersection of line i and column j where we have $V_{ij} = V_i + V_j$. Technically the electrodes are electrically separated into lines and the mirrors into columns. This actuation scheme holds true for actuating a single mirror at a time - exploiting the hysteresis of the electrostatically actuated mirror (see Fig. 2) makes the generation of arbitrary patterns possible: First a hold voltage V_h is set to all the lines. Then a voltage V_δ is set to the first column and V_δ is added to all the lines corresponding to the mirrors to be actuated in the first column, having $V_h + 2V_\delta > V_a$ for the mirrors to be actuated and

1-4244-0641-2/07/$20.00 ©2007 IEEE

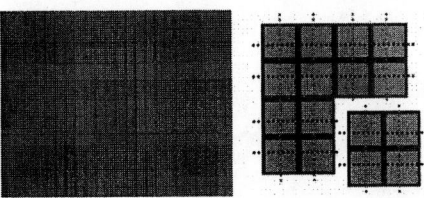

Fig. 1. (a) Optical microscope image of a 200x100 large micromirror array (MMA) currently being fabricated. One micromirror element is 100μm x 200μm in size, the complete array 2cm x 2cm. (b) Mosaicking of these MMA can be used to cover arbitrary large surfaces.

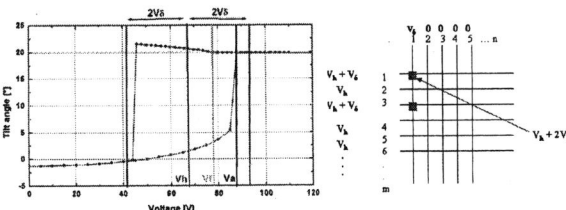

Fig. 2. Electrostatic hysteresis of one micromirror. The column-line addressing scheme is used to generate random patterns on large arrays – requiring only n+m voltages exploiting the electrostatic hysteresis of the micromirrors.

$V_h, V_h + V_\delta < V_a$ for all the others (Fig. 2). Next the voltage of the first column is set to 0 and all the lines again to V_h. This cycle is then effectuated for the columns 2 through n, yielding the desired actuation pattern. At the end, V_h is raised to a value $V_f < V_h < V_a$, the flat zone in the tilt-angle versus voltage hysteresis, ensuring a uniform tilt angle over the whole array [2]. The large arrays currently being fabricated will be utilized to demonstrate this algorithm.

III. CRYOGENIC COMPATIBILITY

The cryogenic compatibility is crucial for the application in an infrared (IR) MOS. The operating temperature must be below 100K for near and mid IR and 30K for far IR. Our MMA is conceived such that all structural elements have a matched coefficient of thermal expansion (CTE) in order to avoid deformation or even flaking within the device when cooling down to the operating temperature. The mirrors themselves must be covered with a gold layer for IR operation, gold having a different CTE than silicon. However we estimate that the induced deformation is small, as the silicon mirror is 10μm thick and the coating 100nm thin.

We did a preliminary characterization of the micromirror arrays from the first generation. A cryogenic setup within a Phillips ESEM was used to study the structural behavior of packaged and unpackaged, coated and uncoated micromirror chips when cooling down from room temperature (RT) to 120K. The characterization included SEM imaging between RT and 120K, actuation tests and interferometric surface characterization before and after the cycling RT-120K-RT. The results are very promising. SEM imaging of the critical structural parts of the device, that is the interface polysilicon-silicon, the interface of the mirror and the electrode chip,

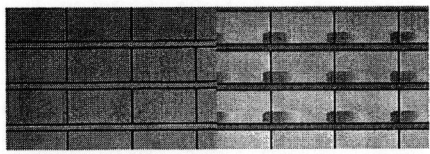

Fig. 3. SEM images of the optical side of a first generation micromirror array. Left side is at 300K and right side is at 120K. The white zones on the mirrors at 120K indicate charging of the mirrors. One micromirror element is 100μm x 200μm in size.

the gold coating on the mirrors and the bonding bumps of the packaged device, showed strictly no structural degradation between RT and 120K. This result was confirmed by the actuation tests before and after the temperature cycle: the device was functional and tilt-angle and actuation voltage where identical before and after cycling (20° mechanical tilt at 89V). The uncoated and the gold coated mirrors showed no surface degradation due to the temperature cycling, as the phase shift interferometric measurement with the Veeco/Wyko 1100NT showed: the peak-to-valley deformation (PTV) of the coated mirrors was 7nm and the PTV of the gold coated mirrors was 15nm, both before and after the cycling. Whereas no structural degradation could be detected at 120K, we saw a possible change of the electrical properties of the polysilicon and/or the interface silicon-polysilicon. Figure 3 shows a SEM of the optical side of the 5x5 micromirror array, left at RT and right at 120K. The white zones on the mirrors at 120K indicate a charging of the material due to bad grounding - these zones are seen on the mirrors but not on the frame, the latter being grounded correctly. As the micromirrors are connected electrically (and mechanically) to the frame by the polysilicon flexion hinges we think that the said polysilicon structures have inferior conductivity at 120K than at RT. How strong this effect is and if it affects the functionality of the device must be examined in further experiments.

A cryogenic chamber for the use on a interferometric setup is currently being realized. This setup will allow interferometric characterization of the mirror surface at 100K and in-situ actuation of the micromirrors and tilt-angle measurement. Conductivity measurements of the silicon-polysilicon-silicon system in function of the temperature will also be effectuated on this setup.

ACKNOWLEDGMENT

The authors would like to thank the SAMLAB and the Service for Micro- and Nanoscopy staffs at IMT and the European FP6 OPTICON program (JRA smart focal planes) for the financial support.

REFERENCES

[1] S. H. Moseley, R. Arednt, R. A. Boucarut, M. Jhabvala, T. King, G. Kletetschka, A. S. Kutyrev, M. Li, S. Meyer, D. Rapchun, and R. S. Silverberg, "Microshutters arrays for the JWST near infrared spectrograph," in *Proc. SPIE*, vol. 5487, 2004, pp. 645–652.

[2] S. Waldis, F. Zamkotsian, P.-A. Clerc, W. Noell, M. Zickar, and N. de Rooij, "Arrays of high tilt-angle micromirrors for multiobject spectroscopy," *IEEE Journal of Selected Topics in Quantum Electronics*, vol. 13, pp. 168–176, 2007.

TuP18
14:30 – 17:30

Improved Control of the Vertical Axis Scan for MEMS Projection Displays

Veljko Milanović

Adriatic Research Institute

828 San Pablo Ave., Ste. 109, Berkeley, CA 94706

veljko@adriaticresearch.org

Abstract - We demonstrate a MEMS projection display system with high speed, broadband, open-loop driving of the vertical axis which improves displayed image quality, increases the refresh rate capability and image brightness/efficiency. High speed capability in the vertical axis allows precise parallel line scanning as well as arbitrary line placement for features such as interlacing.

INTRODUCTION

Raster scanning for a video application presents a difficult high-speed requirement for the horizontal axis [1], even at lower resolution standards such as VGA (640x480 at 60Hz.) In a display application at VGA specifications for example, the fast axis should have a ~31.5kHz scan frequency, although using techniques such as displaying on the forward and backward trajectory can halve that frequency requirement. Often, the horizontal axis specification of a device is the determining factor for its overall performance due to the fact that it is difficult to achieve such high scanning rates (>15kHz,) large angles of deflection, relatively large mirrors, and minimal dynamic deformation during the scan. Ideally, motion of the horizontal axis scan would have uniform velocity across the active video region, i.e. a "triangle" waveform, requiring bandwidth far beyond the above scan rate. If we neglect this latter requirement however, and allow the horizontal axis to move with sinusoidal velocity, the problem becomes significantly simpler. In this regime of operation, raster scanners leverage the quality-factor (Q) of devices and utilize devices capable of only narrow-band sinusoidal driving [1]-[3]. Scanning at/near resonance allows design with very stiff suspensions to be used and in turn provides sufficiently high frequencies and angles.

At the same time, the frame refresh rate requirement creates a vertical axis period specification of 1/60Hz. Seemingly, this would be a relatively easy specification, which is exemplified in the frequently used naming of this axis as the "slow axis." However, the actual waveform requires a significantly higher bandwidth (at least 20X the refresh rate,) due to the fact that a highly linear scan is necessary, and that the retrace time (during which the display is off) should be as short as possible. In other words, we require a sawtooth waveform with at least 10:1 duty cycle. Achieving a high duty cycle is critical to obtaining high refresh rates as well as high brightness/efficiency, by minimizing the laser-off time. In Fig. 1 we show an example case where the ratio of active lines to retrace lines is 10:1, therefore forming a highly asymmetric sawtooth waveform. Fig. 1a shows the frequency content of an ideal waveform with these specifications, compared to a 1:1 triangle wave. Fig. 1b is the measured actual scan of a mirror where the vertical axis bandwidth is limited to only 400Hz. Though 400Hz is significantly higher than the 60Hz refresh rate, the waveform obviously suffers significant distortion from desired sawtooth, showing that such a bandwidth is inadequate for the application.

IMPROVED APPROACH

In order to improve the image quality, refresh rate, and brightness, in our work we implement the vertical scan with devices that have significantly higher bandwidths and also utilize custom feed-forward filters to increase the bandwith beyond first resonance, while preventing ringing. A comparison

example of a measured device waveform is shown in Fig. 1b. Our methodology allows arbitrary line placement (in applications like interlacing,) and purely horizontal and parallel scan lines (instead of continuously increasing vertical scan.) Namely, a vertical axis waveform is actually made up of flat regions and small steps to provide purely horizontal scanning.

In the demonstrations, we utilize gimbal-less two-axis (2D) MEMS optical scanners [3] (Fig. 2b) based on monolithic, vertical combdrive actuators. Normally, they are designed for ultra-fast two-axis beam steering with large optical deflections of >20° over the entire device bandwidth. For video projection application at VGA and SVGA specifications, we design the horizontal axis (X-axis) significantly stiffer to provide high resonant frequency and larger angles, and drive it in resonance. Bode plot of a device with a 0.8mm diameter mirror is shown in Fig. 1c.

It is possible to assemble different size mirrors onto the actuators, which enables a controlled tradeoff between desired aperture and speed. Our gimbal-less two-axis devices have horizontal axis resonant frequencies ranging from 10.0 kHz to ~21.5kHz, for mirror sizes ranging from 1.2mm to 0.8mm diameter. The power consumption of these scanners (< 10mW) is significantly lower than comparably performing scanners.

DISPLAY SYSTEM

We have integrated the MEMS scanner into a compact projection display system (Fig. 2). Video inputs are digitized based on input selection/settings controlled by a PC. Digitized video with sync information is processed by the FPGA DSP unit. This unit also controls the timing and drive signals for the MEMS scanner. It synchronizes the output video data (to laser modulation output) with the mirror position which is governed by the D/A channels and high voltage amplifiers (HVA). Specific X-axis and Y-axis drive signals are prepared in the PC for a specific device and video specifications (refresh rate, etc,) and downloaded to the FPGA memory for continuous readout.

While the X-axis data does not require any filtering or special treatment, the vertical (Y-axis) data is prepared based on the knowledge of the Y-axis capability of the device and in such a way to optimize speed and positional precision, while avoiding ringing. The simplest filtering scheme applies a digital approximation of a Bessel low pass filter while the feed-forward filter [5] uses information about the device resonant frequency and Q to compute a nearly optimal input waveform. A comparison of measured vertical axis waveforms for those two cases is shown in Fig. 1b. The prepared X-axis and Y-axis drive data is then downloaded to the FPGA processing board.

CONCLUSIONS

An ultra-low power portable laser vector display system was demonstrated using MEMS gimbal-less micromirror scanners.

1-4244-0641-2/07/$20.00 ©2007 IEEE

[1] Hakan Urey, "High performance resonant MEMS scanners for display and imaging applications," Proc. SPIE, Vol. 5604, Philadelphia, Pennsylvania, Oct. 2004

[2] A. Yalcinkaya, *et al*, "Two-Axis Electromagnetic Microscanner for High Resolution Displays," J. of MEMS, vol. 15, no. 4, Aug. 06.

[3] Y-C Ko, *et al*, "Eye-type scanning mirror with dual vertical combs for laser display," Proc. of SPIE, vol. 5721, 2005.

[4] V. Milanović, *et al*, "Gimbal-less Monolithic Silicon Actuators For Tip-Tilt-Piston Micromirror Applications," *IEEE J. of Select Topics in Quantum Electronics*, vol. 10, no. 3, Jun 2004.

[5] V. Milanović, K. Castelino, "Sub-100 µs Settling Time for Gimbal-less Two-Axis Scanners", Optical MEMS 2004, Takamatsu, Japan, Aug. 2004.

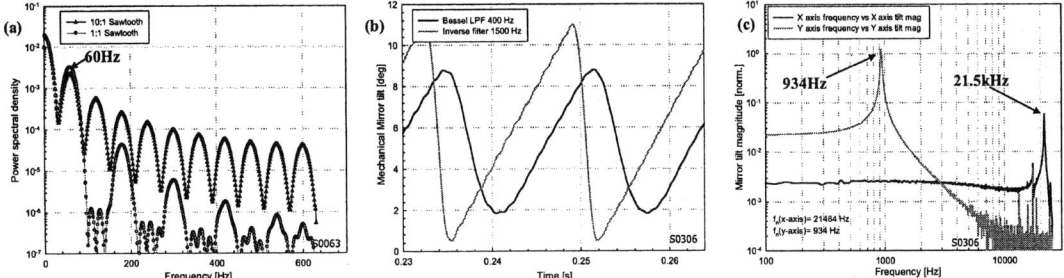

Figure 1. (a) Power spectral density (PSD) of a sawtooth waveform for the vertical scan. The 10:1 ratio sawtooth has a broad power spectrum well beyond the 60Hz refresh rate. (b) Amplitude of the small signal freq. response of a device used to demonstrate SVGA-specification display. (c) vertical axis scan improvement by using a customized inverse filter.

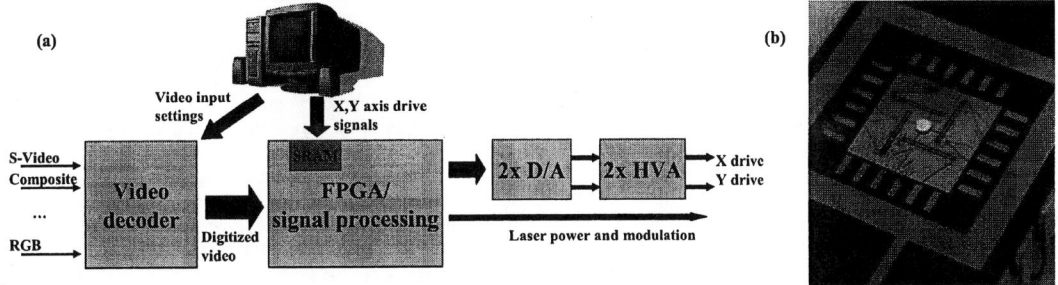

Figure 2. (a) Schematic diagram of the MEMS projection display hardware setup. Video inputs are digitized based on input selection/settings controlled by a PC. Digitized video with sync information is processed by the FPGA DSP unit. This processing unit also controls the timing and drive signals for the MEMS scanning mirror, synchronizing the output video data (laser modulation) with the mirror position which is governed by the D/A channels and high voltage amplifiers (HVA). Specific X-axis and Y-axis drive signals are prepared in the PC for a specific device and video specifications (refresh rate, etc,) and downloaded to the FPGA memory for continuous readout. (b) A two-axis scanning device with a 0.8mm mirror.

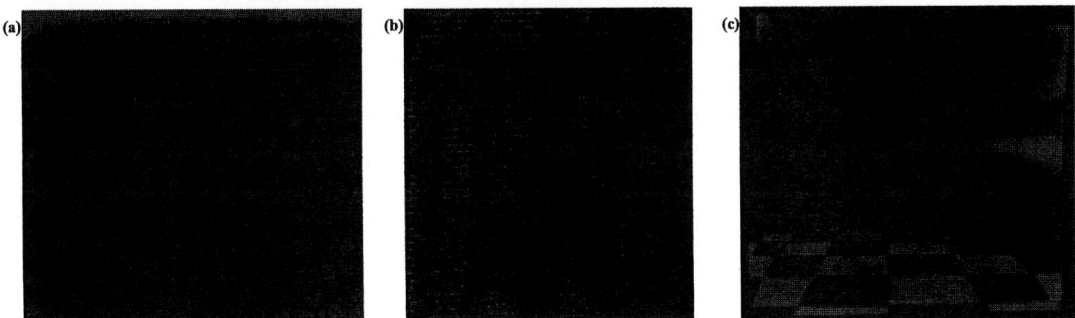

Figure 3. Examples of images displayed by the projection system. (a) Portion of a 320x240 image with a 635nm laser showing 32 grayscale levels. (b) Close inspection of a small segment of an image displayed with a 405nm laser shows horizontal, parallel lines. This is due to flat segments in the vertical scan followed by small steps which move the Y-axis from line to line. (c) a VGA (640x480) demo showing the computer's 2nd monitor through the projection display. Image has 14 grayscale levels due to speed limitation of our 635 nm laser. Full field can be displayed on each refresh, or half a field with interlacing.

TuP19
14:30 – 17:30

High Temperature Operation of Gimbal-less Two Axis Micromirrors

Andrew Miner and Veljko Milanović
Adriatic Research Institute
828 San Pablo Ave., Ste. 109, Berkeley, CA 94706
miner@adriaticresearch.org

Abstract – We demonstrate seamless operation of gimbal-less two axis micromirror devices at high temperatures up to 200°C by characterizing the temperature stability of the mirror tilt angle and first resonant mode. The ability to provide a repeatable, high temperature vector or raster scanning micromirror opens up opportunities for ruggedized products in extreme environments. This capability is also relevant in high incident optical power applications, where energy is absorbed at the mirror. A first order thermal model is presented that allows the temperature rise of mirrors under high optical power operation to be estimated.

INTRODUCTION

Two axis micromirror devices can provide point-to-point, vector, and raster optical beam steering capability for various applications including projection displays, optical data storage, ranging, and bio-medical imaging. The capability of device operation at high temperatures is of particular interest when developing for extreme environment applications; when developing micromirrors for high optical power applications, and finally for robustness in product reliability tests. The two axis micromirror devices in this work are gimbal-less, monolithic single-crystal silicon actuators that may include an integrated mirror, or a mirror that is fabricated separately and subsequently bonded [1],[2]. The actuators are based on electrostatic attractive forces in vertical combdrives that are mechanically coupled to the mirror by bi-axial linkages. Micromirror devices with integrated and bonded mirrors are shown in Figs. 1a and 1b, respectively.

OPERATION AT ELEVATED TEMPERATURES

In order to study the performance of the devices at elevated temperature, a resistively heated stage was constructed that would accept packaged micromirror devices. A schematic of the experimental setup is shown in Fig. 1c. The devices were mounted in commercially available 24 pin dual in line packages using a thermal epoxy. In order to monitor the operating temperature of the die, a K-type thermocouple was mounted on the package as close to the die as possible. The thermocouple was mounted with thermal epoxy, and was typically within 1 mm from the edge of the micromirror die. This allowed the temperature of the device to be monitored and controlled.

Static voltage to angle transfer function as well as a small-signal frequency response of each device was recorded at various temperatures. In each case, the system was allowed to stabilize at a temperature set point for longer than 10 minutes, before performing the measurements using a position-sensitive detector (PSD) and a programmed data-acquisition system.

Static tilt angle for 69V and 81V of actuator voltage for both the x and y axis at temperatures from 25 °C to 200°C is shown in Figure 2a. The tilt angle increases slightly with increasing device temperature. For the data sets shown in Fig. 2a, starting from the top of the plot, linear regression lines show slopes from of 1.2, 0.9, 0.9, and 0.6 milli-degree of mechanical tilt/°C. From frequency response measurements we extracted the frequency of the first resonant mode, which is of interest for both point to point and resonant scanning. In point to point mode the device can be precisely controlled at frequencies that approach this first resonant mode. In other words the useable bandwidth is defined by the first resonant frequency. In resonant scanning, the device oscillates at or near this first resonant mode frequency, so its stability is highly relevant. Figure 2b shows the temperature dependence of the frequency of the first resonant mode for temperatures from 25°C to 200°C. The dependence of this frequency on temperature, based on linear regression of the data, is -73 mHz/°C for the x axis and -44 mHz/°C for the y axis.

Overall, the performance of the devices was highly stable at all tested temperatures, and the upper limit was only defined by our experimental setup. This result is attributed to the monolithic single-crystal silicon construction of the devices and the electrostatic nature of actuation, which both have very low sensitivity to elevated temperatures.

THERMAL MODELING

The above tests apply directly to applications in a high temperature ambient environment. On the other hand, even at room-temperature, the temperature of a device can also increase in applications where the mirror is illuminated with high optical power. In these conditions, the temperature of the mirror and actuator structure will not be uniform. The magnitude of the temperature rise of the mirror is predominately a function of the reflectivity of the mirror, the mirror size, the actuator geometry, and the ambient environment. In order to study the temperature rise of micromirror devices, we conducted numerical studies which showed that when operated in ambient pressure environments, thermal conduction through the gas was typically the dominant mode of heat transfer from the mirrors. This finding is in line with similar work on micromirror heating [3]. This was despite the high thermal conductivity of silicon beams in the actuator, as the relatively large area of the mirror as well as the large area of the actuator itself (including combfingers) contributes to the dominance of gas-conduction.

A representative result for a square mirror on a two axis actuator is shown in Fig. 3a. Based on these numerical studies, a simple analytical model for the temperature rise of the mirrors was developed that considers only two paths of heat conduction operating in parallel: (1) a conduction path through the ambient gas from the mirror to the die and (2) a solid conduction path through the mirror and the silicon actuator to the die. The estimated temperature rise of the mirror for 1 Watt of incident optical illumination is shown in the Fig. 3b. The temperature rise is shown for mirror reflectivities of 90% (characteristic of Aluminum for visible wavelengths) and of 99% (characteristic of thin film multilayer reflective coatings). The data in the table illustrates how gas conduction from a larger mirror to the die and higher reflectivity coatings can significantly reduce temperature rise in this type of micromirror device.

1-4244-0641-2/07/$20.00 ©2007 IEEE

CONCLUSIONS

The static tilt angle and first resonant mode of monolithic, gimbal-less micromirror devices tested in this work show only a slight sensitivity to elevated ambient temperature. For applications where the temperature rise is due to high power optical illumination, any performance degradation can be mitigated by geometric design of the device and the use of highly reflective mirror coatings.

[1] V. Milanović, et al, "Gimbal-less Monolithic Silicon Actuators For Tip-Tilt-Piston Micromirror Applications," *IEEE J. of Select Topics in Quantum Electronics*, vol. 10, no. 3, Jun. 2004.

[2] V. Milanović, K. Castelino, "Sub-100 μs Settling Time for Gimbal-less Two-Axis Scanners", Optical MEMS 2004, Takamatsu, Japan, Aug. 2004.

[3] J. Zhang, Y. C. Lee, A. Tuantranont, and V. M. Bright, "Thermal Analysis of Micromirrors for High-Energy Applications", *IEEE Trans. On Advanced Packaging*. Vol. 26, no. 3, Aug. 2003.

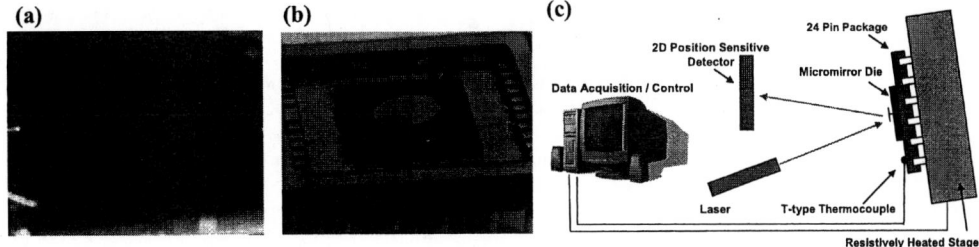

Figure 1. (a) A gimbal-less two axis silicon micromirror device with an integrated 800μm mirror. (b) A similar device with a bonded mirror (3600 μm diameter.) (c) A schematic diagram showing the experimental setup for testing micromirror devices at elevated temperatures. A resistively heated stage was computer controlled using a thermocouple that was attached to a 24 pin package near the device die. Mirror motion was detected using a low power laser and a position sensitive detector.

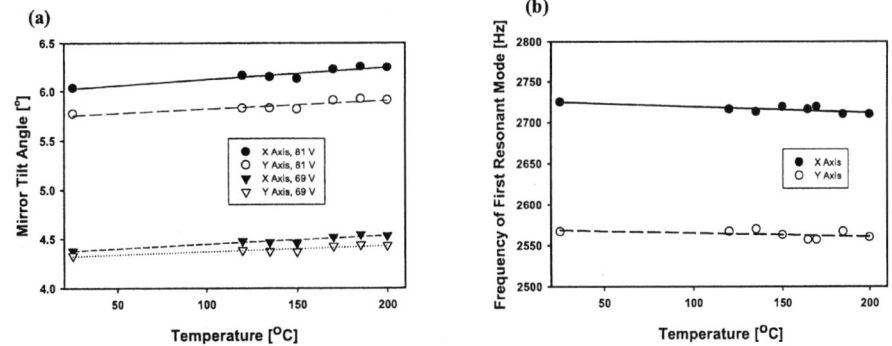

Figure 2. (a) Mirror tilt angles at two select voltages are shown as a function of the device temperature. Both x and y axis were tested and perform similarly. The tilt angle shows a slight increase as a function of temperature. (b) The frequency of the first resonant mode for both x and y axis is shown as a function of device temperature. The resonant mode frequency shows a slight decrease with increasing temperature.

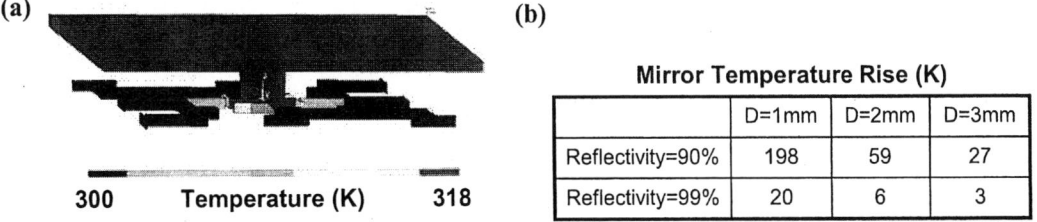

Mirror Temperature Rise (K)

	D=1mm	D=2mm	D=3mm
Reflectivity=90%	198	59	27
Reflectivity=99%	20	6	3

Figure 3. (a) A representative false-color image of numerical simulations performed in order to study the dominant paths of heat conduction in gimbal-less micromirrors. (b) Summarized results from a simple thermal model that considers heat flow from the mirror through the ambient air as well as heat flow through the silicon structure of the device. Temperature rise of the mirror is shown for two values of mirror reflectivity and for three mirror diameters. All simulations assume 1 Watt of optical power incident on the mirror and uniformly distributed over each mirror's surface.

TuP20
14:30 – 17:30

Mechanical-Contact-Based Submicron-Si-Waveguide Optical Microswitch at Telecommunication Wavelengths

E. Bulgan, Y. Kanamori, K. Hane
Department of Nanomechanics, Graduate School of Engineering, Tohoku University,
6-6-01 Aoba, Aramaki, Aoba-ku, Sendai, 980-8579, Japan
Tel +81-22-795-6965, Fax +81-22-795-6963, E-mail bulgan@hane.mech.tohoku.ac.jp

Abstract

We report a mechanical-contact-based single-input single-output normally-closed optical microswitch. The switch comprises an input and an identical output waveguide, and a movable low-loss elliptical waveguide driven by a miniature electrostatic comb drive. Movable waveguide closes the lateral gap between input and output waveguides. Mechanical contact of waveguide tip surfaces enables light propagation from the input to the output waveguide through movable elliptical waveguide. Optical property of the switch is studied by Finite Difference Time Domain (FDTD) Analysis. An SOI wafer with 260 nm-thick device-Si and 2 μm-thick buried-oxide layers is utilized in the fabrication. TE-polarized near infrared light at 1.55 μm wavelength is coupled into the input waveguide through an IR objective lens in free-space, and switch is observed with the help of a vidicon camera. Approximately 30±2 dB switching contrast is obtained experimentally at 540 nm lateral switch motion between on and off states.

Keywords: Optical microswitch, submicron waveguide, mechanical contact, telecommunication, single mode

1 INTRODUCTION

Growing use of information technology and distributed applications for geographically dispersed users have expanded the need for faster and compact optical devices at telecommunications wavelength around 1.55 μm.

Several micro-optical devices are already reported. These devices utilize one of; relative motion of fiber optic cables [1], refractive index modulation of planar optical waveguides through temperature changes [2], employment of arrayed mirrors in free-space [3], or reflective surface creation via forming bubbles in a index-matching-oil filled channel environment [4].

Owing to its high refractive index, 2.45, single crystalline Si can confine light in very compact cross-sectional regions. Besides, Si is transparent in the wavelength range from 1120 nm to 1610 nm, where telecommunications wavelengths fall in. Hence, Si has been a critical material for applications in the submicron-waveguide and photonic crystal fields. Since fabrication of submicron Si is technically feasible today, very compact and faster optical devices are technically feasible today.

In this report, we present an approach towards realization of a novel mechanical-contact-based submicron-Si-waveguide optical microswitch.

2 PRINCIPLE

The optical switch, as illustrated in Figure 1, consists of an input and an identical output waveguide, and a movable waveguide actuated by an electrostatic comb drive. When the comb drive is activated, movable waveguide closes the gap between input and output waveguides. Due to mechanical contact of waveguide tips, light can propagate from the input to the output waveguide through the movable waveguide. When retracted, light propagation ends.

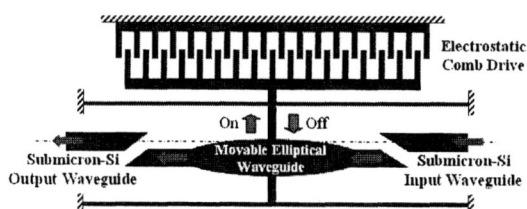

Figure 1. Schematic view of optical microswitch

3 DESIGN AND THEORETICAL CALCULATION

For single mode light propagation at 1.55 μm wavelength, Si waveguides are designed to be 500 nm wide and 260 nm height.

Optical transmission as a function of lateral in-plane motion

1-4244-0641-2/07/$20.00 ©2007 IEEE

at 45°-cut-waveguide-tip geometry is studied by FDTD analysis. Result of the numerical study, as shown in Figure 2, reveals that transmission in the switch is sharply decaying with the increasing lateral gap. Due to use of two-cascade tip pairs, only a lateral gap of 0.4 μm is calculated to yield 96.69 % switching contrast between on and off states. In addition, a minimum longitudinal gap of 1.6 μm is calculated to be necessary to limit free-space transmission leak between input and output waveguides within 1 %.

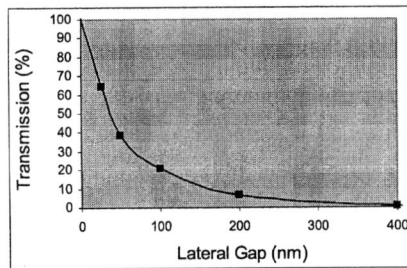

Figure 2. Optical transmission as a function of lateral gap

In order to support the suspended movable waveguide mechanically with minimum optical transmission leak, a low-loss elliptical intersection structure [5] is deployed with slight modifications in order to minimize back-reflection as well.

4 FABRICATION AND EXPERIMENT

The switch is fabricated on an SOI wafer with 260 nm-thick device-Si and 2 μm-thick buried-oxide layers. First, switch is nanopatterned in the Electron Beam Lithography. Then, device-Si is dry-etched by Fast Atom Beam (FAB) etching in full. Figure 3 shows SEM image of the switch after FAB Etching. Next, switch is released by Vapor HF Etching in order to prevent substrate stiction.

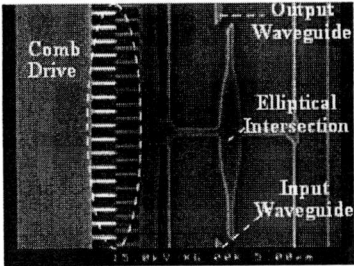

Figure 3. Fabricated optical microswitch

Because active switching region is as small as about 40x60 μm², displacement characteristic of the actuator with respect to applied bias voltage is studied under scanning electron microscope.

TE-like polarized laser light at 1.55 μm wavelength is coupled into the input waveguide through an IR objective lens in free-space, and switch is observed by a vidicon camera. The vidicon camera is calibrated and ensured to be linearly responding with a reference light power. Optical losses and output are evaluated with the help of a custom-written software, which calculates brightness at a desired spot on the switch.

5 RESULTS AND DISCUSSION

Figure 4 depicts the optical response of the switch at both off and on states. Switching region in these figures includes contact-tip pairs of input/movable waveguides and output/movable waveguides, and the comb drive.

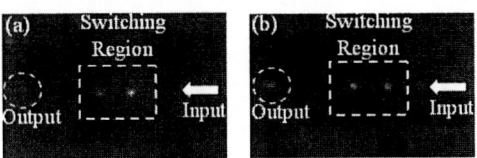

Figure 4. IR camera images of the microswitch at;
(a) Off State
(b) On State at 65 VDC, ~540 nm switch motion

In optical tests, 30±2 dB transmission contrast is obtained at 540 nm lateral switch motion between on and off states. Only about 40 % transmission is observed because of high scattering losses at contact tips due to nanolithography imperfections.

6 CONCLUSION

This study presents numerical study, design, fabrication and testing of mechanical-contact-based single-mode submicron-Si-waveguide optical microswitch. Optical tests have yielded significantly high, approximately 30±2 dB, switching contrast.

REFERENCES

[1] Y. Kanamori, Y. Aoki, M. Sasaki, H. Hosoya, A. Wada and K. Hane, "Fiber-optical switch using cam-micromotor driven by scratch drive actuators," J. Micromech. Microeng., vol.15, pp.118-123, 2005.

[2] T. Watanabe, N. Ooba, S. Hayashida, T. Kurihara and S. Imamura, "Polymeric optical waveguide circuits formed using silicone resin," J. Lightwave Technol., vol.16, pp.1049-1055, 1998.

[3] L. Lin, E. Goldstein and R. Tkach, "Free-space micromachined optical switches for optical networking," IEEE J. Select. Topics Quantum Electron., vol.5, pp.4-9, 1999.

[4] A. Zhang, et al., "Integrated liquid crystal optical switch based on total internal reflection," Applied Physics Letters, vol. 86, 211108, 2005.

[5] T. Fukazawa, T. Hirano, F. Ohno and T. Baba, "Low loss intersection of Si photonic wire waveguides," Jpn. J. of Appl. Phys., vol. 43, no.2, pp.646-647, 2004.

TuP21
14:30 – 17:30

Two-wavelength Grating Interferometry for Extended Range MEMS Metrology

M. F. Toy, O. Ferhanoglu, and H. Urey
Koç University, Department of Electrical Engineering,
Rumelifeneri Yolu, 34450, Istanbul, Turkey
Tel +90-212-338-1772, Fax +90-212-338-1548, E-mail hurey@ku.edu.tr

Abstract
Diffraction gratings integrated with MEMS has many applications as they can offer shot noise limited sub-nm displacement detection sensitivities but are limited in range. A two-wavelength readout method is developed that maintains high sensitivity while increasing the detection range from 105nm to 1.7um assuming sensitivity is maintained at >50% of the maximum sensitivity.

Keywords: Optical Sensing, Grating interferometry, MEMS metrology

1 INTRODUCTION

Diffraction Grating Interferometry is an attractive method for MEMS devices for detecting sub-nm displacements due to its high sensitivity with shot noise level detection ability [1]. The maximum detectable range is limited to $\lambda/4$ of the readout wavelength. In this paper we present 2-wavelength readout method, which offer high sensitivity and long operation range, extending the capabilities of MEMS grating based optical sensors.

Main advantages of the grating based optical readout are that gratings can be micromachined and integrated with single or array of MEMS devices for AFM and ultrasonic sensor applications [1,2], thermo-mechanical IR detector array applications [3], and can serve as comb actuators for Fourier transform spectroscopy applications [4], or other MEMS sensing and actuation applications.

2 EXPERIMENTAL RESULTS

The theory was tested on a MEMS Fourier transform spectrometer shown in Figure 1 [4]. Comb fingers serve the purpose of both actuation and movable diffraction grating. To achieve low-frequency non-resonant mode operation, some tests were conducted by using part of the gratings and a separate moving platform as illustrated in Fig. 1. The device was illuminated using lasers with 632.8, 655.7 nm wavelengths and the reflected first diffraction orders were focused onto two photodetectors.

Sensitivity is given by the intensity change per deflection. The sensitivity for the method is taken as the maximum sensitivity for each wavelength at a given gap. The maximum sensitivity reported with this method is $2\text{x}10^{-4}$

Å/Hz$^{1/2}$ at 20 KHz, that corresponds to <5pm measurement ability at 20KHz [1]. The maximum sensitivity curve corresponding to the above wavelengths are calculated and shown in Fig. 2.

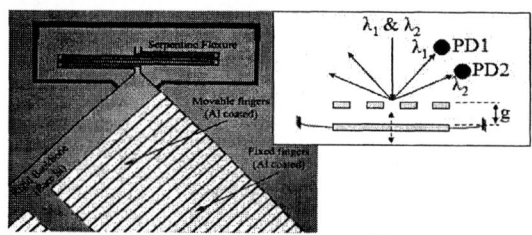

Figure 1. Left: MEMS Spectrometer used in the experiment; Right: side view of setup with moving platform.

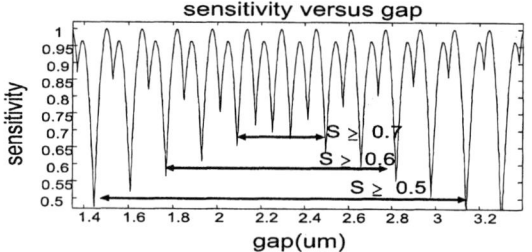

Figure 2. Normalized Sensitivity (S) versus gap.

The sensitivity curve is periodic with D, which is given as:

$$D = \frac{\lambda_1 \lambda_2}{4|\lambda_1 - \lambda_2|} \qquad (1)$$

For the given wavelength pairs $D = 4.6$ um. The curve is symmetric around D/2 as seen in Fig. 2. Range versus sensitivity values for one and two wavelength grating interferometry are summarized in Table 1.

1-4244-0641-2/07/$20.00 ©2007 IEEE

Range values may be extended with different sources with smaller wavelength differences. The actuator in Fig. 1 was actuated sinusoidally at 60 Hz. The data obtained from the photodetectors are shown in Fig. 3. Since the gap was modulated sinusoidally, the PD signals are chirped sinusoids.

Table 1. Ranges for different sensitivities for 1 wave readout (633 nm) and 2-wave readout (633nm & 656 nm) Full range gives the period of the sensitivity curve

	S > 0.7	S > 0.6	S > 0.5	Full range
1-wave readout	80 nm	95 nm	105nm	160 nm
2-wave readout	0.4 um	1 um	1.7 um	4.6 um

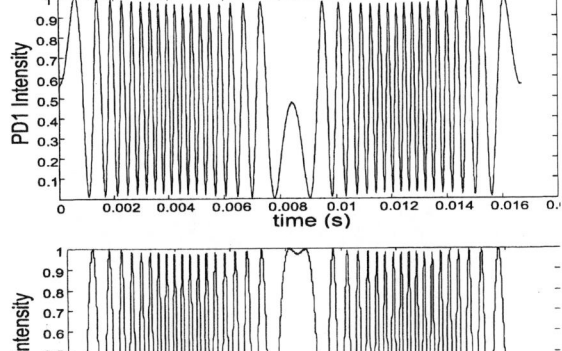

Figure 3. Data collected from two detectors (λ_1= 656 nm in the top figure and λ_2=633 nm in the bottom figure).

The corresponding displacement was calculated to be ±3um using the normalized intensities and is shown in Fig. 4. Each photodiode intensity value corresponds to several possible solutions within the range of operation with a period of half the wavelength. The solution space using PD1 and PD2 signals are compared and the closest possible solution is chosen to find the absolute position in the range. The solution has two gap values calculated from PD1 and PD2 signals, the one with higher sensitivity near the particular operating point should be used for the sensitivity calculations and to determine the absolute position. Similar to single wavelength interferometers, a calibration measurement has to be taken for each sensor to determine the maximum and the minimum intensities corresponding to each photodiode output. When one is dealing with deflections smaller than $\lambda/4$, similar procedure should apply, and the peak intensities determined prior to the experiment would be needed. Laser noise reduction and active calibration can be performed by simultaneously monitoring the 0th order light for each wavelength, which gives the best sensitivity results.

3 APPLICATIONS

The wide range capability enables this technique to be used in dynamic measurements such as the experiment presented here. This technique is able to measure very small deflections at both high frequencies as well as at low frequencies and DC, which is a limitation in Laser Doppler Vibrometer measurement devices that measure velocity.

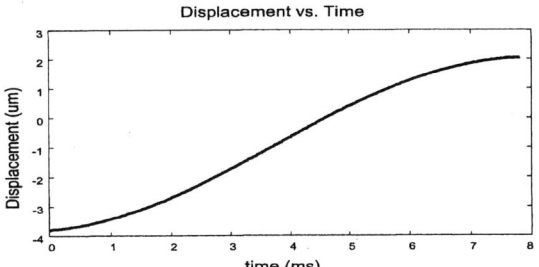

Figure 4. Calculated displacement using PD1 and PD2 data

Another important application is illustrated in Fig. 5. The nanoimaging and biomolecular mechanics measurement devices have shot-noise limited deflection measurement ability with integrated probe and grating interferometer but limited measurement range due to single wave readout [2]. Two-wave readout technique improves the range to several microns, which relaxes requirement on gap fabrication and active control of the gap.

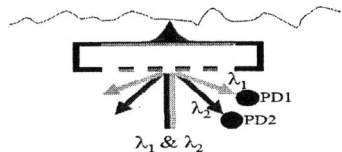

Figure 5. Illustration of AFM application

We thank Caglar Ataman for help with MEMS IR Spectrometer experiments and Prof. Degertekin's group for providing CMUT devices for tests. This work is partly sponsored by ASELSAN Inc.

REFERENCES

[1] W. Lee, N. A. Hall, Z. Zhou, and F. L. Degertekin, "Fabrication and characterization of a micromachined acoustic sensor with integrated optical readout", *IEEE JSTQE.* **10**, pp.643-651, 2004.

[2] H.Torun, J Sutanto, K K Sarangapani1, P.Joseph, F.L.Degertekin, C. Zhu1,"A micromachined membrane-based active probe for biomolecular mechanics measurement", *Nanotechnology* **18**, 165303, 2007.

[3] H. Torun, O. Ferhanoglu, H. Urey, "Thermo-Mechanical Detector Array with Optical Readout," IEEE-LEOS Optical MEMS Conference, P18, Big Sky, Montana, USA, 2006.

[4] C. Ataman, H. Urey, A. Wolter, "MEMS-based Fourier Transform Spectrometer," *J.Micromech.Microeng.* **16**, p.2516-2523, 2006.

TuP22
14:30 – 17:30

A New Fast Infrared Tracking System With Thermopile Array Implementation

Po-Hsiang Chang, Chih-Hsiung Shen

National Changhua University of Education

No.101, Chongde 2nd St., Tainan City, Taiwan, R.O.C

Tel + 886963227971, Fax + 8864-7211149, E-mail ben97971@yahoo.com.tw

Department of Mechatronic Engineering,

National Changhua University of Education, Changhua, Taiwan

Abstract

This paper proposes a novel design and implementation of tracking infrared system with a 12x2 CMOS thermopile linear array. Without complicated image processing circuits, the tracking system compares four signal outputs from the linear thermopile array and the location of infrared target is determined very fast and efficiently. A CMOS compatible thermopile array is realized using 1.2 um standard CMOS process and two low offset operational amplifiers (OPA) is included to transfer the analog signals of thermopile output to digital signals. The system incorporated with a Germanium lens is realized and measured with an infrared blackbody carefully and the results are quite accurate and successful.

Keywords: CMOS, Thermopile, Operation amplifier, IR Sensor, Thermal image, Thermal tracking system.

Ⅰ. INTRODUCTION

The technology of infrared image tracking is powerful in daily, and it could be applied in security system, heat tracking, and infrared image tracking system. This system is made of thermopile array and chopper amplifier. The chopper amplifier is very powerful with it's low noise and offset, so we don't need to worry that the signal of thermopile would be interfered. The offset of chopper amplifier could less than 1μv. Using thermopile and chopper amplifier could achieve this system, and get the digital signal easily.

Ⅱ. THE SYSTEM: Thermopile

Each one pixel of thermopile [1]—[3] array with 24 pairs of thermoelectric infrared sensors can be fabricated by using a standard CMOS 1.2μm process, thus it's capable of integrating the interface circuit on-chip. So that we can save the chip size, increase the reliability, and cost down. The device consists of a thermally insulated absorbing area and a thermopile with the "hot" junctions in the absorbing area and the "cold" junctions on a heat sink (i.e. the silicon bulk). As the length of thermocouple lead increased, the sensitivity of thermopile sensor will increase too. Scheme of thermopile array was shown in Fig.1 and the shape of v-groove window was observed.

Two low offset chopper amplifies are used and connected to four pixels of thermopile linear array and output two digital signals which are easily to identify the location of target.

Fig. 1 Photograph of array thermopile

Ⅲ. ARCHITECTURE IN TWO PARTS

Because of the thermopile characteristic, we use thermopile to produce small voltage when temperature changed. Through system we can get the signal by the small voltage is enough to distinguish the exactly position and direction of infrared image without readout circuit immediately. Besides the operation amplifier (OPA) can read analog signal, and it can transfer to a digital one. The system also includes efficacy of A/D converter and readout circuit. Due to directly receive a digital signal, the IR sensor needn't use complex computer soft to identify position and direction of movement. So we have to discuss about the circuit and optics.

A) Part of circuit

We use the chopper amplifier [4] to amplify the small voltage caused by the thermopile, because it has excellent characteristic that is low offset voltage (less than 1μv), high noise reject ratio (CMRR/PSRR) and have more than 100dB gain. The output signal of system is the positive limiting

1-4244-0641-2/07/$20.00 ©2007 IEEE   97

level voltage or negative limiting level one. That could be regarded as a digital signal (High/Low). The system block diagram is shown in Fig. 2.

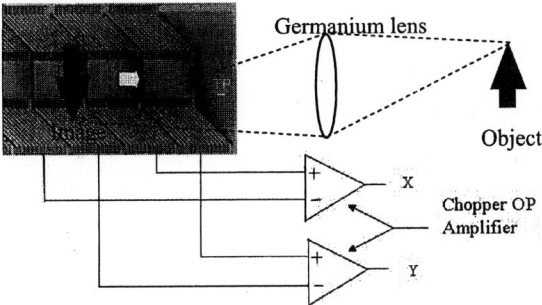

Fig.2 simplified block diagram of infrared image tracking system

Using chopper amplifier it could export digital signal. The thermopile would produce voltage to two negative of OPA when infrared image was located at the first pixel and the second one. Because negative of OPA imports voltage from thermopile, the output signal can be regarded Low and Low. When the system output was Low/Low signal, we know the blackbody is located on the right of system. The system has three situations (Low/Low, High/Low, High/High). According to the output we can distinguish that the position of blackbody was located on the right or left or middle of system. The true table which is shown in table II

Table II. True table of output

The position of image	Signal of x	Signal of y
Between 1 and 2	L	L
Between 2 and 3	H	L
Between 3 and 4	H	H

B) Part of optics

The system consists of array thermopile, OPA, and germanium lens. The system is shown in Fig.3. The array sensor includes twenty four pixels (shown in Fig. 1). The system uses four pixels to trace infrared image. Using blackbody produces infrared image which is focused on two pixels by lens. Then, by using OPA, we utilize the voltage from thermopile to export digital signal. We have to figure out each distance between lens and thermopile and infrared image.

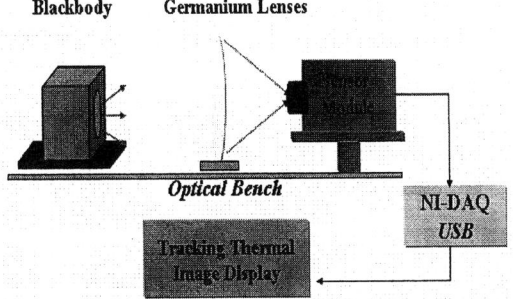

Fig.3 Experimental setup for infrared tracking

IV. EXPERIMENTS AND MEASUREMENT

The system used AD8551, array thermopile (standard CMOS 1.2μm process), and germanium lens shown in Fig. 2 (f=25.4mm). According to experimental results, we can verify the reliability. We can find the results in table III. The result is totally matching with our supposition.

Table III. The output voltage of system when different position

Position / Temperature	2cm		4cm		6cm		8cm		10cm		12cm		14cm		16cm		18cm	
90°C	0	0	5.6	6.02	5.7	6.03	5.83	6.03	5.98	0	0	0	0	0	0	0	0	0
80°C	0	0	5.61	6.03	5.73	6.03	5.73	6.03	5.73	0	0	0	0	0	0	0	0	0
70°C	0	0	5.57	6.02	5.75	6.03	6.03	6.03	5.95	0	0	0	0	0	0	0	0	0
60°C	0	0	5.7	6.03	5.7	6	6.04	6.04	5.95	0	0	0	0	0	0	0	0	0
50°C	0	0	5.7	6.03	5.7	6.03	6.04	6.03	5.87	0	0	0	0	0	0	0	0	0
40°C	0	0	5.8	6.03	5.8	6.02	6.04	6.03	5.82	0	0	0	0	0	0	0	0	0
30°C	0	0	5.8	6.03	5.8	6.0	5.8	6.01	5.7	0	0	0	0	0	0	0	0	0
20°C	0	0	0	0	0	0	0	0	6	5.98	6.03	5.7	6.03	5.7	6.02	0	0	
15°C	0	0	0	0	0	0	0	0	6	5.8	6.0	6	6	6	6	0	0	
10°C	0	0	0	0	0	0	0	0	6	5.7	6.0	5.7	6.02	5.7	6.0	0	0	
5°C	0	0	0	0	0	0	0	0	6	5.7	6.0	5.7	6.02	5.8	6.03	0	0	

V. CONCULUSIONS

By experimenting and measuring, it is successful that we prove the design of system. We can trace the infrared image easily, and reduce complex circuit (readout circuit and A/D converter). Then we could distinguish the movement of infrared image, witch is from the output of system without computer soft. It's powerful technique in SOC, because it can save the area of chip. By using CMOS standard process, we could manufacture thermopile and circuit. Therefore this architecture can be achieved by system-on-chip (SOC). There will be more such researches in the future

REFERENCES

[1] J. A. Huang, and C. H. Shen. "Novel Improvement of Thermal Conductance for Microstructure of CMOS Compatible Thermopile" The 3nd Conference on Precision Machinery and Manufacturing Technology pp.19-25 2005

[2] Chin-Shown Sheen, Sien Chi, "CMOS compatible thermoelectric infrared sensors", Electronics Letters pp:1117-1118 2000

[3] Bakker, A., Thiele, K., and Huijsing, J.H. "A CMOS Nested-Chopper Instrumentation Amplifier with 100-nV Offset "Solid-State Circuits, IEEE Journal vol:35 no:12 pp: 1877-1883 Dec 2000

[4] Dzahini D., and Ghazlane H "Auto-zero stabilized CMOS amplifiers for very low voltage or current offset" IEEE Nuclear Science Symposium Conference Record vol.1,no. pp 19-25 Oct. 2003

TuP23
14:30 – 17:30

Self-Supported Pitch-Variable Guided-Mode Resonant Grating Filters at Telecom Wavelengths

Y. Kanamori, N. Matsuyama, J.-S. Ye and K. Hane
Tohoku University
6-6-01 Aoba, Aramaki, Aoba-ku, Sendai 980-8579, Japan
Tel +81-22-795-6965, Fax +81-22-795-6963, E-mail kanamori@hane.mech.tohoku.ac.jp

Abstract

We fabricate a pitch-variable guided-mode resonant grating filter. The grating is a Si self-supported structure and expanded by opposed comb-drive actuators. The initial grating period of 860 nm is designed for a wavelength-selective filter at telecom wavelengths. By controlling the grating period, the resonant wavelength can be tuned.

Keywords: Guided-mode resonant grating filters, subwavelength gratings, wavelength-selective filters, comb-drive actuators

1 INTRODUCTION

A subwavelength grating, which has a smaller period than the wavelength of an incident light, has gotten a lot of attention as one of novel optical elements. A guided-mode resonant grating (GMRG) filter, which is one of the subwavelength grating filters, works as a high efficiency (100% in theory) band-stop filter[1]. Since the band-stop filters with both narrow (< 1 nm) and broad bandwidths can be designed with silicon as the GMRG material, telecom applications with GMRG filters are attractive. The optical characteristics are sensitive to the grating structure. So, mechanical change of the grating structure causes large optical modulation. We have investigated pitch-variable subwavelength gratings at visible wavelengths[2-4]. Using a GMRG filter with the pitch-variable mechanics, we can realize a tunable wavelength-selective filter at telecom wavelengths. In this paper, we design and fabricate a pitch-variable GMRG filter at telecom wavelengths for the first time. The filter is fabricated on a silicon-on-insulator (SOI) substrate, and estimated the optical characteristics.

2 PRINCIPLE AND DESIGN

Figure 1 shows concept of wavelength-selective filters based on pitch-variable GMRG filters. The pitch-variable grating consists of a self-supported Si structure into the air. It is satisfied guided-mode resonant condition at the incident wavelength λ, which impinges normally on the grating. When the grating is initial state with the period $\Lambda(0)$ as shown in Fig.1(a), the 100% of incident light at a resonant wavelength is reflected causing by the strong coupling with the grating. When the grating period is increased by expanding the groove width from d to d+x as shown in Fig.1(b), the resonant wavelength is changed depending mainly on the grating period. Therefore, by changing the grating period $\Lambda(x)$ mechanically, the selective wavelength can be controlled as shown in Fig.1(c).

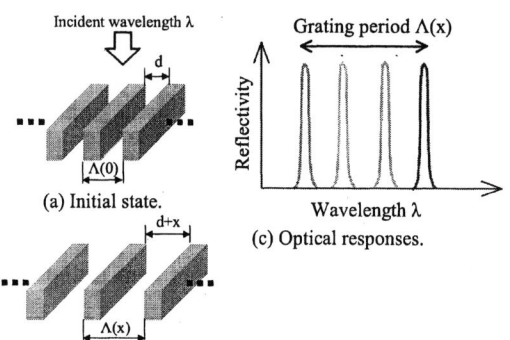

Figure 1. Concept of wavelength selective filters based on pitch-variable GMRG filters.

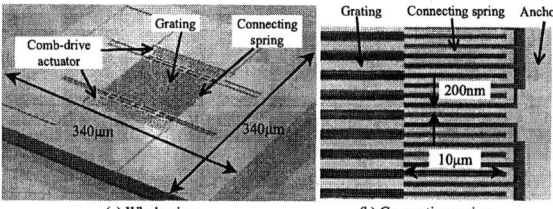

Figure 2. Schematic of the proposed wavelength-selective filter.

Figure 2 shows the schematic of the proposed wavelength-selective filter. The filter mainly consists of a grating, connecting springs and comb-drive actuators on a substrate. These materials are silicon. The grating and connecting springs are self-supported structures and the grating beam is connected next to each other by the connecting springs as shown in Fig.2(b). The width and length of the connecting springs are 200 nm and 10 μm, respectively. The opposed two comb-drive actuators are connected to the grating. By expanding the grating with the comb-drive actuators, the grating period can be changed.

1-4244-0641-2/07/$20.00 ©2007 IEEE   99

The reflectivity of the grating as shown in Fig.3(a) has been calculated using rigorous coupled-wave analysis. The incident light is assumed to impinge normally on the grating with TE polarization. The dimension of the grating structure has been designed to obtain the resonant wavelength around 1.55 µm. Figure 3(b) shows the calculated reflectivity as a function of the incident wavelength and the grating period. Fill factor, which is the ratio of grating width (= 430 nm) to the Λ, is also shown. The white line shows a resonant wavelength line. As you can see, 100% reflectivity is obtained around the line, although the resonant-wavelength widths are different depending on the grating period. By controlling the grating period using the comb-drive actuators, the wavelength of the reflection peak can be changed.

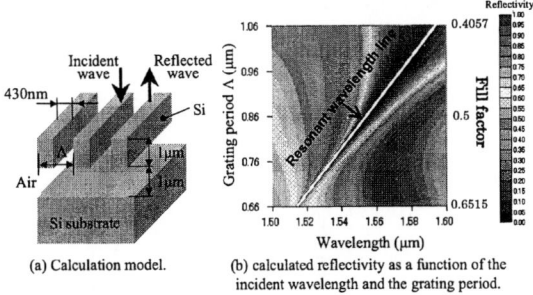

(a) Calculation model.

(b) calculated reflectivity as a function of the incident wavelength and the grating period.

Figure 3. Optical design.

3 FABRICATION

In the fabrication, we used a SOI substrate. The filter pattern was drawn by electron beam lithography. To etch vertically a Si device layer with 1µm thickness, three-steps etching technique using fast atom beam and reactive ion etching machines was carried out. Finally, SiO_2 sacrificial-layer etching was carried out using HF gas.

Figure 4 shows the optical micrograph of the fabricated filter. The initial grating period is 860 nm. As you can see, the filter is fabricated well. The grating area looks dusky-red resulting from structural color under the white light illumination.

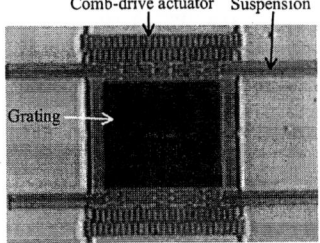

Figure 4. Optical micrograph of the fabricated filter.

4 RESULTS AND DISCUSSION

To measure the changing of the grating period, the pitch-variable operation was performed into a scanning electron microscope (SEM) chamber. The magnified views

of the grating and connecting spring parts at the initial and expanding states are shown in Figs.5(a) and (b), respectively. By driving the comb-drive actuators, the grating period is changed from 860 to 875 nm.

Figure 6 shows the measured reflectivity as a function of the incident wavelength. As the grating is expanded by increasing the applied voltage (0, 40 and 80 V) to the comb-drive actuators, the wavelength of the reflection peak shifts to the long-wavelength side as expected by the calculation in Fig.3(b).

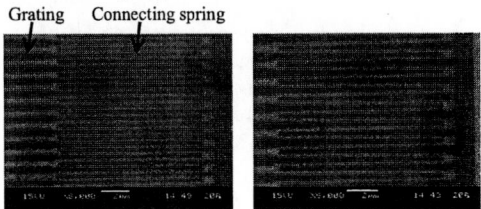

(a) Initial state (Λ=860nm).   (b) Expanding state (Λ=875nm).

Figure 5. Magnified views of the grating and connecting spring parts.

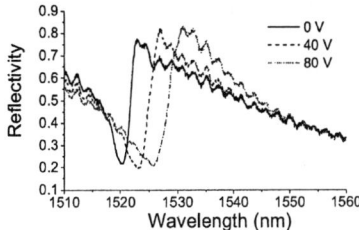

Figure 6. Measured reflectivity as a function of the incident wavelength.

5 CONCLUSIONS

We fabricated a pitch-variable GMRG filter. The grating was a Si self-supported structure and designed at telecom wavelengths. Using a SOI substrate, the proposed filter was fabricated well. The pitch-variation was estimated precisely under the SEM observation. The resonant wavelength shifted to the long-wavelength side as expected by the calculation.

REFERENCES

[1] S. Tibuleac and R. Magnusson, "Reflection and transmission guided-mode resonance filters," *J. Opt. Soc. Am. A*, vol. 14, pp.1617-1626, 1997.

[2] T. Kobayashi *et al.*, "Pitch-variable Subwavelength Gratings Driven by Comb Actuators," Proc. of IEEE LEOS Optical MEMS 2004, pp.198-199, 2004.

[3] K. Hane *et al.*, "Variable optical reflectance of a self-supported Si grating," *Appl. Phys. Lett.* vol. 88, pp.141109-1-3, 2006.

[4] J.-S. Ye *et al.*, "Self-supported subwavelength gratings with a broad band of high reflectance analysed by the rigorous coupled-wave method," *J. Mod. Opt.* vol. 53, pp.1995-2004, 2006.

TuP24
14:30 – 17:30

An optical MEMS pressure sensor based on phase demodulation

Yixian Ge, Ming Wang, Xuxing Chen, Haitao Yan , Hua Rong
Jiangsu Key Lab of Opto-electronic Technology, School of Physical Science
and Technology, Nanjing Normal University, 210097, Nanjing China
Tel +86-25-83598685, Fax +86-25-83598685, E-mail geyixian820925@163.com

Abstract

A novel optical fiber pressure sensor based on Fabry-Perot (FP) interferometer and phase demodulation method is described. MEMS techniques and the common communicational components are used to fabricate the sensor. The principles of pressure measurement and sensor design have been introduced. Phase demodulation method based on Fourier transformation is explored, which can reduce errors resulting from intensity variation of light source. Experimental results demonstrate that the sensor has reasonable linearity, sensitivity and a wide pressure measurement range from 0.1MPa to 3MPa.

Keywords: optical fiber sensor; F-P interference; MEMS; phase demodulation

1 INTRODUCTION

By employing the MEMS technology, a variety of optical MEMS pressure sensors based on Fabry-Perot interferometry have recently been proposed and fabricated [1-3]. In the previous works, different configurations of optical MEMS Fabry-Perot pressure sensors have been proposed, such as the ones by anodic bonding a silicon diaphragm onto a glass substrate with a previously etched cavity [4]. This kind of sensor is usually demodulated by detecting intensity changes in the interference. In this paper, we designed and fabricated a kind of optical MEMS pressure sensor based on phase demodulation method, which provides a wider measurement range, better detection linearity and sensitivity.

2 SENSOR DESIGN

Fig.1 shows a sketch of the optical MEMS pressure sensor. Light is introduced into the sensor from an optical fiber. One part of the incident light will be directly reflected by the fiber plug end face, and the other will come into the sensor and be reflected by the air-silicon interface. When pressure is loaded onto the silicon diaphragm, the Fabry–Perot cavity depth will change. By measuring the phase shift of the reflection spectrum, it is expected that one can easily know the loaded pressure.

The intensity of the reflected light is given as [5]:

$$I_r = \frac{r_{12}^2 + r_{23}^2 + 2r_{12}r_{23}\cos\varphi}{1 + r_{12}^2 r_{23}^2 + 2r_{12}r_{23}\cos\varphi} I_0 \qquad (1)$$

The deflection of the center position of the diaphragm y due to a loaded pressure P is given by [6]:

$$\frac{Pr^4}{Eh^4} = \frac{16}{3(1-u^2)}(\frac{y}{h}) + \frac{7-u}{3(1-u)}(\frac{y}{h})^3 \qquad (2)$$

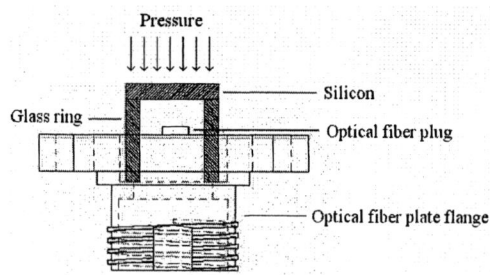

Fig.1 Sketch of optical fibre MEMS pressure sensor

Figure 2 shows the relation between the deflection and the pressure under the different thickness of silicon diaphragm. We initially designed pressure sensors to respond over the pressure range 0 to 3 MPa, to choose r=1.65mm, h=180μm for fabrication.

Another key design parameter is the depth of the F-P cavity. The visibility is shown as following:

$$V = \frac{\sqrt{R_1[1 + (Ad/2\pi)^2]}}{1 + R_1[1 + (Ad/2\pi)^2]} \qquad (3)$$

For this sensor, $R_1 = 0.04$, λ=1.55μm , $v=2.094, a=5\mu m$, $A=0.0458$. The relation between the visibility and the F-P cavity depth is shown in Figure 3. From it, we could see when cavity depth is 6mm, the visibility is also 0.1, viz. the depth could be long.

3 FABRICATION PROCESS

In our work, surface and bulk MEMS techniques are used to fabricate the sensing elements. Fig.4 shows the processing steps of fabrication. The thickness of silicon wafer is 340μm and its orientation is <100>.

1-4244-0641-2/07/$20.00 ©2007 IEEE   101

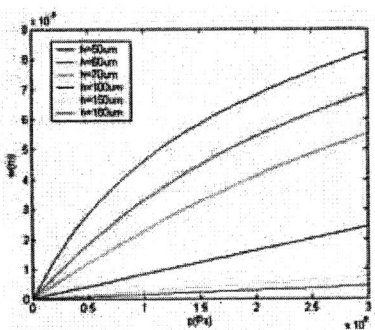

Fig.2 the relation between deflection and the pressure under the different thickness of the silicon diaphragm

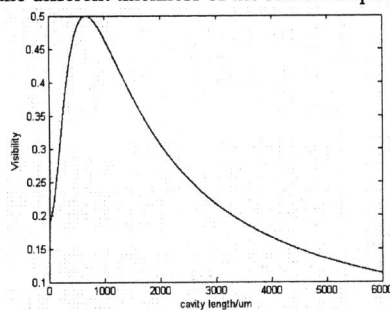

Fig.3 The relation between the visibility and the F-P cavity depth

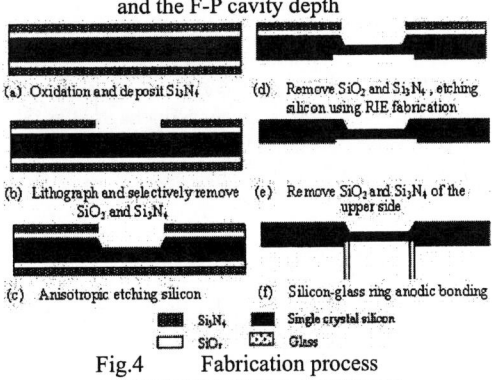

(a) Oxidation and deposit Si₃N₄

(d) Remove SiO₂ and Si₃N₄, etching silicon using RIE fabrication

(b) Lithograph and selectively remove SiO₂ and Si₃N₄

(e) Remove SiO₂ and Si₃N₄ of the upper side

(c) Anisotropic etching silicon

(f) Silicon-glass ring anodic bonding

Si₃N₄   Single crystal silicon
SiO₂   Glass

Fig.4   Fabrication process

4 EXPERIMENTAL RESULTS

Pressure tests have been carried out by employing a set-up illustrated in Fig.5. We apply the Fourier transformation method [7] to the reflected spectrum and find out maximum Ω_L, the cavity length L can be calculated through

Eq.4: $L = \dfrac{c \cdot \Omega_L}{4\pi}$ (4)

Fig.6 shows the experimental results where the cavity of the sensor is plotted as a function of pressure. The fitting line is L=322.98-3.65*P, the linearity is 0.9%, and the sensitivity (i.e. change in cavity/loaded pressure) is 3.50μm/MPa.

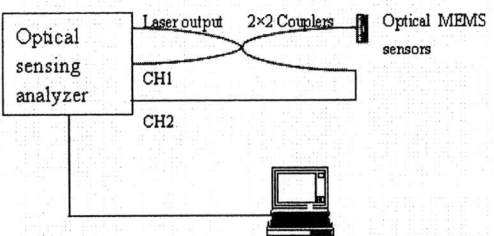

Fig.5 Experimental setup for the optical MEMS Pressure sensor

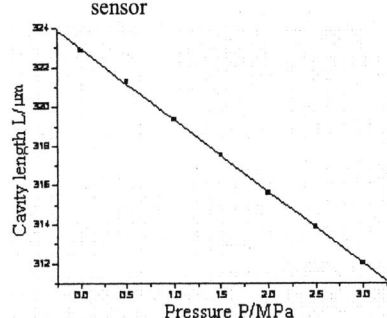

Fig.6 Relation between pressure and cavity length of the sensor

Acknowledgements

This work was supported by the National Natural Science Foundation of China, the Specialized Research Fund for the Doctoral Program of Higher Education (20050319007) and Jiangsu province high-novel technology project（BG2003024）

References

[1] Don C.Abeysinghe, A Novel MEMS Pressure Sensor Fabricated on an Optical Fiber, IEEE photonics technology letter, Vol.13, No.9, September, 2001:993-995.

[2] W.J. Wang, R.M. Lin, D.G. Guo, T.T. Sun, Development of a novel Fabry–Perot pressure microsensor, Sens. Actuators A 116 (2004):59-65.

[3] Pawel Niewczas, Lukasz Dziuda, dt al.. Interrogation of Extrinsic Fabry–Pérot Interferometric Sensors Using Arrayed Waveguide Grating Devices, Vol.52, No.4, August, 2003:1092-1096.

[4] Jie Zhou, Samhita Dasgupta, et al.. Optically interrogated MEMS pressure sensors for propulsion applications, Optical Engineering, Vol.40, No.4, April, 2001:598-604.

[5]J.P.Mathien. OPTICS[M]. Beijing:Science Press, 1987: 251-253.

[6] Wei Jin, Shuangshen Ruan, et al.. Science Press,2005: 268-269.

[7]Gregory   t.a.kovacs,   Micromachined   Transducers Sourcebook[M].Beijing: Science Press,2003:180-181

TuP25
14:30 – 17:30

A MEMS-based Organic Deformable Mirror with Tunable Focal Length

Tyng-Yow Chen, Chao-Hu Li, Jen-Liang Wang, Chen-Wei E. Chiu, Guo-Dung J. Su

National Taiwan University, Graduate Institute of Electro-Optical Engineering & Department of Electrical Engineering

No.1, Sec. 4, Roosevelt Road, Taipei 10617, Taiwan

Tel +886-2-3366-3652, Fax +886-2-2367-7467, E-mail:gdjsu@cc.ee.ntu.edu.tw

Abstract

In this paper, we present a deformable mirror with circular membrane combined with single transparent electrode which can used as concave or convex mirrors. The membrane is made by organic material to achieve large deformation is actuated by electrostatic force. The mirror's focal length is adjustable by tuning the applied voltage. The beam diameter can be shrunk to 1/3 of original diameter when the focal length is adjusted to 87mm. The fabrication process, mirror's tunable range, and experiment result will be discussed in this article.

Keywords: Optical MEMS, deformable mirror, focal length, polyimide

1 INTRODUCTION

Currently, deformable mirror (DM) is a critical device for astronomy observation [1], and laser beam shaping it's also applying on biomedical field [2]. Deformable mirror for adaptive optics usually drives by several actuators which are made from PZT[3], PWN or using electrostatic force. One can control the mirror contour to desired shape by driving the specific actuators. As the progress in micromachining process, we fabricated the deformable mirrors with relatively small-size (diameter=3.5mm) and low surface roughness, which is potentially can be applied to optical image system. Moreover, we avoid the snap-down effect by spinning pillar-like photoresist. It can also solve the air-trapping problem when we actuate the deformable mirror. By using the batch fabrication, we made the low-cost deformable mirror chips for many applications, such as optical switches, variable optical attenuator, and imaging systems.

The focal length of the deformable mirror can be calculated from two geometric analyses. The first is use the formula shown below [4]

$$\delta = A \tan(\frac{\sin^{-1}(A/R)}{2}) \qquad (1)$$

where δ is the center displacement, A is the effective

aperture radius of the mirror, and R is the radius of curvature. According to Gaussian optics, the focal length of a curved mirror can be approximately expressed by f = 0.5R. The second way is illustrated in Fig1. When assuming the center displacement of the mirror is small compared to focal length, we have

$$\frac{f - f'}{f} = \frac{d}{D} \qquad (2)$$

where f is the focal length of the mirror, f' is the distance between mirror and detector, D is the beam diameter before focused by the DM, and d is the beam diameter after focused by DM. Rearrange equation (2) and we can obtain the focal length, f = f'D / (D-d) = (1-d/D)⁻¹ f'.

2 DEVICE FABRICATION

Our tunable-focal-length organic deformable mirror is fabricated using MEMS process which is suitable for batch fabrication. We started with growing thermal oxide 600nm on two-sides of (100) silicon wafer, and patterned the wafer with 3.6(mm) x 3.6(mm) opening. After thermal oxide is removed, we etch the opening silicon by tetramethyl ammonium hydroxide, TMAH, at the temperature of 90C. The etching will be hold until the thickness of remaining silicon about 20um. Due to thin film thermal mismatch concern, we deposit chromium and aluminum orderly. The organic thin film, polyimide, is spin-coat around 2um and hard bake at 300C in the oven. The conducting film is deposited at last using aluminum, and dry-etching is performed to etch through remaining silicon. Thermal oxide and chromium is also sequentially etched away. All the process steps are shown in Fig2.

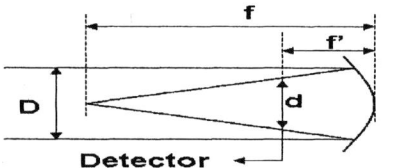

Fig.1. Schematic drawing of geometric optics for DM.

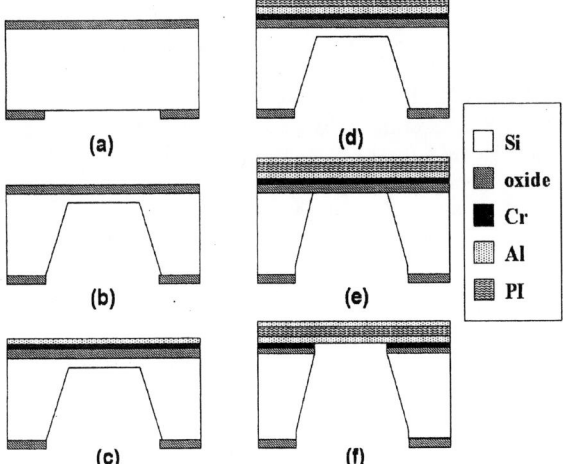

Fig.2. (a) Oxidation for 6000A, pattern and etch backside oxide. (b) Etching silicon with a thin Si membrane left. (c) Deposit Cr/Al on front-side oxide. (d) Spin-coat polyimide. (e) Deposit Al 1200A as electrostatic layer, then using dry etching to release remaining Si membrane. (f) Removing oxide, and then etching away Cr.

The fabricated membrane had excellent mechanical properties. Relatively low young's modulus to 15GPa and residual stress down to 3MPa make this chip adequate for specific applications. Lower residual stress can make us achieve smaller radius of curvature easier, due to this, we reduce the operation voltage and enhance the deformable mirror's performance in tunable range. The deformable mirror chip is composed of top-side reflective mirror and bottom-side electrode which is fabricated by coating an aluminum thin film. We actuated the deformable mirror by producing the potential difference between two parts. By decreasing the gap we can reduce the driving voltage even lower, but it may limit the tunable range, either. Here we choose 55(um) as our gap, it can allow more than about 9(um)'s deflection or 11.5 diopter (1/f).

3 EXPERIMENT RESULTS

The reflected laser beam profile for our deformable mirror is shown in Fig3. The ratio of diameter for (b) over (a) is d/D = 0.33, which gives f = 87mm for f' =58mm from equation (2). Since the laser beam is incident to the center of the DM, we can use the paraxial approximation to calculate radius of curvature, R = 2f, which gives R=174mm. In addition, the measured center displacement of the DM is δ = 0.009mm. Substitute δ and A=1.75mm to equation (1), we have R=170mm, which is well matched with above calculation calculated by equation (2).

We also flip our chip and the concave DM becomes a

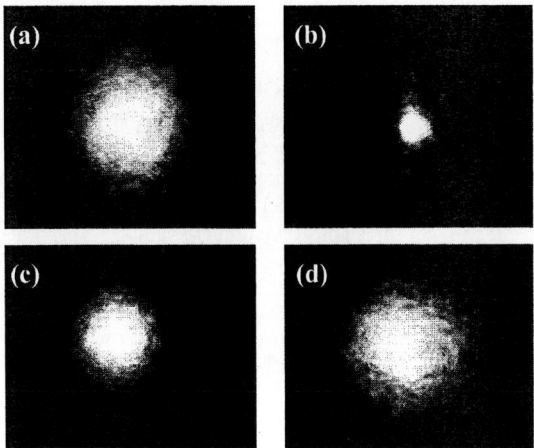

Fig.3. The beam profile for DM with (a) flat, (b) curved surface at concave side, and (c) flat, (d) curved surface at convex side.

convex mirror. The beam profile is shown in Fig3 (c) and (d). The laser beam is indeed diverged by the convex mirror and the beam diameter is 1.4 times larger than the original laser beam diameter.

4 CONCLUSIONS

We reported a tunable-focal-length deformable organic mirror fabricated by MEMS technology. The main structure of the mirror is consisted of an organic membrane, which provides a lower residual stress. The focal length of the mirror can be tuned from infinity to about 87(mm). Two geometric analyses are also calculated and have a good match with each other.

REFERENCES

[1] Andrei Tokovinin, San drine Thomas, Gleb Vdovin, "Using 50-mm electrostatic membrane deformable in astronomical adaptive optics." Proc. Of SPIE Vol.5490, pp.580-585

[2] Nathan Doble and David R. Williams, "The Application of MEMS Technology for Adaptive Optics in Vision Science." IEEE J. Sel. Top. Quantum electron. Vol. 10, No. 3, May/June 2004.

[3] Isaku Kanno, Takaaki Kunisawa, Takaaki Suzuki, and Hidetoshi Kotera, "Development of Deformable Mirror Composed of Piezoelectric Thin Films for Adaptive Optics" IEEE J. Sel. Top. Quantum electron. Vol. 13, No.2, 2007, PP. 155-161.

[4] Y. W. Yeh, CW. E. Chiu, and GD. J. Su, "Organic amorphous fluoropolymer membrane for variable optical attenuator applications." J. Opt. A: Pure and Appl. Opt., Vol. 8, No. 7, July 2006, pp. S377-S383

TuP26
14:30 – 17:30

Novel Adaptive Optics System with an Electrostatically-driven Deformable Mirror and Wavefront Compensation Algorithm

Akio Kobayashi[1], Hiroyuki Kawashima[1], Noriko Saito[1], Masayuki Momiuchi[1],
Akihiro Koga[2], Ryo Furukawa[2], and Kei Masunishi[2]

[1] Nano-Opto Laboratory, Corporate R & D Center, Topcon Corporation (Japan)
Phone : +81-3-3558-2562, Facsimile : +81-3-5966-5054, E-mail: a.kobayashi@topcon.co.jp
[2] Mechanical System Laboratory, Corporate R & D Center, Toshiba Corporation (Japan)

Abstract

We have fabricated a membrane deformable mirror (DM) for ophthalmologic adaptive optics. An algorithm is demonstrated to compensate wavefront aberration of eyes, effectively. We have developed analysis systems that can predict the DM characteristics.

Keywords: Adaptive Optics, Deformable Mirror

1. Introduction

Adaptive optics (AO) has been applied to medical and laser engineering fields as well as astronomy. Wavefront aberration correction of human eyes has been studied in ophthalmology and vision science to obtain high-resolution retinal images [1]. This paper describes novel adaptive optics system with DM and wavefront compensation algorithm. The deformation behavior of DM was modeled by general-purpose solvers.

2. Retinal imaging system with AO

2.1. Optical system with the DM

Retinal imaging system with AO is illustrated in Fig. 1. Sensing-beacon(λ=840nm) and imaging-illumination lights (λ=633nm) are scattered by the retina and transmitted through the eye lens to DM, which can arbitrarily be deformed by electrostatic force to compensate the eye-lens aberration. The lights reflected by DM are splitted into two sensors, those are, an imaging CCD and a Shack-Hartmann wavefront sensor, which determines the residual aberration. The wavefront is then calculated by a computer(PC), and the imaging aberration is minimized by adaptively modifying the mirror deformation. Therefore, the system can obtain high-resolution retinal images, even if the eye has significant aberration. The DM is the key device of the system.

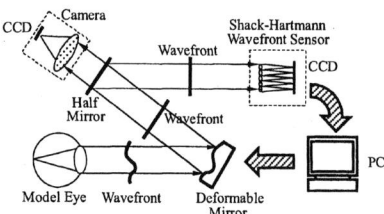

Fig. 1. The retinal imaging system with AO.

2.2. Structure of the DM

The structure of the DM is schematically shown in Fig. 2. The DM reflection mirror is a 12mm in diameter, 6μm thick silicon membrane that was etched out from an SOI wafer and coated aluminum film. The mirror is mounted with 50um-spacers on a multi-layer ceramic substrate, which has 85 concentric electrodes. The electrode pattern is shown in Fig.3. DM deforms up to 15 μm when 180 V is simultaneously applied to the 85 electrodes. DM is hermetically packaged with nitrogen or vacuum sealing.

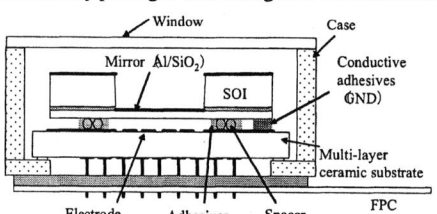

Fig. 2. Structure of the DM.

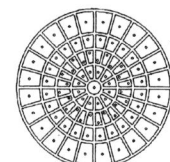

Fig. 3 An electrode pattern of the DM

3. Wavefront compensation algorithm

The wavefront aberration can be generally expanded by Zernike polynomials. We determined voltage "templates" for unit deformation corresponding to each of Zernike modes up to the 6th order. The aberration was compensated by superimpose of template voltages, weighted by each coefficients which appeared in the Zernike expansion[2]. Although this method is simple and fast, the aberration was not completely cancelled, due to the limited number of the electrodes, their discrete configuration and voltage resolution. In this study, the residual aberration was

1-4244-0641-2/07/$20.00 ©2007 IEEE   105

dramatically reduced by optimizing the superimposition coefficients using the least-square method.

As shown in Fig. 4, aberration compensation has been successfully demonstrated for a model eye. In this case, the system incorporates Zernike modes up to the 10th order for sensing and up to the 6th order for the correction. The residual aberration shown in Fig. 4(b) is due to Zernike modes higher than the 6th order, but it is suppressed sufficiently to obtain an essentially diffraction-limited image. Thus, in order to obtain higher resolution retinal images, Zernike modes should be higher than the 6th order.

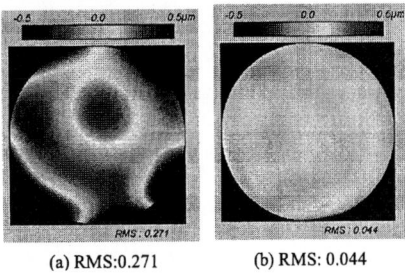

(a) RMS:0.271   (b) RMS: 0.044

Fig. 4. Aberration maps. (a) Initial. (b) Compensated.

4.Static analysis for the DM

The 85 electrode configuration does not take account of Zernike modes higher than the 6th order. In order to design a new electrode configuration that deals with higher Zernike modes, we simulated static deformation of DM with general-purpose solvers, ANSYS Muliphysics, by sequentially coupling the structural and electrostatic analyses (coupled analysis). The modeling was conducted for the region defined by the electrodes, reflection mirror and space between them. The stratified shell model was assumed due to the small thickness of the mirror. In order to fit to an actual deformation, we introduced internal stress to DM by deforming the frame of the membrane. The stress was determined to be 4.4MPa(tensile) [3].

The DM shape modeled by the coupled analysis for the (4, -4) voltage template is shown in Fig. 5 with Zernike expansion coefficients, in comparison with the measured shape. The modeled shape exhibits a good agreement, including the Zernike modes other than (4, -4). This coupled analysis can be used to optimize the electrode configuration.

When voltage is applied to only two electrodes, stress dependence of DM deformation is shown in Fig. 6. In the absence of the internal tension stress, the analysis predicts a bath-tub shape although the two separated electrodes pull the membrane. In the presence of the stress, DM exhibits two deformation cusps at the two electrodes. The analysis indicates that the deformation behavior of DM has high sensitivity to the internal stress.

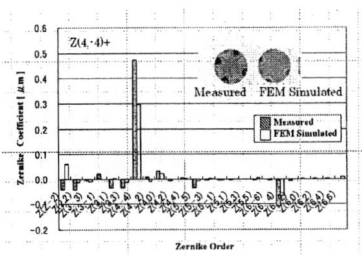

Fig. 5. Analyzed and measured DM shapes that expressed in Zernike mode coefficients.

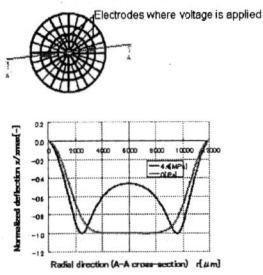

Fig. 6. Membrane deformations for various internal stress.

5. Conclusions

We developed the wavefront compensation system with membrane-type electrostatic DM. The compensation algorithm effectively suppresses the aberration. The static analysis is useful in predicting the deformation behavior of DM. Further improvement of the system is underway by incorporating Zernike modes higher than the 6th order.

Acknowledgment

This research was supported by Japan Science and Technology Agency under Project "SENTAN".

References

[1] for example: H. Hofer, L. Chen, G. Y. Yoon, B. Singer, Y. Yamauchi, and D. R. Williams, "Improvement in retinal image quality with dynamic correction of the eye's aberrations," Opt. Express, Vol. 8, pp. 631-643 (2001).
[2] H. Kawashima, M. Nakanishi, N. Takeda, I. Minegishi, and A. Kobayashi, "Compensation of model eye's aberration by using deformable mirror," Proceedings of SPIE, Vol. 5717, No.28, pp. 219-229 (2005).
[3] K. Masunishi, A. Koga, R. Furukawa, M. Momiuchi, H. Kawashima, and A. Kobayashi, "Development of an electrostatic driven DM (Deformable Mirror) (2): Numerical analysis of the static characteristics," Proceeding of Conference on Information, Intelligence and Precision Equipment (IIP2007), pp. 60-65 (2007) (in Japanese).

TuP27
14:30 – 17:30

Reflectance Study of Nano-Scaled Textured Surfaces

Cheng-Chung Chen[1], Peichen Yu[2], Hao-Chung Kuo[2]
1. Department of Photonics and Display Institute,
2. Department of Photonics and Institute of Electro-Optical Engineering,
National Chiao Tung University, 1001, Ta Hsueh Road, Hsinchu 300,Taiwan R.O.C.
Tel: 886-3-571-2121 ext-56354, FAX: 886-3-571-6631, E-mail: yup@faculty.nctu.edu.tw

Abstract

An algorithm based on rigorous coupled wave analysis is used to calculate the reflectance of two kinds of nano-scaled textured surfaces, periodic pyramid and random pillar structures. The device dimension and angular dependence are investigated.

Keywords: nanostructure, sub-wavelength structure, SWS, anti-reflection, omni-directional reflector, ODR

1 INTRODUCTION

The ability to suppress or alter the Fresnel reflection of a material over a broadband range offers new prospects for optoelectronic devices, where the figure of merit relies on the transmission efficiency or refractive index contrast. Some apparent examples include light emitting diodes, solar cells, distributed feedback (DFB) laser, and distributed Bragg reflectors (DBR). Several reports have been made on anti-reflective surfaces employing periodic sub-wavelength structures(SWS)[1]. Silicon-based nano-pyramids fabricated by various techniques are of particular interests due to dramatic increase in the demand of solar cells. Recently, the interests in the reflectance of a periodic SWS surface have been extent to that of a disordered textured surface due to several advantages. First, the fabrication cost is significantly reduced as the e-beam lithography is no longer required. Second, a variety of materials, including Si[2], GaN[3], ZnO [4], TiO$_2$/SiO$_2$[5] and indium tin oxide (ITO)[6], are available for the applications, offering versatile device possibilities.

The textured surface most commonly consists of straight or tilted nano-pillar structures involving either deposition or etching. The modified refractive index, and hence the reflectance of the structure are believed to be highly correlated to the density and geometry of the nano-pillar structures. However, because of the nature of disorder, numerical simulation for the reflectivity of the nano-pillar structures becomes a challenge, hindering further process in device design and analysis. Here, we report our calculations of the reflectance of nano-pyramid and nano-pillar structures using an algorithm based on rigorous coupled wave analysis (RCWA) and transmission line method. The device dimension and angular dependence of the incident field are studied without considering side-wall roughness and dispersion effects.

2 CALCULATION METHOD

The RCWA method is often employed to solve the diffraction and transmission efficiency of optical diffractive elements, where vectorial Maxwell's equations are solved in the Fourier domain. A three-dimensional periodic structure is first decomposed into layers of periodic structures along the vertical z axis. In each layer, the transverse field components of the Maxwell's equations can be simplified to a set of linear functions using Bloch Theorem. Transmission line method is then used to solve the boundary problems for the entire structure [7]. Reflectance is obtained as a sum of the reflection diffraction efficiencies of different diffraction orders.

The first SWS surface consists of nano-pyramids with a base width, W of 150 nm, and a height, L of 350 nm. The schematic is shown in the inset of Fig. 1. The incident field is assumed to be He-Ne laser with a wavelength of 633 nm. The material is assumed to be silicon with a refractive index of 3.9, and also lossless and non-dispersive. The transverse grid size was chosen to be 5 nm, and the vertical calculation step is 2nm. The simulation result will be verified with the reported device of Ref. [1].

To calculate the random nano-pillar structures, a unit cell consisting of multiple pillars is defined, where the dimension of the unit cell is at least an order of magnitude larger than the incident wavelength, in order to minimize the calculation errors arising from periodicity. Each pillar has a height, L of 100nm and diameter, D of 150nm. The density of nano-pillar structure is $\sim 10^9$ cm^{-2}.

3 RESULTS AND DISCUSSION

To verify the calculation algorithm, the reflectance of a Si-substrate is computed as a function of the incident angle for both transverse-electric (TE) and transverse-magnetic (TM) modes. As seen in Fig. 1, the reflectance of TM mode

1-4244-0641-2/07/$20.00 ©2007 IEEE   107

vanishes at Brewster angle. The reflectance for nano-pyramid structures with W= 350 nm and L= 150 nm are also calculated for both polarizations. The results are nearly identical to that of Ref. [1]. It is evident that the nano-scaled textured film exhibits anti-reflective characteristics with little reflectance, corresponding to nearly unity index of reflection. It is worth to note that TM polarizations show less angular dependence of the incident field, up to an angle of 60^0. The vanishing of the Brewster angle can be explained as a result of coupled transverse field components in the inhomogeneous structures.

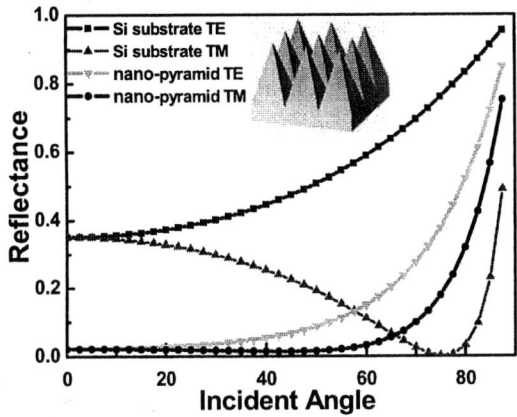

Figure 1. Calculated reflectance of a Si substrate and nano-pyramid structures as a function of incident angle for both TE and TM modes. The calculated results agree with that of Ref. [1].

We next investigate the dependence of the reflectance on the device dimensions. The reflectance as a function of the height of the pyramid is plotted with varying base widths for

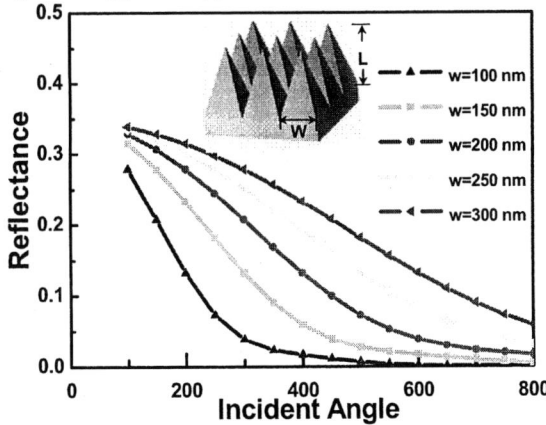

Figure 2. Reflectance is plotted as a function of the height and the base width of the nano-pyramids for TE polarization.

TE polarization. It is found that the reflectance decreases significantly when the aspect ratio of the nano-pyramid increases. In general, the reflectance of <0.05 can be obtained as the aspect ratio is >3.

Fig.3 shows a preliminary calculation result of the reflectance of a nano-pillar structure versus the incident angle for both TE and TM polarizations. The calculation results show that the random nano-pillar structure can easily achieve nearly-perfect anti-reflection with an aspect ratio less than unity. It also suggests that the antireflection can be omni-direction for TM polarization. Further investigation on the density of nano-pillar structure will be conducted. Other nano-pillar structures involved tilted angles will also be presented.

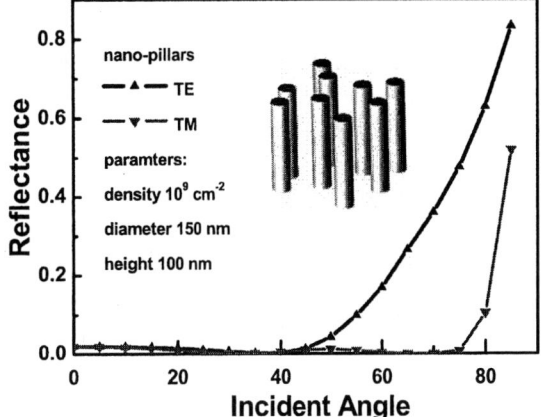

Figure 3. Calculated reflectance of a nano-pillar structure as a function of incident angle for both TE and TM modes.

REFERENCES

[1] Y. Kanamori, M. Sasaki, and K. Hane, Opt. Lett. **24**, 1422 (1999).

[2] G. R. Lin, H. C. Kuo, H. S. Lin, and C. C. Kao, Appl. Phys. Lett. **89**, 073108 (2006).

[3] C. L. Tseng, M. J. Youh, G. P. Moore, M. A. Hopkins, R. Stevens and W. N. Wang, Appl.Phys. Lett. **83**, pp. 3677-3679 (2003).

[4] T. Fujibayashi, T. Matsui, and M. Kondo, Appl. Phys. Lett. **88**, 183508 (2006).

[5] J. Q. Xi, M. F. Schubert, J. K. Kim, E. F. Schubert, M. Chen, S. Y. Lin, W. Liu and J. A. Smart, Nature Photonics Lett. **1**, pp. 176-179 (2007).

[6] S. W. Kim, D. S. Bae, and H. Shina, Appl. Phys. Lett. **96**, pp. 6766-6771 (2006).

[7] C. H. Lin, K. M. Leung, and T. Tamir, Opt. Soc. Am. A **19**, pp. 2005-2017 (2005).

TuP28
14:30 – 17:30

Experimental Observation of Self-Propelled Cavity Soliton-like Evolutions in VCSELs with Photonic-Crystal Micro-Structures

Tsin-Dong Lee[1], Chia-Yu Chang[1], Meng-Hong Wu[1], Jhao-Ren Jhang[1], Te-ho. Wu[1] and Ray-Kuang Lee[2,*]

[1]Graduate School of Optoelectronics, National Yunlin University of Science & Technology
123 Sec. 3, University Rd. Douliu, Yunlin, Taiwan 64002, R.O.C.
[2]Institute of Photonics Technologies, National Tsing-Hua University, Hsinchu, Taiwan 300, R.O.C.
Tel +886-5-534-2601 Ext: 4342, Fax +886-5-534-2063, E-mail leetd@yuntech.edu.tw

Abstract

We report experimental observation of transverse optical pattern distribution and its evolution in photonic-crystal-structured VCSELs by using near-field scanning optical microscope. With fixed driving current but different distances above the emitting surface of the cavity, self-propelled soliton-like cavity patterns are demonstrated.

Keywords: VCSEL, near field scanning microscope, optical pattern

1 INTRODUCTION

Vertical Cavity Surface Emitting Lasers (VCSELs) are recognized as powerful semiconductor lasers that have played a significant role in high-speed laser printing, optical storage and long-wavelength telecommunications. With optical output vertically emitted from the surface, VCSELs also act as an interesting platform for studying optical pattern formation in mesoscopic system. In this scenario, self-organized linear and nonlinear optical modes have been demonstrated in semiconductor microresonators, such as scar modes [1, 2] and cavity solitons [3].

In recent years, with the advance of new fabrication technologies, it becomes more feasible to actually utilize one- or higher-dimensional periodic dielectric structures (or especially the photonic bandgap crystals) to modify the resonance modes in semiconductor lasers. In this work, a two-dimensional photonic crystal micro-structure is fabricated on a VCSEL surface [4] to investigate the transverse optical pattern formation and its evolution by using near-field scanning optical microscope (NSOM) technology. Localized soliton-like structures in the output surface are formed and observed. Due to the confinement of the surrounding photonic crystal cavity, experimental evidence of self-propelled evolutions at different distances above the emitting surface for a fixed driving current are demonstrated directly from the NSOM images. The nonlinear interaction potentials inside the semiconductor cavity make a single-peaked soliton-like transverse pattern split into two well-defined patterns. The experimental investigations in this paper provide new results for studying

the dynamics between nonlinear dissipative system and bandgap effect in semiconductor microstructures.

2 PHOTONIC-CRYSTAL-STRUCTURED VCSELS

The SEM image of the microstructured VCSEL with photonic crystal pattern on the surface used in our experiments is shown in the insert of Figure 1. The epitaxial layers of these VCSELs are grown by MOCVD on n^+-GaAs substrate, with graded-index separate confinement heterostructure (GRINSCH) active region formed by undoped triple-$Al_{0.12}Ga_{0.88}As$-$Al_{0.3}Ga_{0.7}As$ quantum wells placed in one lambda cavity. The p- and n-doped distributed

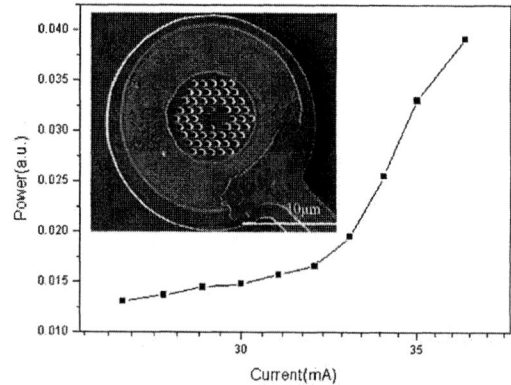

Figure 1: The L-I curve, light versus current, of the photonic-crystal-structured VCSELs and the corresponding SEM image for the microresonator (insert).

1-4244-0641-2/07/$20.00 ©2007 IEEE

Bragg reflectors (DBR) in the vertical cavity consist of 24 and 40.5 pairs of $Al_{0.3}Ga_{0.7}As/Al_{0.9}Ga_{0.1}As$ layers, respectively. We perform reactive ion etch (RIE) to define mesas with diameters of 22 μm, where the $Al_{0.98}Ga_{0.02}As$ layer within the $Al_{0.9}Ga_{0.1}As$ confinement layers is selectively oxidized to AlOx. The oxidation depth is about 3 μm towards the center from the mesa edge so that the resulting oxide aperture is 16 μm in diameter. We also introduce an oxide aperture to reduce the lateral optical loss and the leakage current. The p-contact ring with a diameter of 20 μm and a width of 5μm is formed on the top of the p-contact layer. The n-contact is formed at the bottom of the n-GaAs substrate. The micro-structured surface pattern is defined by the focused ion beam. In this work, hexagonal lattice patterns with seven-defects are fabricated within the p-contact ring to introduce photonic crystal structures. Moreover two small extra holes are introduced in core region to break the polarization symmetry of the cavity. The lattice constant of the hexagonal lattice for this photonic-crystal-structured VCSEL is 1 μm and the diameters for the surrounding and two extra holes are 500 nm and 100 nm, respectively.

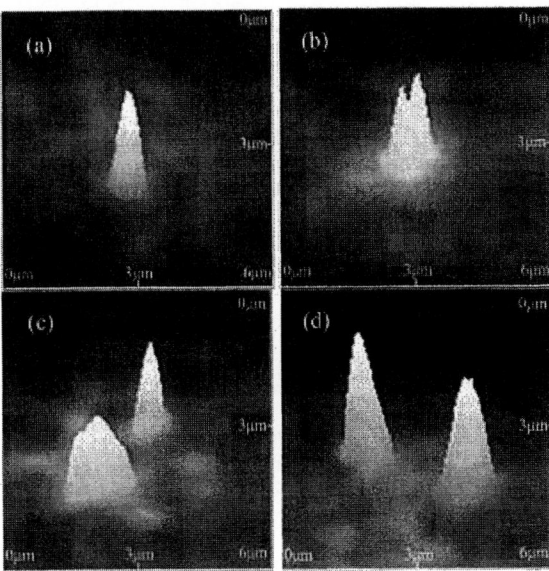

Figure 2: For a fixed driving current, 34 mA, the intensity distributions of the output optical pattern collected by NSOM measurement at different distances above the cavity surface are shown for (a) z = 0, (b) z = 1.5, (c) z = 3, and (d) z = 4.5μm, respectively.

3 EXPERIMENTAL RESULTS AND DISCUSSIONS

Figure 1 shows the CW light-current relation, L-I curve, of our AlGaAs-based photonic-crystal-structured VCSEL. The threshold current for lasing condition of our VCSEL is about 27 mA. After the VCSEL is turned on, we using near-field scanning optical microscope (NSOM) operated at collection mode to measure the electromagnetic intensity distribution on the aperture surface. For a fixed driving current we perform the NSOM measurement at different distances at the emitting window above the cavity surface. The intensity distributions at the near-field for the operation current at 34 mA are shown in Figure 2, with estimated distance about 0, 1.5, 3 and 4.5 μm, respectively. In Figure 2(a), it is clearly seen that a single-peaked self-localized pattern is formed on the emitting surface at z = 0μm. Due to that the operation current is well-above the threshold current, the nonlinear effects are believed to play an important role in the formation of the transverse optical patterns [5]. Moreover, after leaving the cavity surface, z = 1.5μm, the single-humped transverse pattern is split into a double-humped one observed in Figure 2(b). Then the two well-defined soliton-like optical patterns are demonstrated in Figure 2(c) for the distance of 3μm. And finally the nonlinear interaction potential between the two soliton-like pattern repels each other and makes them separated farther and farther, as shown in Figure 2(d).

4 CONCLUSIONS

With microstructure patterns and NSOM technologies, we investigate the formation and evolution of the transverse optical patterns in AlGaAs-based photonic-crystal-structured VCSELs. Near-field images of soliton-like patterns and self-propelled evolutions for such localized patterns at different distance above the surface cavity are observed. More detail modeling of such novel semiconductor microresonators are under investigated for possible all-optical information processing.

REFERENCES

[1] K. F. Huang, Y. F. Chen, H. C. Lai, and Y. P. Lan, *Phys. Rev. Lett.* **89**, 224102 (2002).

[2] T. Gensty, K. Becker, I. Fischer, W. Elsaßer, C. Degen, P. Debernardi, and G. P. Bava, *Phys. Rev. Lett.* **94**, 233901 (2005).

[3] S. Barland, J. Tredicce, M. Brambilla, L. Lugiato, S. Balle, M. Giuduci, T. Maggipinto, L. Spinelli, G. Tissoni, and T. Knoedel, *Nature* (London) **419**, 699 (2002).

[4] N. Yokouchi, A. J. Danner, K. D. Choquette, *Appl. Phys. Lett.* **82**, 1344 (2003).

[5] Y. Menesguen, S. Barby, X. Hachair, L. Leroy, I. Sagnes, and R. Kuszelewicz, *Phys. Rev. A* **74**, 023818 (2006).

TuP29
14:30 – 17:30

Elastic-like Collision of Gap Solitons in Nonlinear Nonlocal Photonic Crystals

YuanYao Lin, I-Hung Chen, Ray-Kuang Lee*
TEL: +886-3-5742439, Email: rklee@ee.nthu.edu.tw
Institute of Photonics, National Tsing-Hua University, 101, Section 2, Kuang-Fu Road, Hsinchu City 300, Taiwan

Abstract

We analyze the existence, stability, and mobility of gap solitons in photonic crystals with diffusion mechanism of the nonlinearity numerically. For the bands of Bragg gap, solitons with nonlocal effects are more stabilized and become more movable due to the combinations of non-locality effect and the oscillation tails of the wave packets. We show that gap solitons can revive an elastic-like collision even in the photonic systems due to non-locality.

Keywords: solitons, photonic crystals, nonlocal effect

1. Introduction

Recently experimental observations of nonlocal response have been demonstrated in various systems, such as photorefractive crystals, nematic liquid crystals, and thermo-optical materials. Nonlocal effect comes to play an important role as the characteristic response function of the medium is comparable to the transverse content of the wave packet. Meanwhile photonic crystals - artificial periodic structures with a high index contrast - have provided an efficient control of wave transmission and localization, making it possible to tailor dispersion, diffraction, and emission of electromagnetic waves [1]. The study of photonic crystals made of a Kerr-type nonlinear material, the so-called nonlinear photonic crystals, has revealed a wealth of nonlinear optical phenomena and, in particular, they can support self-trapped nonlinear localized modes of the electromagnetic field in the form of so called *gap solitons*. The existence of gap solitons is a unique property of nonlinear periodic systems. The study of nonlocal nonlinearity brings new features in solitons [2], such as modification of modulation instability and azimuthal instability, suppression of collapse in multidimensional solitons, change of the soliton interaction, and formation of soliton bound states.

For *nonlocal* nonlinear medium, the non-locality is known to improve the stabilization of solitons due to the diffusion mechanism of the nonlinearity. For Kerr-type nonlinear photonic crystals, it has been predicted that with non-locality Peierls-Nabarro potential barrier for solitons moving across the lattice is drastically reduced in the total reflection band [3]. In stead of the total reflection band, in this work we study the existence of gap solitons in the Bragg gaps of nonlinear nonlocal photonic crystals. Moreover in addition to stabilize soliton solutions in the Bragg gaps, the non-locality effect helps to reduce the Peierls-Nabarro potential barrier further more due to the oscillation tails of the gap solitons.

2. Nonlocal solitons in Bragg gaps

We consider a wave packet propagating along the ξ axis in the photonic crystals with Kerr-type nonlinearity and diffusion-type non-locality,

$$i\frac{\partial U}{\partial \xi} + \frac{1}{2}\frac{\partial^2 U}{\partial \eta^2} + V(\eta)U + n(\xi,\eta)U = 0, \quad (1)$$

$$n - d\frac{\partial^2 n}{\partial \eta^2} = |U|^2, \quad (2)$$

where η is the transverse coordinates, and $n(\xi,\eta)$ is the refractive index profile induced by the exponential-type diffusion kernel function responding to the intensity soliton intensity [4]. The coefficient d stands for the degree

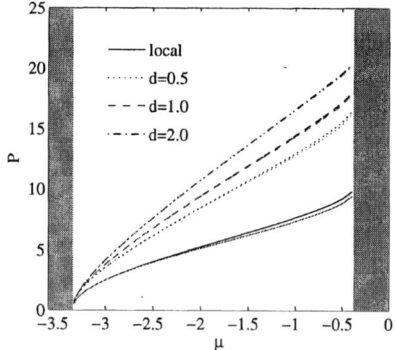

Fig. 1. Bifurcation curves for gap-soliton of even-mode and odd-mode in first gap region are plotted in red and black, respectively. The shadow regions represent linear bands of photonic crystals. The solid lines represents "local" nonlinearity. The dotted lines, dashed lines, and dotted-dashed lines show the degree of non-locality of $d = 0.5, 1.0$ and 2.0, respectively.

1-4244-0641-2/07/$20.00 ©2007 IEEE

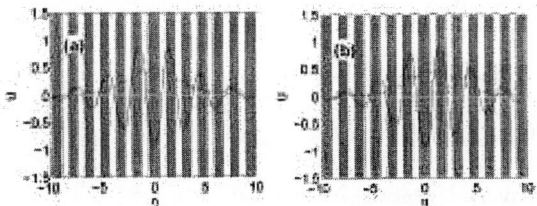

Fig. 2. Solutions of gap soliton of odd mode (a) and even mode (b) in the first gap region for the degree of non-locality $d = 0.5$. The solid lines in black and red are the field and the corresponding refractive index, respectively.

of non-locality which governs the diffusion strength of refractive index. The stationary of Equations (1-2) is given by the solution $U(\xi, \eta) = u(\eta)e^{i\mu\xi}$. Figure 1 shows the bifurcation curves for gap-soliton of even-mode and odd-mode in first gap region through the relations of μ and soliton power P, respectively. In the simulations, the periodic potential is set as $V(\eta) = -3\cos(4\eta)$. Compared to the local nonlinear medium, $d = 0$ in Fig. 1, non-locality effect increases the formation power for gap solitons. Solutions of these gap solitons for odd and even modes in the first gap region for the degree of non-locality $d = 0.5$ are shown in Fig. 2. In addition to the localized wave packet, gap solitons have similar oscillation tails as linear Bloch modes in the bands. The stability of these gap soliton solutions is calculated through *linear stability analysis*, with the perturbed gap soliton solution,

$$U = [u + (v - w)e^{i\lambda\xi} + (v^* + w^*)e^{i\lambda^*\xi}, \qquad (3)$$

where $Im\{\lambda\}$ indicates the growth rate of the small perturbation. Moreover we study the mobility of these gap soliton solutions by calculating the Peierls-Nabarro potential barrier in the periodic systems, i.e. the power differences δH between the even and odd modes. It is clearly seen that in the first Bragg gap the growth rate of modulation instability in Fig. 3(a) as well as the Peierls-Nabarro potential barrier in Fig. (b) are both drastically reduced with the comparisons of the local nonlinearity, $d = 0$. But with the oscillation tails similar to the linear Bloch-waves, we find that nonlocal solitons in the Bragg gaps are more stable and more movable than those in the internal reflection band [3], .

Based on the results of reduction of the Peierls-Nabarro potential barrier of gap soliton by non-locality, we demonstrate a potential-free collisions between two gap solitons in the photonic crystals. It is well known that without periodic potentials, solitons experience elastic collisions when the relative phases between them are zero. But such elastic collision between two solitons is destructed in the photonic systems due to the confinement of the Peierls-Nabarro potential barrier, as shown in Fig 4 (a) for $d = 0$. But with non-locality $d \neq 0$, in Figure 4(b) we show that gap solitons can revive an elastic-like collision even in the

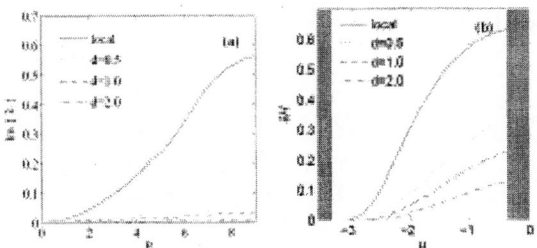

Fig. 3. (a): Modulation instability of gap soliton. The growth rate of the small perturbation, $Im\{\lambda\}$ versus soliton power is given. (b): The Peierls-Nabarro potential barrier, δH, of gap soliton versus μ. The solid, dotted, dashed, and dotted-dashed lines represent the cases for $d = 0, 0.5, 1.0$ and 2.0, respectively.

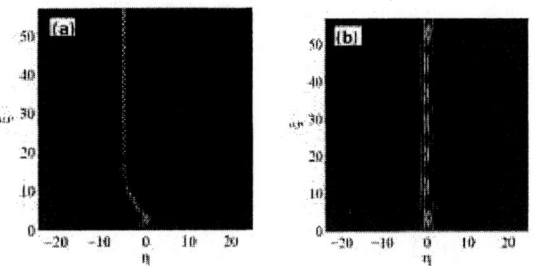

Fig. 4. Collision of two soliton in photonic crystals with local (a) and nonlocal (b) nonlinearities. The degree of non-locality is set as $d = 0.5$.

photonic systems due to nonlocal effect.

In conclusion, we demonstrate the existence of solitons in nonlinear nonlocal photonic crystals in the Bragg gaps. The stability and mobility of such novel gap solitons are calculated through the linear stability method and the Peierls-Nabarro potential barrier. Compared to the internal reflection band, nonlocal gap solitons in the Bragg gaps are more stabilized and becomes more movable due to the oscillation tails of wave packets. Moreover we demonstrate that it is possible to have elastic-like collisions between gap solitons with the help of nonlocal effects.

References

[1] C. M. de Sterke and J. E. Sipe, "Gap solitons," in *Progress in Optics*, E. Wolf, ed., (North- Holland, Am-sterdam, 1994), Vol. XXXIII, pp. 203V260.

[2] A. W. Synder and D. J. Mitchell, *Science* **276**, 1538 (1997).

[3] Z. Xu, Y. V. Kartashov, and L. Torner, *Phys. Rev. Lett.* **95**, 113901 (2005).

[4] W. Królikowski, O. Bang, J. J. Rasmussen, and J. Wyller, *Phys. Rev. E* **64**, 016612 (2001).

TuP30
14:30 – 17:30

Wide-Angle Low-Loss 1×2 Multimode Interference Optical Power Divider with Tilted Input and Output Waveguides

Yih-Bin Lin, Tien-Lun Ting*, and Way-Seen Wang*

Department of Electrical Engineering, Lunghwa University of Science and Technology
No. 300, Sec.1, Wanshou Rd., Guishan, Taoyuan County 33306, Taiwan
Tel +886-2-8209-3211 ext. 5520, Fax +886-2-8209-4650, E-mail lebbb@mail.lhu.edu.tw
* Graduate Institute of Electro-Optical Engineering, National Taiwan University
No. 1, Sec. 4, Roosevelt Road, Taipei, 10617 Taiwan
Tel +886-2-3366-3671, Fax +886-2-23621950, E-mail wswang@cc.ee.ntu.edu.tw

Abstract

Multimode interference (MMI) coupler with tilted input/output waveguides is proposed. By the self-imaging theory of MMI couplers, the output fields at the end of the multimode region have tilted phase fronts and automatically propagate toward the branching waveguides. The waveguides with layered structure and vertical sidewalls are used to simulate the proposed device. The excess loss is only 0.24 dB even though the branch angle is 20 degrees.

Keywords: multimode interference, optical waveguide, power divider

1 INTRODUCTION

Optical power divider is often used to separate the power of input light into two or more beams. These output beams have to be apart from each other at a certain distance large enough to prevent from coupling. Conventional 1×2 optical power dividers can be realized by using y-branches, directional couplers, or multimode interference (MMI) couplers [1]. However, the output branch waveguides of the optical power divider usually need to be apart furthermore for waveguide elements of next stage or for coupling to fibers. Bending waveguides often are used to separate the output waveguides further, but these bending waveguides cause large excess losses.

In this paper, we proposed an MMI optical power divider with the input waveguide being tilted. According to the self-imaging property [1] of MMI devices, the output field is a two-fold image of the input field. One portion of the output image is the same as the image of the input field and the other portion of the output image is the mirror image of the input field. Therefore, while the input field is tilted, the two output field are automatically tilted and directed into the output waveguides as shown in Fig.1. The two portions of the optical fields do not feel the bending structures while they propagate to the output waveguides. As additional bending structures are not required, the propagation loss of the proposed power divider is lower than that of the conventional one.

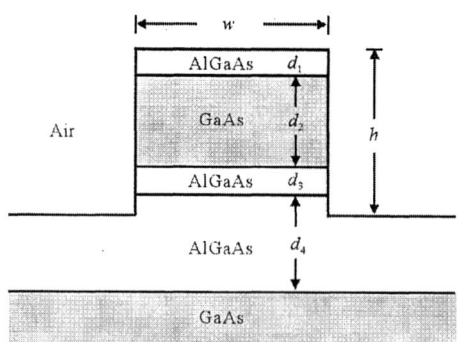

Fig.1 Top view of the proposed optical power divider.

Fig.2 Cross-sectional view of the waveguide

1-4244-0641-2/07/$20.00 ©2007 IEEE   113

2 SIMULATION AND RESULTS

The cross-section of the waveguide is a multilayered ridge waveguide with vertical sidewalls on GaAs substrate [2] as shown in Fig.2. In our simulation, the thickness of the layers, d_1, d_2, d_3, and d_4 are 0.3, 1.5, 0.3, and 3.1 μm, respectively, and the corresponding refractive indices are 3.3676, 3.4804, 3.3676, and 3.4519. The etch depth h is 2.2 μm and the operational wavelength is 1.064 μm. The single-mode condition for this waveguide is satisfied when the waveguide width w is 2.6 μm at the input and output sections. The width, W_{mmi}, of the multimode region is set to be 6 μm for supporting enough guiding modes and compact-sized considerations. The output field is a two-fold image while the length of the multimode region $L_{mmi} = 1.5L_\pi$ according to the self-imaging theory [1] under the excitation condition of general interference.

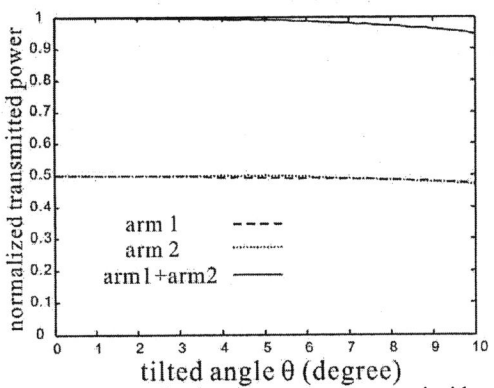

Fig.3 Normalized output power versus incident angle when $L_{mmi} = 244$μm.

The length of the multimode region, L_{mmi}, should be optimized by simulating a conventional MMI divider with straight input and output waveguides, that is, by setting the tilted angle θ to zero. The length of the MMI coupling region is found to be 244μm and the simulation results show that the total loss is 0.006dB when $\theta = 0°$. As θ is increased up to $10°$ and the input and output waveguides are tilted to the same angle θ. The normalized transmitted power of the two output arms and the total power are shown in Fig.3. The results show that the transmitted powers of the two output arms are decreased slightly as θ is increased. The propagating field distribution is shown in Fig. 4 when $\theta = 5°$ and the total loss is only 0.033dB. The total loss is as low as 0.24dB even when $\theta = 10°$, corresponding to a branch angle of $20°$. Obviously, the performance of the proposed device is very close to that of the conventional one when $\theta = 0°$.

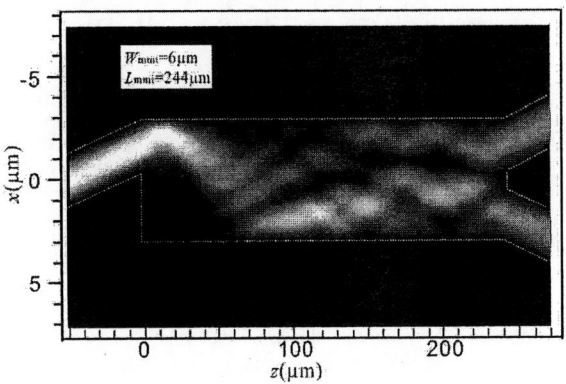

Fig.4 The propagating field distribution along z axis when $\theta = 5°$ and $L_{mmi} = 244$μm.

For the excitation condition of paired-interference in the self-imaging theory [1], only one-third of the length of a general-interference MMI coupling region is needed to produce the imaging. However, when the input waveguide is tilted, the transmission efficiency falls rapidly with the increasing angle. Simulation results show that no complete imaging replication appears at the output. A paired-interference MMI power divider requires not only a position of input field to be at one-third of the width of MMI coupling region, but also a symmetric input field to meet the zero inner products with the characteristic modes (mode number = 2, 5, 8,...). When the input waveguide is tilted, the input field is no longer symmetric, the inner products with those characteristic modes are then not zero anymore. Therefore, no imaging duplication appears at the output. That is, the paired-interference excitation is not suitable for this design of tilted input and output MMI configuration.

In conclusion, a wide-angle and low-loss 1×2 MMI optical power divider by tilting the input and output waveguides with the same angle is proposed. The total loss is as low as 0.24dB even though the tilted angle is $10°$. Details of the application will be of great interest for future study.

REFERENCES

[1] L. B. Soldano and E. C. M. Pennings, "Optical multi-mode interference devices based on self-imaging: principles and applications, " J. Lightwave Technol., vol.13, pp. 615-627 1995.

[2] J. M. Heaton, M. M. Bourke, S. B. Jones, B. H. Smith, K. P. Hilton, G.W. Smith, J. C. H. Birbeck, G. Berry, S. V. Dewar, and D. R. Wight, "Optimization of deep-etched, single-mode GaAs/AlGaAs optical waveguides using controlled leakage into the substrate," J. Lightwave Technol., vol.17, pp. 267–281 1999.

TuP31
14:30 – 17:30

Double Reflection in the Blazed Grating

Cheng-Hao Ko[1]*, Wei-Chih Liu[1], Nien-Po Chen[1], Pao-Ting Cheng[1] and Jian-Shian Lin[2]

1 Department of Electrical Engineering, Yuan Ze University, Chung-Li, 32003, Taiwan
2 Advanced Manufacturing Core Technology Division, Mechanical and System Research Laboratories,
Industrial Technology Research Institute, Chu-Tung, 31040, Taiwan
*Corresponding author: ko.chenghao@gmail.com

Abstract

The blazed grating with the Rowland structure has the advantage of self-focusing and is chosen as the major component in the spectrometer chip. In simulations for the blaze angle design in visible spectrum, we discover the phenomenon of the double reflection diffraction. Its cause and parameter space are discussed. The spectrometer utilizing the phenomenon has similar performance to the grating with the ordinary diffraction and is easier to manufacture in microelectromechanical systems (MEMS) technology. The discovery will greatly ease the design of the spectrometer chip.

Keywords: Gratings; Reflection, Spectrometers and spectroscopic instrumentation, Microstructure devices

1 INTRODUCTION

An ordinary grating does not have great efficiency other than the zeroth order. In order to enhance the efficiency of the chosen diffraction order m, a technique called blazed grating was developed [1]. The principle is to have the blazed oblique reflection surface with the blaze angle such that the angle between the incident and the reflected light is equal to that between the zeroth and the m-th order of the diffraction in the multiple-slit grating diffraction. The resultant diffraction has the optimal efficiency at the m-th order.

Besides, the blazed grating can be arranged on to the concave surface to converge the diffracted light. The design can reduce the lens system for focusing purpose in an ordinary blazed grating. This is a welcome reduction when it is implemented as a spectrometer chip.

In order to find the optimal blazed angles in the concave reflective surface design, we use electromagnetic theory of gratings to simulate the efficiency of the gratings [2,3], where the Rowland circle structure is adopted as the concave reflective surface [4], and the non-polarized spherical incident light in the visible spectrum ($\lambda = 550$ nm) is chosen as the light source.

2 THE PARAMETERS OF THE GRATINGS

In Fig. 1, the concave reflective surface is cylindrical, with the Meridional radius R (also the diameter of the Rowland circle) of 44.404 mm, the cylindrical arc length of 7.750 mm, the cylindrical height of 0.125 mm, and the grating pitch d of 3 μm. The first way to arrange the blazed grating on to the concave surface is to arrange the baseline of the blazed grating in the parallel fashion; the second way is to arrange the baseline as the tangent of the arc. We denote them as

"parallel" and "non-parallel" gratings, respectively. In the present study, we will focus on the case of non-parallel grating. Next, the normal line of grating is defined as the line where the Meridional radius of the concave surface and the diameter of the Rowland circle coincide. The incident arm length r and the incident angle α are defined from the light source to the middle of the concave surface, with r = 22.202 mm and $\alpha = 60°$. Furthermore, the desired diffraction order is the third. The reflective coating is chosen to be gold, because of its difficulty to oxide and its excellent reflectivity in the range of 550-nm wavelength. The coating thickness tolerance is practically set to 20 nm in the simulation.

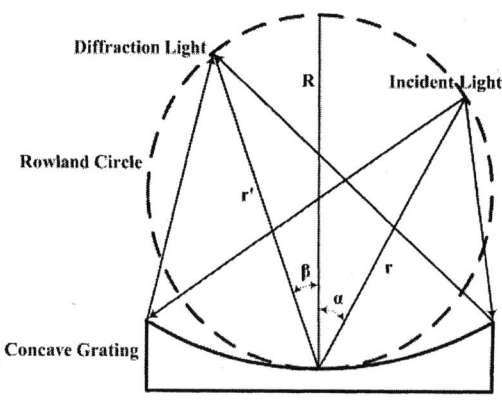

Fig. 1. Rowland circle structure.

3 RESULTS

We use the application software PCGrate [5] to perform the diffraction intensity simulation. It is a general-purpose grating efficiency simulator, which is based on the coupled

1-4244-0641-2/07/$20.00 ©2007 IEEE

electromagnetic wave theory and the modified integral method. To verify the validity of the angle ϕ (the right blaze angle), we use PCGrate to simulate the efficiency of the third order diffraction with variable ϕ, with a fixed θ (the left blaze angle). The simulation is performed against θ of 45°, 50°, 55°, 60°, and 65°. The results are plotted in Fig. 2. All of the curves have a peak at $\phi = 20.79°$.

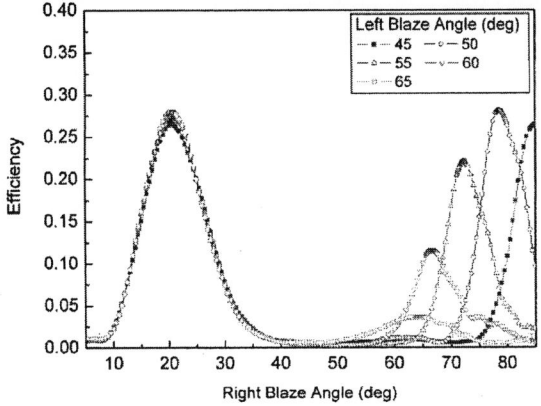

Fig. 2. The diffraction efficiency as the function of the right (ϕ) and the left (θ) blaze angles.

On the other hand, there is the second peak in each curve, whose location (ϕ) depends on the value of the left blaze angle θ. The second peaks are located at about 84.7°, 78.6°, 72.2°, 66.6°, and 64.6°, respectively. Their corresponding efficiencies vary, with the highest efficiency for $\theta = 50°$ and $\phi = 78.6°$. This is an interesting phenomenon and has not been reported. It results from the multiple reflections between the right and the left blazed surfaces. Fig. 3 shows the geometric explanation of the multiple reflections (double reflection).

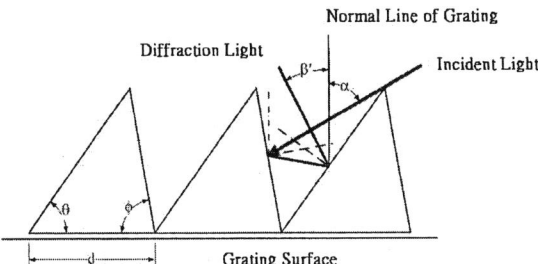

Fig. 3. The schematic of the double reflection diffraction.

The diffraction angle (of the double reflection) is denoted as β' and given as

$$\beta' = 2(\phi + \theta) - \alpha - \pi \qquad (1)$$

The condition to have the double reflection diffraction at the identical diffraction angle as for the regular diffraction ($\beta' = \beta = 18.42°$) can be obtained from Eq. (1) which yields $\theta + \phi = 129.21°$. This is consistent with the simulation result, with the corresponding values of 129.7°, 128.6°, 127.2°, 126.6°, and 129.6°, respectively. In addition, the geometry imposes further restriction on the ranges of ϕ and θ. That is, both angles must be acute and the left blaze angle θ cannot exceed the incident angle α in order to have the double reflection. The former restriction results in the lower bound for the left blaze angle θ to be 40°; the latter restriction yields the upper bound to be 60°.

Therefore, we can use Eq. (1) to obtain the requirements for the left blaze angle θ in the double reflection diffraction, provided that the incident angle α and the diffraction order m (consequently the diffraction angle β') are given.

4 DISCUSSION

The efficiency of the double reflection diffraction could be as high as that of the ordinary diffraction, as shown by the curve of $\theta = 50°$ in Fig. 2. However, in practical applications, the surface roughness can be worse than that in our simulations (20 nm).

The discovery of the double reflection diffraction phenomenon can greatly reduce the process difficulty in manufacturing. In the design of the blazed grating with the ordinary diffraction, the blaze angle for the desired diffraction order could be small. It results in a short side that is too shallow to etch in the chip processes (smaller than the critical dimension) [6].

REFERENCES AND LINKS

[1] E. Hecht, *Optics*, 4th ed. (Addison Wesley, 2002).
[2] M. C. Hutley, *Diffraction Grating* (Academic Press, 1982).
[3] G. J. Swanson, "Binary optics technology: the theory and design of multilevel diffractive optical elements," MIT Lincoln Laboratory Report p.854 (1989).
[4] W. B. Peatman, *Gratings, Mirrors, and Slits* (Gordon and Breach Science Publishers, 1997).
[5] International Intellectual Group, Inc., *PCGrate*. http://www.pcgrate.com.
[6] T. J. Suleski and D. C. O'Shea, "Gray-scale masks for diffractive-optics fabrication: I. Commercial slide imagers," Appl. Opt., vol. 34, pp.7507–7517, 1995.

TuP32
14:30 – 17:30

Effect of a Vertical Stack of Aligned Subwavelength Metal Hole Arrays on Extraordinary Transmission Spectra

[1]J Provine, [1]Rishi Kant, [2]David A. Horsley, and [1]Roger T. Howe
[1]Department of Electrical Engineering, Stanford University
[2]Department of Mechanical and Aeronautical Engineering, University of California at Davis
127X Allen CIS Building, Stanford, CA 94305
Tel +1.510.717.5952, Fax +1.650.644.0464, E-mail jprovine@stanford.edu

Abstract

We report on the effect of stacking a pair of metallic subwavelength hole arrays on the extraordinary transmission spectrum. A novel fabrication process was developed to create a self-aligned vertical stack of subwavelength metal gratings with precise control of the separation between the two metal films. For separations within the coupling region of the evanescent surface modes of the metal films, minor sharpening of the transmission filter characteristics was observed.

Keywords: Optical MEMS, Plasmonics, IR Spectroscopy, Subwavelength Gratings

1 INTRODUCTION

Subwavelength diffraction gratings composed of a metal film patterned with a regular array of holes exhibit some interesting optical properties, in particular extraordinary transmission due to surface plasmon (SP) coupling[1]. SP theory predicts the EM field on the metal film surface is affected by the complex permittivity of the surrounding media. Previous work has shown the transmission of hole arrays can be actively tuned by utilizing MEMS actuation to control the proximity of the metal to a dielectric media[2]. Simulations predict that the coupling between a pair of vertically stacked photonic crystal (PC) slabs will produce many strong resonance phenomena[3]. In this work, we study the coupling effects in an analogous system of vertically stacked metallic gratings.

2 DESIGN & FABRICATION

Hole arrays of various permutations of pitch spacing (a) and hole diameter (d) were included on a single contact lithography mask to explore the dependency of transmission on array and hole geometry. In general the hole arrays were designed for the mid-IR regime ($a \approx \lambda \approx 10\mu m$). The central wavelength of surface resonance modes can be estimated from

$$\lambda_p = \frac{a}{\sqrt{i^2 + j^2}} \sqrt{\frac{\epsilon_d \epsilon_m}{\epsilon_d + \epsilon_m}} \quad (1)$$

where ϵ_d and ϵ_m are the dielectric constants of the surrounding media and the metal, respectively, and i and j are indexing integers. Self-aligned identical gratings were fabricated with a controlled vertical separation (g).

Starting with a (100) Si wafer, a 750nm thermal oxide layers was grown at 1000°C. An undoped poly-Si film of thickness g was deposited by LPCVD at 580°C. The gap between the stacked metal films is determined by this film.

With the front side of the wafer protected by hard baked photoresist, through wafer holes were lithographically defined then etched through the backside poly-Si film, the backside buried SiO₂, and the full thickness of the original wafer stopping on the frontside buried oxide (BOX). The oxide etching was performed by a magnetically enhanced reactive ion etch in an Applied Materials P5000 etcher using CF_3 and Ar, and the Si etching was performed by a Bosch-process DRIE of alternating passivation with CH_4 and etching with SF_6 in a STS Advanced Silicon Etch (ASE) System. After stripping the photoresist, the front side BOX exposed by the through wafer holes is chemically etched by a dip in 6:1 BOE. Next, 200nm films of Al were sputtered on both sides of the wafer. The sputter occurred in an Ar ambient with pressure of ~6mTorr yielding a film stress of 92MPa tensile. The frontside of the wafer was lithographically patterned with the hole arrays aligned to the through wafer holes. A 1.6μm thick photoresist was used to transfer the grating patterns through the top Al film using a Cl_2 RIE in a Lam etcher, then the poly-Si layer via DRIE in the STS ASE, and finally the bottom Al layer again using Cl_2 RIE in a Lam etcher. The metal membranes were released through isotropic RIE in the STS ASE by etching with SF_6 without passivation cycles. Once the photoresist was stripped by O₂ plasma etch, the devices were complete.

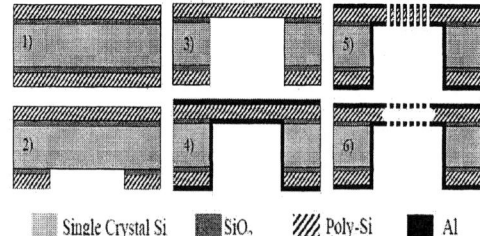

Figure 1. Cross-section of the fabrication process flow.

1-4244-0641-2/07/$20.00 ©2007 IEEE

The fabrication process is shown as a cross-sectional schematic in Figure 1. It should be noted that this fabrication process allows for actuation of the Al membranes relative to each other by utilizing the poly-Si gap level as a mechanical structural layer.

Figure 2 shows SEM images of a grating with 2.4μm gap. The left image shows the top view of the device, while the right shows a magnified image of a region where part of the upper hole array was removed by ion milling to expose the lower hole array.

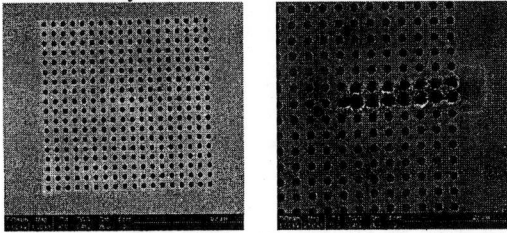

Figure 2. (left) SEM of a 150μm x 150μm grating with a=8μm, d=5μm, and g=2.4μm. (right) SEM of the same device showing the lower grating through a region where the upper grating was removed by focus ion beam milling.

3 OPTICAL ANALYSIS

The fabricated grating stacks were spectrographically analyzed using a Nicolet 6700 FT-IR system connected to a Nicolet Continuum Infrared Mircroscope. The transmission spectra of gratings with g=2.4μm and g=3.6μm are shown in Figure 3 (a=10μm and d=6μm) and Figure 4 (a=8μm and d=5μm). In both figures the spectra for a single grating of 400nm thick Al is included with the transmission squared to account for the double filtering that occurs for two gratings.

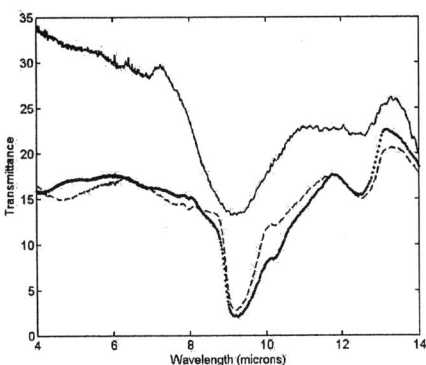

Figure 3. Transmission spectra gratings with a=10μm and d=6μm for a single grating (solid), g=3.6μm (dashed), and g=2.4μm (dotted).

For both gratings, the main spectrographic features of the single grating are preserved in the double grating structures. The evanescent surface wave caused by SP is known to extend ~λ/2 from the metal surface[1], so it is expected that a second metal film within this range would effect the filter performance beyond a simple appended filter. In both hole arrays, the interaction of the second Al film causes the maximum transmission mode to be sharpened. Additionally, the secondary minima and maxima are narrowed and enhanced. The effects are only a few percent in each case.

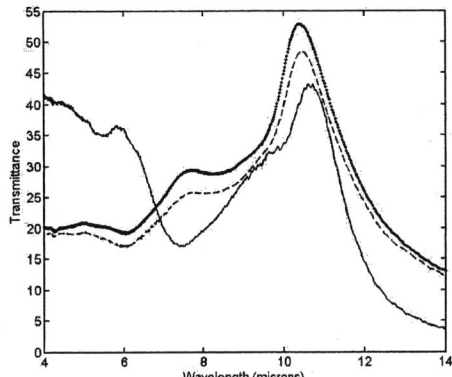

Figure 4. Transmission spectra gratings with a=8μm and d=5μm for a single grating (solid), g=3.6μm (dashed), and g=2.4μm (dotted).

4 CONCLUSIONS & FUTURE WORK

We report on the effect of a double stack of self-aligned metal subwavelength gratings on extraordinary transmission spectra. The transmission maxima and minima modes exhibit are mildly enhanced compared to a single metal film by the interaction of the two metal films. The fabrication process developed for this study is being extended to allow electro-static actuation of metal films via the poly-Si spacer film as a mechanical layer. The metal films could be able to actuate with both vertical and lateral relative motion.

REFERENCES

[1] W.L. Barnes, A. Dereux, and T.W. Ebbesen, "Surface plasmon sub-wavelength optics," *Nature*, 424, 824-830, 2003.

[2] J. Provine, J.L. Skinner, and D.A. Horsley, "*Subwavelength Metal Grating Tunable Filter,*" Proc. of IEEE MEMS, Istanbul, Turkey, January 23-27, 2006.

[3] W. Suh, M.F. Yanik, O. Solgaard, and S. Fan, "Mechanically switchable photonic crystal structures based on coupled photonic crystal slabs," Proc. of SPIE, 5360, 299-306, 2004.

TuP33
14:30 – 17:30

The Measurement of Liquid Refractive Index by D-shaped Fiber Bragg Grating

Hong-Wei Chen[1], Chuen-Lin Tien[2], Wen Fung Liu[2], Shane-Wen Lin[1]

[1]Graduate Institute of Electrical and Communications Engineering, Feng Chia University, Taichung, Taiwan
[2]Department of Electrical Engineering, Feng Chia University, Taichung, Taiwan
No. 100 Wenhwa Rd., Seatwen, Taichung, Taiwan 40724, R.O.C.

Abstract- **The purpose of removing a part of fiber cladding to form a D-shaped FBG is to improve the sensing sensitivity. By using different types of liquids with different refractive index for the experiment, the Bragg wavelength shift of side-polished FBG is measured.**

I. INTRODUCTION

Optical fiber sensors can be used in the harsh environments for measuring the required parameters, since the optical fiber is mainly made of silicon materials that have high resistance to electromagnetic interference, thermal shock and corrosion. The fiber sensors based on fiber Bragg grating (FBG) have been potentially developed due to their many inherent advantages such as the small size, immunity to electromagnetic interference, wavelength multiplexing, and distributed sensing possibilities[1-5].

In general, the fiber Bragg grating is often used to measurement the temperature, pressure or stress. It is scarcely used to measure the refractive index variation of a test material. The sensitivity of the ordinary fiber sensors can be improved by the method of etching or heat-fusion fiber. But the disadvantage of this way is that it easily breaks and not good durability. For enhancing the disadvantage, we proposed a novel refractive index sensing device by using side-polished FBG. The technique of side-polished fiber is used for removing the fiber side-cladding to make the gratings to detect the Bragg wavelength shift caused by the refract index change of the ambient materials. The technique of side-polished fibers provides an interesting structure of D-shape fiber Bragg gratings for expecting to increase the sensor sensitivity. The side-polished FBG exhibits an asymmetric geometry fiber structure with the characteristics both of transmission optical-power loss and reflective power reduction. In this study, we present the experimental results by using different refractive-index ambient materials, such as alcohol, water, acetone, methyl alcohol and isopropanol at room temperature.

II. PRINCIPLE

A fiber Bragg grating consists of a periodic modulation of the refractive index in the core of a single-mode optical fiber. When the phase-matching condition is satisfied, the contributions of reflected light from each grating plane add constructively in the backward direction to form a back-reflected peak with a center wavelength defined by the grating parameters. The first-order Bragg condition is given by

$$\lambda_B = 2n\Lambda \tag{1}$$

where the Bragg wavelength, λ_B, is the center wavelength of the input light that will be back-reflected from the Bragg grating, and n is the effective index of the fiber core and Λ is the grating period. As the fiber cladding is polished to expose the core to air, the effective index of the fiber core can be changed to result in the shift of reflected Bragg wavelength and bandwidth expansion. Differentiating equation (1) with respect to the reflective Bragg wavelength, λ_B, can be expressed as

$$\frac{\Delta\lambda_B}{\lambda_B} = \frac{\Delta n_{eff}}{n_{eff}} + \frac{\Delta\Lambda}{\Lambda} \cong \frac{\Delta n_{eff}}{n_{eff}} \tag{2}$$

where $\Delta\lambda_B$ is the Bragg wavelength shift. Δn_{eff} and $\Delta\Lambda$ are the change of grating effective refractive index and the grating period, respectively. For the side-polished fiber grating, the index change Δn_{eff} will dominate the Bragg wavelength shift in comparison with the effect of grating period change ($\Delta\Lambda$). This is due to that the absence of strain and the constant temperature exist in the grating. Therefore, the effect of grating period variation ($\Delta\Lambda$) can be neglected. The shift of Bragg wavelength in Eq. (2) can be simply expressed as

$$\Delta\lambda_B \cong 2\Delta n_{eff}\Lambda \tag{3}$$

Thus this FBG sensor for detection of the environmental refractive index change is based on the Bragg wavelength shift.

FBG sensors have been proposed for different sensing functions, for example strain, temperature, dynamic magnetic field, etc. It is dependent on the grating center wavelength shift when the sensing parameters cause grating effective index or grating period variation. The sensing mechanism of this device is based on refractive index variation that is induced by the different liquids such as, alcohol, water, acetone, methyl alcohol and isopropanol. When the D-shaped sensor is inserted into different liquids, the variation of the ambient refractive index leads to change the effective index and then causes Bragg wavelength shift. The value of Bragg wavelength shift can be measured by using an optical spectrum analyzer (OSA) for estimating which type of liquid to be tested.

III. EXPERIMENTAL RESULTS AND DISCUSSIONS

In our experiment, the used FBG is written in a hydrogen-loaded single-mode fiber (SMF-28) by using the phase mask writing technique of a KrF excimer laser (248 nm). The grating length is 10 mm with the reflectivity of 90 %. Annealing the fiber grating takes 8 hours at 150℃. For fiber polishing it is glued in silicon V-groove with the setup similar to that shown in Ref. 8. The fibers were glued in the Si wafer

1-4244-0641-2/07/$20.00 ©2007 IEEE 119

with V grooves. The length of the side-polished interaction section is about 10mm. In the fiber-optic sensing region, the fiber diameter is polished down to 66.3 μm to be measured by an optical microscope. The proposed sensing device and the experiment setup are schematically illustrated in Fig. 1. Different types of liquids including the water (n=1.33), acetone (n=1.36), isopropanol (n=1.38), methyl alcohol (n=1.44), and alcohol (n=1.46) at 20℃ are dropped into the sensor. The reflection spectra of before and after side-polished FBG was observed by an OSA as shown in Fig.2. The sensing mechanism of this device is based on the change of the effective refractive index of fiber Bragg grating. The results indicate that the reflection spectrum red-shifts as ambient refractive index increases. The FBG wavelength as a function of the ambient refractive index is shown in Fig. 3.

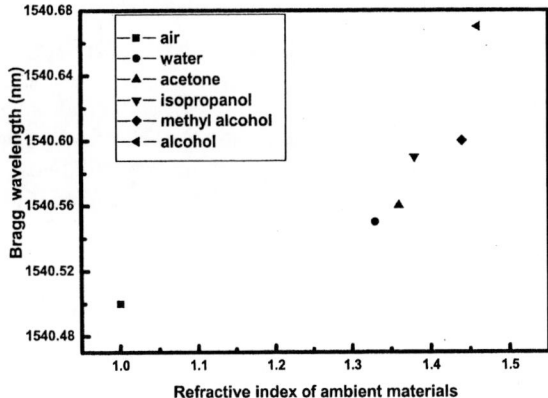

Fig. 3 Refractive index of ambient materials versus Bragg wavelength.

IV. CONCLUSIONS

The proposed sensing device has been experimentally demonstrated to be applied for detecting different liquid concentrations and the variation of refractive index. There are still some parameters requiring to be further investigated, such as interaction length, the diameter of side-polished fiber and the flatness of polished surface for improving this sensor sensitivity.

V. REFERENCES

[1]. K.O. Hill and G Meltz, "Fiber Bragg Grating Technology Fundamentals and Overview," *J. Lightwave Technol.*, Vol.15, pp.1263-1276, August 1997.

[2]. A. D. Kersey, M.A. Davis, H. J. Patrick, M. LeBlance, K. P. Koo, C.G. Askins, M. A. Putnam,and E. J. Friebele, "Fiber Grating Sensors," *J. Lightwave Technol.*, Vol. 15, pp.1442-1463, August 1997.

[3]. A. Othonos, "Fiber Bragg Gratings," *Rev. Sci. Instrum.*, Vol. 68, pp. 4309-4341, December 1997.

[4]. K. O. Hill, Y.Fujii, D.C. Johnson, and B. S. Kawasaki, "Photosensitivity in Optical Fiber Waveguides: Application to Reflection Filter Fabrication," *Appl. Phys. Lett.*, Vol. 32, pp. 647-649, May 1978.

[5]. A. T. Andreev, B. S. Zafirova, and E. I. Karakoleva, "Single-mode Fiber Polished into the Core as a Sensor Element," *Sensor and Actuators A*, Vol. 64, pp. 209-212, January 1998.

[6]. Joel Villatoro, Antonio Diez, Jose L. Cruz, and Miguel V. Andres, "In-line Highly Sensitive Hydrogen Sensor Based on Palladium-coated Single-mode Tapered fibers," *IEEE Sensor Journal*, Vol. 3, No. 4, August 2003

[7]. Wei Liang, Yanyi Huang, Yong Xu, Reginald K. Lee and Amnon Yarive "Highly Sensitive Fiber Bragg Grating Refractive Index Sensors," *Applied Physics Letters* 86, 141122, 2005.

[8]. Shiao-Min Tseng and Chin-Lin Chen, "Side-polished Fibers," *Applied Optics*, Vol.31, No.18/20, June 1992.

[9]. Alberto Alvarez-He'rrero, Hector Guerrero, and David Levy "High-Sensitivity Sensor of Low Relative Humidity Based on Overlay on Side-Polished Fiber," *IEEE Sensors Journal*, Vol.4, No.1, February 2004.

[10]. Jaejoong Kwon, Younghee Jeon, Byoungho Lee*, "Tunable Dispersion Compensation with Fixed Center Wavelength and Bandwidth Using a Side-polished Linearly Chirped Fiber Bragg Grating," *Optical Fiber*, pp. 159-166, Technology 11, 2005.

[11]. M. J. F. Digonnet, J. R. Feth, L. F. Stokes and H. J. Shaw, "Measurement of Core Proximity in Polished Fiber Substrates and Couplers," *Optics Letters*, Vol. 10, No.9, September 1985.

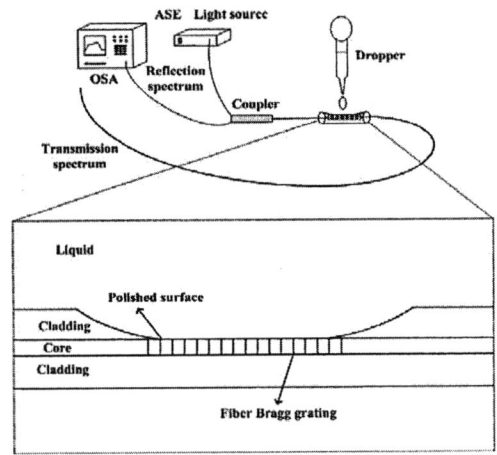

Fig. 1 Experiment set-up used to measure liquid refraction

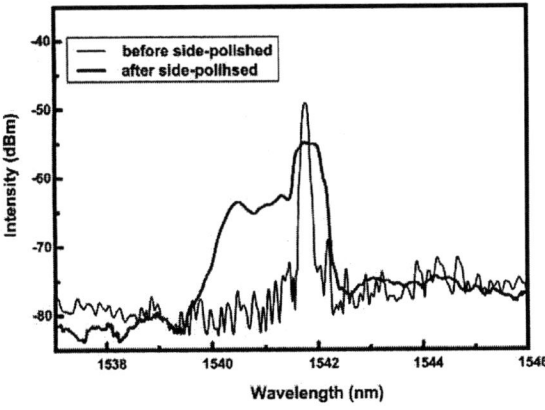

Fig. 2 Reflection spectra of before and after side-polished FBG sensor

TuP34
14:30 – 17:30

A Compact Silicon-on-Insulator MMI-based Polarization Splitter

Yao-Feng Ma, and Ding-Wei Huang*

Graduation Institute of Electro-Optical Engineering, National Taiwan University
1 Roosevelt Rd. Sec.4, Taipei, 106 Taiwan, R.O.C.
Tel/Fax +886-2-33663664, *dwhuang@cc.ee.ntu.edu.tw

Abstract

A compact MMI-based polarization splitter is designed using SOI channel waveguides. Even though silicon does not have any material birefringence, the slab-like geometry of MMI makes it possible to induce significant birefringence which allows to achieve polarization splitting with a suitable design. In this work, we demonstrate the polarization splitting in an SOI waveguide device as short as 100 μm designed by utilization of BPM technique.

Keywords: Integrated optics, multi-mode interferometer (MMI), silicon-on-insulator (SOI), polarization splitter, BPM.

1 INTRODUCTION

Polarization splitters are widely used in optical systems, where polarization states of light should be taken into serious considerations, such as communications, sensing, data storage, imaging, and signal processing [1], [2]. For many birefringent optical devices, separation of orthogonal polarization states is a straightforward solution where polarization splitters can be used.

In this paper, we demonstrate a compact silicon-on-insulator (SOI) multimode interference (MMI) waveguide device that exhibits excellent polarization splitting behavior. The design and fabrication SOI MMI devices have already been studied [3], [4]. However, there is still not any SOI MMI devices reported for polarization splitting applications. In our study, we carefully manipulate the self-imaging effect and the geometry-induced birefringence in SOI MMI devices so that a significant difference of the MMI self-imaging length between TE and TM modes can be obtained.

In Section 2, we show how to design and utilize this feature to achieve a polarization splitter. Simulation results and analysis follow.

2 DESIGN AND SIMULATION

The polarization-dependent behavior of integrated optical waveguides has two main sources. One source is the intrinsic material birefringence which can be due to stress in the waveguides. The other source is due to waveguide geometry or cross-sectional profile. However, as the cross-sectional dimensions reduce, the birefringence can increase and such waveguides can be designed for a specific

birefringence value. Analysis of SOI waveguides by three-dimensional (3-D) vector beam propagation (BPM) simulations confirm the evolution of geometrical birefringence in single-mode (SM) SOI waveguides as their size is reduced. The slab-like region of MMI exhibits the geometry-induced birefringence in a similar way.

Within the length of the MMI device, the number of the occurrences of the self-imaging effect for TE and TM modes can be different. When the number of the occurrences of the self-imaging effect for TE and TM modes differs by 1 within a suitably designed length of the MMI device, the TE and TM components can be successfully separated into different output ports.

In our design, because the silicon core is usually covered by a silica cladding layer above it to avoid the influence of moisture and temperature, the photonic wire can be regarded as a channel waveguide which have core index $n_o = 3.474$ for silicon and cladding index $n_a = 1.444$ for silica at 1.55 μm (Fig. 1). The width w and height h of the SM SOI channel waveguide is chosen to be 0.32 μm and 0.28 μm for satisfying the SM condition for an SOI channel waveguide.

For best efficiency of the self-imaging effect of the MMI device, we set the locations of the input and the two output ports to be approximately equal to one-third of the width of MMI (W_{mmi}) to eliminate some higher modes excited by the input. For different W_{mmi}, due to the geometry-induce birefringence, the self-imaging lengths of TE and TM modes are different and are shown in Fig.2. When the ratio of the self-imaging length of TE mode L_{TE} and the self-imaging length of TE mode L_{TM} equals to the ratio of two consecutive

1-4244-0641-2/07/$20.00 ©2007 IEEE   121

integers m and $m+1$, i.e., $L_{TE}/L_{TM} = 1/2, 2/3, 3/4, \ldots, m/(m+1)$, meanwhile the length of the MMI device is chosen to satisfy $L_{mmi} = mL_{TM} = (m+1)L_{TE}$, then the two modes can be successfully separated into 2 different output ports as shown in Fig.2. The intersects between horizontal lines and the curve of W_{mmi} are the design solutions, we pick W_{mmi} as small as possible to achieve the largest difference of the self-image lengths between TE and TM modes and the shortest length of the entire device. Note that when W_{mmi} is too small, the separated modes may couple to each other between the two output ports. To avoid the coupling, we choose $W_{mmi} = 2.2$ μm and $L_{mmi} = 120.3$ μm to be the dimensions for our design.

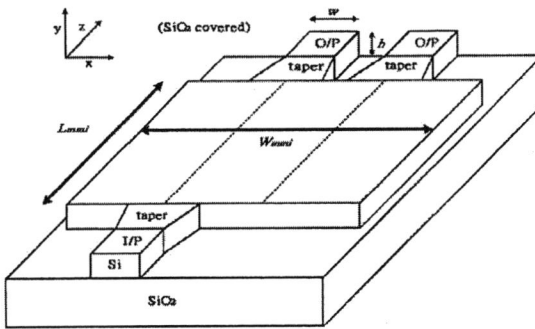

Fig.1 Schematic of MMI-based polarization splitter in a SOI channel waveguide.

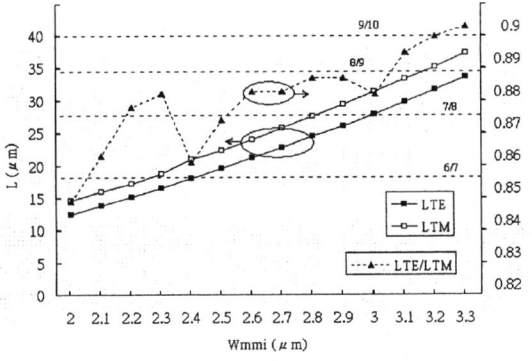

Fig.2 The self-imaging lengths of TE and TM modes and the ratio of the self-imaging length of TE and TM mode, L_{TE}/L_{TM}. The intersects between horizontal lines and the curve of W_{mmi} are the design solutions.

Furthermore, we insert tapers between waveguides and the slab region of MMI to reduce the excess loss. The simulation is conducted by using BPM technique as shown in Fig. 3. The simulation result confirms that, with a suitable design, the proposed SOI MMI exhibits excellent polarization splitting behavior.

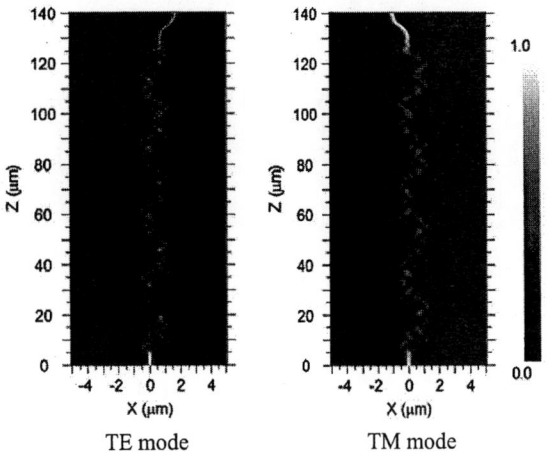

Fig.3 The simulation is conducted by using BPM technique. The left-hand side is the result of TM mode and the right-hand side is the result of TE mode.

3.CONCLUSION

The geometry-induced birefringence in an SOI MMI device becomes apparent as its size reduced. This idea is used to design and fabricate an MMI coupler polarization splitter based on geometrical birefringence. The Length of the device is about 120 μm when the width of the MMI is 2.2 μm and the extinction ratio of the two polarization components can achieve -20dB. We have demonstrated the first compact polarization splitter that manipulates the geometry-induced birefringence of the MMI-based high-index optical channel waveguides on SOI platform.

REFERENCES

[1] A. Miliou, R. Srivastava, and R. V. Ramaswamy, "A 1.3-_m directional coupler polarization splitter by ion exchange," *J. Lightw. Technol.*, vol.11, no. 2, pp. 220–225, Feb. 1993

[2] K. Okamoto, M. Doi, T. Irita, Y. Nakano, and K. Tada, "Fabrication of TE/TM mode splitter using completely buried GaAs/GaAlAs waveguide," *Jpn. J. Appl. Phys.*, vol. 34, no. 1, pp. 114–115, 1990.

[3] E. C. M. Pennings, R. J. Deri, A. Scherer, R. Bhat, T. R. Hayes, N. C. Andreadakis, M. K. Smit, L. B. Soldano,and R. J. Hawkins, Appl. Phys. Lett. **59**, 1926 (1991).

[4] L. B. Soldano and E. C. M. Pennings, J. Lightwave Technol. **13**, 615 (1995).

TuP35
14:30 – 17:30

Near-field images of surface plasmon eigenmodes in gold nanogratings

FU HAN HO*, SHENG YUAN CHEN, YU HSUAN LIN, JIEN YIN SU

Tel +886-3-577-9911 ext 479, Fax +886-3-577-3947,
*E-mail: fuhanho@itrc.org.tw
Nanotechnology Division, Instrument Technology Research Center (ITRC), National Applied Research Laboratories,
20, R&D Rd. VI, Hsinchu Science Park, Hsinchu, Taiwan, R.O.C.

Abstract

In this study, we have fabricated gold nanograting structures by using a focused-ion-beam system. The nano-optical properties were examined by using both the scattering- and transmission-mode near-field scanning optical microscope (NSOM). The surface plasmon (SP) eigenmode in this periodic nanostructure is directly observed by the scattering-mode NSOM. Two SP waves, localized fields and propagating waves, generating in the near field were found by the transmission-mode NSOM, and the dependence of wavelengths was studied. The relations of scattering- and transmission-mode results were discussed as well.

Keywords: Plasmonics ,nano-optics, near-field scanning optical microscope, and focused-ion-beam system

1 INTRODUCTION

Metallic nanostructures are now widely studied and used in the applications of nanophotonics and biosensors for their unique optical properties.   Plasmonics is one of these attractive topics. Surface plasmon (SP) is the quantized oscillation of the collective electrons excited at the interface between a conductor and an insulator. [1] By matching the dispersion-matching condition through different designs of the structures, SP is easily generated in the ways of highly localized field or propagating surface wave. Many kinds of patterns or shapes were proposed for the different functional nanodevices, such as sensors [2], waveguides [3,4], nanolithography mask [5], etc. To realize the behavior of SP, the mechanism of SP excitation modes is an important issue. Many models were recently proposed and predicted [6-8], and theoretical simulations were also shown interesting optical phenomenon. However, the experimental observation is still limited. In this research, we have studied the optical properties of the gold nanograting by using a near-field scanning optical microscope (NSOM). The SP eigenmode in this periodic nanostructure is directly observed by the scattering-mode NSOM.   We also find two SP waves, localized fields and propagating waves, generating in the near field and study the dependence of wavelengths by transmission-mode NSOM. The relations of scattering- and transmission-mode results were discussed as well.

2 Sample preparation

The sample in this study is a gold film with the thickness of 60nm, which the gold is coated by DC sputtering on a glass slide. The nanograting structures are fabricated by using a focus ion beam (FIB) on the gold film. The results are shown in Figure 1. The pattern is a square with the size of 10×10 μm as Fig 1(a). The width of each slit is about 70 nm, and the distances between the slits were controlled as 140 nm in Fig 1(b).

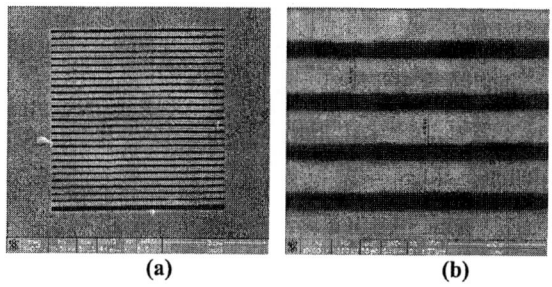

(a)   (b)

Figure 1. (a) SEM micrograph of the nanograting pattern. (b) The zoom-in picture.

3 Experimental setup

The nano-optical property of gold nanograting is investigated by using the near-field scanning optical microscope (Aurora-3, Veeco Instruments Inc.).   Figure 2 shows the experimental setup of NSOM measurement. The NSOM is based on an atomic force microscope (AFM) with a tuning-fork force-sensing device for the feedback control. The fiber tip for NSOM is coated with aluminum in the thickness of 100nm, and the aperture size is about 50nm. A multi-line Ar-Kr laser is used to illuminate the sample either

1-4244-0641-2/07/$20.00 ©2007 IEEE   123

through an objective lens from the bottom side (transmission mode) or the aperture of a fiber tip (scattering mode). The near-field scattered light can be collected by the fiber tip (transmission mode) or an objective lens at the same side of fiber tip (scattering mode). The signal was detected by a photomuliplier tube (PMT). The NSOM fiber tip is scanning in three dimensions for simultaneously imaging the topography and photography of the sample surface.

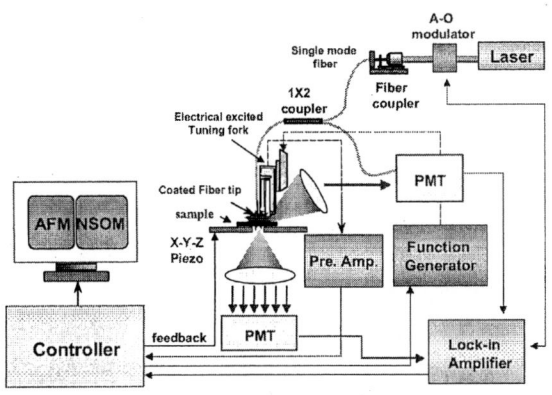

Fig 2. Experimental setup

4 Results and Conclusion

Figure 3 shows the near-field optical results of scattering-mode measurement. The input light is a point-like source generated by the small aperture and illuminated in the near field, and the scattered light is collected by an objective lens in the reflection side. The wavelength is 532nm. The topography (AFM) and NSOM image of the nanograting are simultaneously obtained for the same location with the scanning size of 3×3 μm. The nanostructure of eight slits is clearly shown in the left-hand side of the topography image, and the gold film with no structure is in the right-hand side. In the NSOM image, we find eight dark lines in the left-hand side correctly corresponding to the slits of topography image. The reason of the dark line is that the light is almost transmitting through the nanoslits. In between the dark lines, the bright lines are observed and the position is corresponding to the centers of gold nanowires. We also find optical interference-like patterns in the right-hand side of the NSOM image, which is corresponding to the area of pure gold film. We believe that the bright lines and interference-like patterns are the spatial eigen-mode of SP in this nanograting structure. This means that the optical properties and SP behaviors are already spatially determined when the structure is formed. The local illumination by a point-like light source is one of measurements to spatially resolve the eigen-mode.

Furthermore, the transmission-mode NSOM experiment

is also studied in order to compare with the results of scattering-mode experiment, which the results is not shown here. We find that the SPs generated by a normally incident light from the bottom side were coupled into the grating and propagating over the sample surface. Two kinds of SP waves, localized resonance field and propagating wave, in the near field are observed by the fiber tip, which these two SP waves are in a strong correlation with the eigen-mode results. We also find the behavior of localized resonance fields depend on different laser wavelengths.

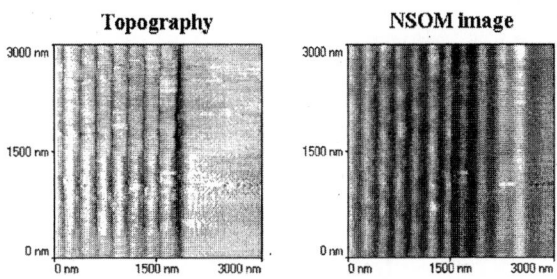

Figure 3. The topography and near-field optical images of scattering-mode measurement.

In conclusion, the optical near fields and eigenmodes of SP in the gold nanogratings were directly examined by using both the scattering- and transmission-mode NSOM. The size of the structures and wavelengths of input light play important roles for SP waves. These results will provide significant information for the further applications, like nanolithography, optical filter, SP generator, and sensors.

Reference

[1] H. Raether, *Surface plasmons on Smooth and Rough Surfaces and on Gratings*, Springer-Verlag, Berlin (1988).

[2] Jirí Homola (Ed.), *Surface Plasmon Resonance Based Sensors*, Springer-Verlag, Berlin (2006).

[3] Yatsui, M. Kourogi, and M. Ohtsu, Appl. Phys. Lett. 79, 4583 (2001)

[4] W. Nomura, M. Ohtsum and T. Tatsui, Appl. Phys. Lett. 86, 181108 (2005)

[5] Z. W. Liu, Q. H. Wei, and X. Zhang, Nano. Lett. 5, 957 (2005).

[6] S.-H. Chang and S. K. Gray, Opt. Exp. 13, 3150 (2005).

[7] S. A. Darmanyan and A. V. Zayats, Phys. Rev. B 67, 035424 (2003).

[8] W. L. Barnes, W. A. Murray, J. Dintinger, E. Devaux, T. W. Ebbesen, Phys. Rev. Lett. 92, 107401 (2004).

TuP36
14:30 – 17:30

Surface Plasmon Leakage in Its Coupling with an InGaN/GaN Quantum Well through an Ohmic Contact

Dong-Ming Yeh, Chi-Feng Huang, Yen-Cheng Lu, Cheng-Yen Chen, Tsung-Yi Tang, Jeng-Jie Huang,
Kun-Ching Shen, Ying-Jay Yang, and C. C. Yang
Graduate Institute of Electro-Optical Engineering and Graduate Institute of Electronics Engineering, National
Taiwan University, Taipei, Taiwan, R.O.C.
(phone) 886-2-23657624   (fax) 886-2-23652637   (e-mail) ccy@cc.ee.ntu.edu.tw

ABSTRACT

We demonstrate the loss of surface plasmon (SP) energy through oscillating electron leakage via the Ohmic contact of either
p-type or n-type GaN layer in the coupling process between SP and an InGaN/GaN quantum well (QW). The observation
implies that in using the SP-QW coupling for enhancing emission in a light-emitting diode, the metals for Ohmic contact and
SP generation must be separated. A thin dielectric interlayer is required in the region for SP-QW coupling to avoid the leakage
of SP energy.

Keywords: surface plasmon, quantum well, Ohmic contact

1. INTRODUCTION

In this paper, we report the loss of SP energy through the
leakage of oscillating electrons via the required impurity
doping in the p- or n-type layer in an LED when the SP-QW
coupling is to be used for enhancing emission efficiency.
The basic idea in applying the SP-QW coupling process to
an LED is to use the required p-, n-type Ohmic contact, or
current spreading metal for generating SPs. However, the
Ohmic contact built between the p- or n-type semiconductor
layer and the coated metals may create a channel of electron
leakage or SP energy loss. We compare the SP-QW coupling
behaviors between the three samples of p-, n-type, and
un-doped GaN layer between an InGaN/GaN QW and the
coated metals. It is found that doped GaN layer can always
lead to SP loss through electron leakage.

2. SAMPLE DESCRIPTIONS

The two used samples contain an InGaN/GaN QW grown on
a 2-μm i-GaN layer (grown at 950 °C), which was deposited
on c-plane sapphire substrate with metalorganic chemical
vapor deposition. The QW, grown at 760 °C, has a width of
3 nm with the indium content around 10 % leading to the
photoluminescence (PL) spectral peak between 450 and 460
nm. Then, a 10-nm undoped GaN layer (grown at 950 °C) is
deposited to serve as the barrier of the QW before
impurity-doped cap layer is grown. In the p-type sample, the
growth of a GaN layer of 10 nm in thickness at 950 °C with
Mg doping of around 10^{20} cm^{-3} impurity concentration is
realized. After the thermal activation process at 750 °C for
30 min, a hole concentration of around 10^{18} cm^{-3} is obtained.
In the n-type sample, Si doping of also around 10^{20} cm^{-3} in
concentration, leading to around the same electron
concentration, is implemented in the 10-nm GaN cap layer

(also grown at 950 °C). The p- and n-type samples are
coated with various metals for generating SPs. Their SP-QW
coupling behaviors are compared.

3. OPTICAL CHARACTERIZATIONS

Fig. 1 shows the PL spectra of the n-type sample with the
conditions of bare-QW (labeled with QW), 100-nm-Al
coating (labeled with Al), 100-nm-Ag coating (labeled with
Ag), and 4-nm-Ni/100 nm-Ag coating (labeled with
Ag). Fig. 2 shows the corresponding TRPL traces. The Al coating
means to fabricate an Ohmic contact on the n-type layer.
One can see that the n-type Ohmic contact leads to a
tremendously decreased PL intensity and slight blue shift in
PL spectral center-of-mass. It is believed that the SP-QW
coupling results in the PL suppression in this case due to the
nonradiative SP energy transfer. Because the signal is weak,
it becomes difficult to obtain the TRPL data in the case of
n-type Ohmic contact. Because of different work functions,
the Ag and Ni/AG coatings on the n-type layer do not
produce any Ohmic contact. Compared with the bare-QW
condition, the PL intensity of the Ag-coated condition is
slightly decreased. Also, the TRPL decay time becomes
shorter. Such behaviors may not strongly support the
existence of SP-QW coupling. However, the PL suppression
and the further decay time reduction in the case of Ni/Ag
coating clearly indicate the SP-QW coupling effects.

Then, Fig. 3 shows the PL spectra of the p-type sample
with the conditions of bare-QW (labeled with QW),
100-nm-Ag coating (labeled with Ag), 4-nm-Ni/100 nm-Ag
coating (labeled with Ni/Ag), and 10-nm-SiN/100-nm-Ag
(labeled with SiN/Ag). Fig. 4 shows the corresponding
TRPL traces. The Ag coating means to create an Ohmic
contact on the p-type GaN. The better attachment of Ni to

1-4244-0641-2/07/$20.00 ©2007 IEEE

the semiconductor leads to a better Ohmic contact with Ni/Ag coating. The coating of SiN means to create an insulating dielectric layer between the p-type semiconductor and metal. From Fig. 4, one can see that the p-type Ohmic contacts result in significant decreases of PL intensity, indicating the SP-QW coupling-induced PL suppression. However, with an insulating layer between the p-type layer and the metal, the PL intensity is significantly enhanced with a clear blue shift of PL spectral feature and a decrease of PL decay time.

The SiN/Ag coating condition for PL enhancement implies that the insulating layer (SiN) plays a key function in reducing the SP nonradiative energy loss. In the cases of creating Ohmic contacts, either n-type or p-type, the SP-QW coupling always leads to a stronger SP energy loss. In this situation, it is believed that the oscillating electrons of SPs may leak via the transport through the Ohmic contact. The SP modes lose their energy nonradiatively through electron leakage. One may think that such electron leakage can occur only in the type case. However, from the data in Fig. 1, one can see that the n-type Ohmic contact can also lead to electron leakage.

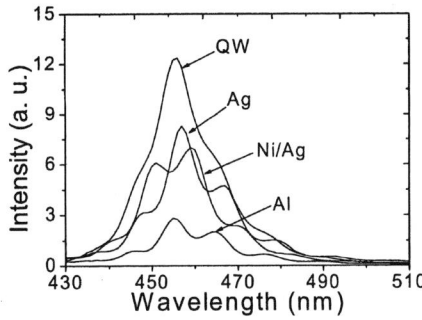

Fig. 1 PL spectra with different coatings on the n-type epitaxial structure in comparison with that of the bare-QW sample.

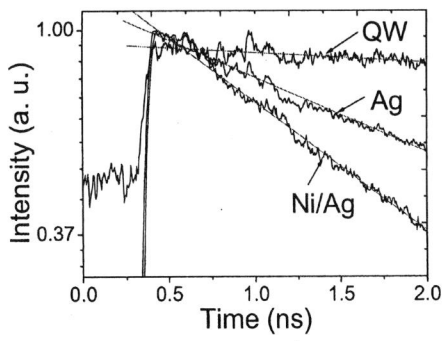

Fig. 2 The corresponding TRPL traces except the Al-coating sample of the samples in Fig. 1.

4. CONCLUSIONS

In summary, we have demonstrated the loss of SP energy through oscillating electron leakage via the Ohmic contact of either p-type or n-type GaN layer in the coupling process between SP and an InGaN/GaN QW. It was shown that the PL intensity was significantly reduced although the TRPL decay time was not significantly decreased when an Ohmic contact was formed, in contrast to the case of PL enhancement when an insulating thin layer existed between the doped semiconductor and metal. The observation implied that in using the SP-QW coupling for enhancing emission in an LED, the metals for Ohmic contact and SP generation must be separated. A thin dielectric interlayer is required in the region for SP-QW coupling to avoid the leakage of SP energy.

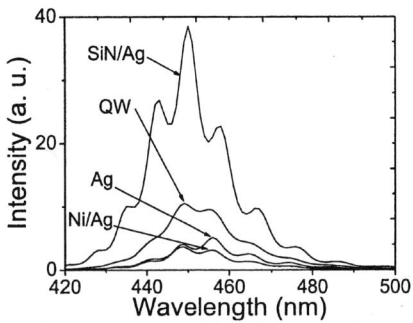

Fig. 3 PL spectra with different coatings on the p-type epitaxial structure in comparison with that of the bare-QW sample.

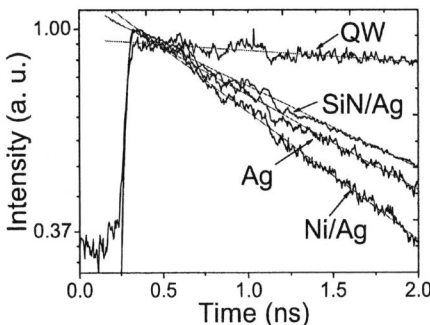

Fig. 4 The corresponding TRPL traces of the samples in Fig. 3.

ACKNOWLEDGEMENT

This research was supported by National Science Council, The Republic of China, under the grant of NSC 95-2120-M-002-012 and NSC 95-2221-E-002-287, and by US Air Force Scientific Research Office under the contract AOARD-06-4052.

TuP37
14:30 – 17:30

Temperature-dependent Behaviors of the Surface Plasmon Coupling with an InGaN/GaN Quantum Well

Yen-Cheng Lu, Cheng-Yen Chen, Dong-Ming Yeh, Chi-Feng Huang, Tsung-Yi Tang,
Jeng-Jie Huang, and C. C. Yang
Graduate Institute of Electro-Optical Engineering and Department of Electrical Engineering,
National Taiwan University, 1, Roosevelt Road, Section 4, Taipei, Taiwan, R.O.C.
(phone) 886-2-23657624   (fax) 886-2-23652637   (e-mail) ccy@cc.ee.ntu.edu.tw

Abstract

We demonstrate the temperature dependent behavior of the surface plasmon (SP) coupling with an InGaN/GaN quantum well (QW). The SP coupling efficiency relies on the availability of carriers with sufficient momentum for transferring the energy and momentum into the SP modes. At low temperatures, the carriers are trapped by the potential minima in the QW and the SP coupling is weak. As temperature increases, more and more carriers escape from the potential minima leading to the stronger and stronger SP coupling. When the temperature is close to the room condition, the SP coupling strength saturates because most carriers have escaped from the potential minima. The three temperature ranges of different SP coupling behaviors can be clearly identified from the data of photoluminescence (PL) enhancement ratio and PL intensity decay rate.

Keywords: surface plasmon, temperature-dependence, InGaN/GaN, quantum well

1 INTRODUCTION

The revisit to surface plasmon (SP) in the past several years has brought us with many potential applications, including the enhancement of light emission efficiency through the coupling between an SP mode and a radiating dipole such as those in a quantum well (QW) [1,2]. In the SP-QW coupling process, the energy and momentum of an exciton or a pair of carriers are transferred into an SP, particularly a propagating SP or surface plasmon polariton, for effective SP radiation. To provide the SP momentum, which can be several times larger than that of a photon, a carrier pair for SP coupling needs to have a certain momentum.

In this paper, we report the temperature-dependent behaviors of the SP coupling with an InGaN/GaN QW. It is found that the SP coupling efficiency increases with temperature. The possible mechanism behind this behavior, including the carrier delocalization from the potential minima in the QW, is discussed. Although the required momentum matching condition only needs the thermal energy corresponding to a few tens of Kelvin, the carrier delocalization process results in a significantly higher probability of SP-carrier momentum matching and hence SP-QW coupling.

2 SAMPLE PREPARATION

The InGaN/GaN QW sample was grown with metalorganic chemical vapor deposition. It consists of an InGaN/GaN QW sandwiched by a top GaN cap layer of 20 nm in thickness and the lower GaN cladding layer of 2 μm in thickness. This bare-QW sample is coated on top with a 50-nm Ag thin film at room temperature for creating SP coupling (called the Ag-coated sample).

3 EXPERIMENTAL RESULTS

For understanding the temperature dependence of PL spectral feature, we calculate the center-of-mass energies of the two samples at various temperatures and plot the results in Fig. 1 (the left ordinate). Here, one can see that both curves show the S-shape variations with temperature. In the low temperature range (10-90 K in this QW sample), carriers have little thermal energy and are basically trapped in the potential minima. Hence, the PL energy follows the typical temperature-dependent band gap shrinkage behavior. In this temperature range, the trapped carriers have little momenta. However, in the medium temperature range (around 90-180 K), the thermal energy of carrier results in the elevation of their average energy state level and hence the blue shift of the PL spectral feature. Also, in this temperature range, carriers start to escape from the potential minima to become free in motion. In other words, in this range, as temperature increases, more and more carriers gain significant momenta. Finally, in the high-temperature range (180-300 K), most carriers have sufficient kinetic energy (and momentum) for escaping from the potential trapping and becoming free in motion.

In Fig. 1, we also plot the ratio of the integrated PL intensity of the Ag-coated sample over that of the bare-QW sample (see the right ordinate). The ratio values represent the PL enhancement factors due to SP coupling. It is interesting to see that the temperature dependent behavior of the enhancement ratio can also be divided into the three aforementioned temperature ranges. In the low and high

1-4244-0641-2/07/$20.00 ©2007 IEEE   127

temperature ranges, the ratio is not significantly changes in varying temperature. However, in the medium temperature range, the enhancement ratio increases monotonically with temperature (from about 3.5 to 8).

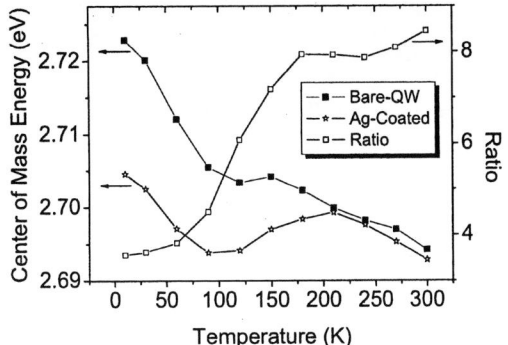

Fig. 1 Temperature dependencies of center-of-mass energy and integrated intensity ratio of the two samples.

Figure 2 shows the calibrated PL decay rates of the two samples as functions of temperature (filled squares and filled circles). Also, their differences are shown with the filled triangles. In the bare-QW sample, the decay rates are almost unchanged with temperature below 90 K due to carrier trapping by the potential minima. Beyond that and up to 210 K, it basically increases linearly with temperature due to the capture of carriers by the non-radiative recombination centers. Beyond 210 K, the increase of the decay rate saturates due to the almost complete carrier escape. On the other hand, in the Ag-coated sample, the temperature-dependent variation of the decay rate also follows the aforementioned pattern of the three temperature ranges. In the low and high temperature ranges, the decay rate changes slowly with temperature. In the medium temperature range, the decay rate increases almost monotonically with temperature. Such a variation trend can also be applied to the decay rate difference. Generally, the decay rate difference increases with temperature indicating that phonon does play an important role in the SP coupling.

4 DISCUSSIONS

For effective SP-QW coupling, the momentum mismatch between SP and carrier can be easily overcome with thermal energy above several tens Kelvin. However, when the carriers are trapped by potential minima, the coupling efficiency can be significantly decreased although their wave functions include certain spatial-frequency components for matching the momenta of SP and carrier. By assuming that the wavenumber of SP, k_{sp}, is three times larger than that of a photon, i.e., $k_{sp} \sim 3 \times (2\pi n/\lambda) \sim 0.1$ nm^{-1}. For an electron to have such a wavenumber, it requires a temperature of only

30K and even lower for a hole. Hence, sufficient momentum can be easily obtained for free carriers. However, for a trapped carrier, only a fraction of spatial-frequency components can provide such a momentum mismatch. Assuming that a trapped carrier has a Gaussian-form wave function of 5 nm in width (standard deviation), we can evaluate the mean value and standard deviation of the wave function in the k-space to give around 0 and 0.1 nm^{-1}, respectively. The probability of finding a carrier with the wavenumber larger than 0.1 nm^{-1} is less than 0.5 in a 2-D structure. Hence, at a certain temperature, although a pair of trapped carriers can couple with an SP mode, its probability is significantly lower than that of a pair of free carriers. In other words, the delocalization of carriers with increasing temperature can enhance the SP-QW coupling rate.

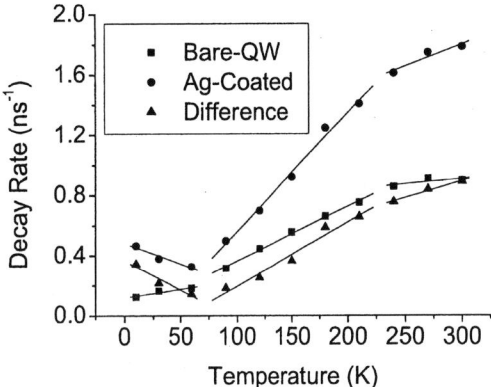

Figure 2 Decay rates and their differences as functions of temperature.

REFERENCES

[1] N. E. Hecker, R. A. Hopfel, N. Sawaki, T. Maier, and G. Strasser, *Appl. Phys. Lett.* **75**, 1577 (1999).
[2] A. Neogi, C. W. Lee, H. O. Everitt, T. Kuroda, A. Tackeuchi, and E. Yablonvitch, *Phys. Rev. B.* **66**, 153305 (2002).

ACKNOWLEDGEMENT

This research was supported by National Science Council, The Republic of China, under the grant of NSC 95-2120-M-002-012 and NSC 95-2221-E-002-287, and by US Air Force Scientific Research Office under the contract AOARD-06-4052.

TuP38
14:30 – 17:30

The Role of the Quantum-confined Stark Effect in an InGaN/GaN Quantum Well During its Coupling with Surface Plasmon for Light Emission Enhancement

Cheng-Yen Chen, Yen-Cheng Lu, Dong-Ming Yeh, and C. C. Yang

Graduate Institute of Electro-Optical Engineering and Department of Electrical Engineering,
National Taiwan University, 1, Roosevelt Road, Section 4, Taipei, Taiwan, R.O.C.
(phone) 886-2-23657624   (fax) 886-2-23652637   (e-mail) ccy@cc.ee.ntu.edu.tw

ABSTRACT

We analyze the contribution of the screening of the quantum-confined Stark effect (QCSE) to the light emission enhancement behavior in the surface plasmon (SP) coupling process with an InGaN/GaN quantum well (QW). From the measurements of excitation power-dependent photoluminescence and time-resolved photoluminescence (TRPL), and the fitting to the TRPL data based on a rate-equation model, it is found that when the excitation level is high, the QCSE screening effect not only contributes significantly to the emission enhancement, but also enforces the SP coupling process because of the blue shift of emission spectrum caused by the screening effect. Therefore, the emission strength from SP radiation, relative to that from QW radiative recombination, increases with the excited carrier density.

Keywords: surface plasmon, quantum-confined Stark effect, InGaN/GaN quantum well, time-resolved photoluminescence

1 INTRODUCTION

It has been shown that the coupling of surface plasmons (SPs) generated at the interface of a metal and a semiconductor with a nearby InGaN/GaN quantum-well (QW) could enhance the photoluminescence (PL) of the QW [1]. In calibrating the SP coupling effect, either the enhancement or the suppression of emission, we usually rely on the variation of PL intensity and the reduction of PL decay time in the time-resolved PL (TRPL) measurement. Besides, the quantum-confined Stark effect (QCSE), particularly in the current consideration of an InGaN/GaN QW, may play a crucial role in the coupling process. Normally, the coupling becomes stronger when the effective band gap of the QW is closer to the SP resonance frequency. Therefore, either PL enhancement or suppression is more significant on the high energy side of the PL spectrum. On the other hand, the existence of the piezoelectric field in an InGaN/GaN QW grown on c-plane sapphire substrate leads to the QCSE. When carriers are injected into a device fabricated with such a QW structure, the carrier screening effect reduces the QCSE leading to the emission blue-shifted and radiation efficiency increased. Hence, both the screening of the QCSE and the SP-QW coupling result in the similar phenomenon of the emission modification on the high-energy side of the PL spectrum. In this research, we intend to identify the contribution of the QCSE screening to PL enhancement in the coupling process of an InGaN/GaN QW with SP modes generated at the interface between a GaN cap layer and an Ag thin film by performing excitation

power dependent PL and TRPL measurements.

2 SAMPLE PREPARATION

The InGaN/GaN single-QW epitaxial structure was grown on (0001) sapphire substrate with MOCVD. After the growth of a buffer layer and a 2-μm GaN layer, a 3-nm InGaN QW layer was deposited. The QW was covered by a 20-nm GaN cap layer. Then, a 50-nm Ag film was then coated on the epitaxial sample for generating the SPs. The internal quantum efficiency of the QW sample was calibrated to give from 32 to 42 % when the excitation power increases from 5 to 25 mW.

3 EXPERIMENTAL RESULTS

The room-temperature PL and TRPL measurements were excited with the second-harmonic of a fs Ti:sapphire laser at 380 nm. Fig. 1 shows the PL spectra of the bare-QW and the Ag-coated samples at two excitation power levels: 5 and 25 mW. Here, one can see that the PL intensity is enhanced when the Ag thin film is coated in either the high or low excitation case. The high-energy side is more enhanced when compared with the low-energy side. Meanwhile, a higher excitation level leads to a larger blue shift range in PL spectrum. Such blue shifts are indicated with the variation of spectral center of mass, as marked with the grey dot on the four spectra of Fig. 1.

In Fig. 2, six curves show the photon energy-dependent PL decay times for the conditions as: curve A: low excitation (5 mW) in the bare-QW; curve B:

high-excitation (25 mW) in the bare-QW; curve C (D): the slow- (fast-) decay component of the Ag-coated sample with low excitation; curve E (F): the slow- (fast-) decay component of the Ag-coated sample with high excitation. In the Ag-coated sample, the non-single-exponential decay becomes clearer at high excitation that can be attributed to the recovery of QCSE from carrier screening. When the excitation is increased to 25 mW, the decay time is generally reduced, particularly significantly in the high-energy range. With SP coupling, the decreasing extent of decay time in increasing the excitation power is not as large as the case of bare-QW, particularly in the low photon energy range. By comparing the levels of curves A, B, and C, one can see that the decrease of decay time due to the Ag coating is significantly larger than that due to the five-time excitation power increase in the bare-QW sample.

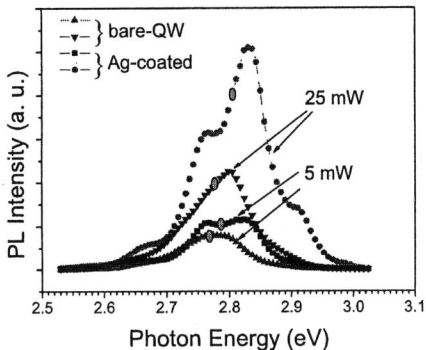

Fig. 1 PL spectra of the two samples at two different excitation power levels.

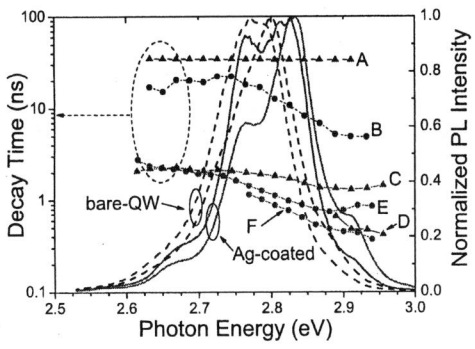

Fig. 2 Decay time and normalized PL intensity as functions of photon energy.

4 DISCUSSIONS

To further identify the role of QCSE screening in the SP-QW coupling process, we improve the rate-equation model we built before for calibrating various decay time constants in the coupling process [2]. The carrier density-dependent PL decay time (in ns) of the bare-QW sample was first calibrated from the excitation-power dependent TRPL measurements. The considerations of the excitation power dependencies of the internal quantum efficiency and QW radiative decay time imply the incorporation of the effect of QCSE screening into our calibration procedure. Besides, we consider the QW-SP coupling rate ($1/\tau_{SP}$) as

$$\frac{1}{\tau_{SP}} \propto \frac{1}{\tau_R} \cdot n_{SP}(t)^{\alpha}.$$

Here, τ_R and n_{SP} represent the QW radiative decay time and SP density, respectively. Also, α is a fitting parameter describing the saturation of SP coupling. The fitting parameters for best curve fitting are tabulated in Table I. Here, $\tau_{QW,min}$ represents the minimum value of τ_{QW}, which is defined as $1/\tau_{QW} = 1/\tau_R + 1/\tau_{NR}$ with τ_{NR} standing for the QW non-radiative decay time. Also, $\tau_{SP,min}$ represents the minimum value of the SP coupling time constant. The decreasing trend of $\tau_{QW,min}$ with increasing excitation level originates from the QCSE screening effect. This result can be partially attributed to the fact that the QCSE screening effect shifts the emission wavelength to become closer to the SP resonance energy, near which the SP density of state is higher. The increase of $R_{R/NR}$ means SPs are more radiative at higher excitation. It increases with the excitation level. The decrease trend of α indicates the saturation effect existing in the coupling process.

Table I

	5mW	15mW	25mW
$\tau_{QW,min}$ (ns)	38.7	23.0	12.5
$\tau_{SP,min}$ (ns)	10	5	3.6
$R_{R/NR}$	0.35	0.56	1.39
α	0.83	0.73	0.57
$R_{SP/QW}$	1.87	2.6	2.94

REFERENCES

[1] K. Okamoto, I. Niki, A. Shvartser, Y. Narukawa, T. Mukai, and A. Scherer, Nat. Mater. **3**, 601 (2004).

[2] C. Y. Chen, D. M. Yeh, Y. C. Lu, and C. C. Yang, Appl. Phys. Lett. **89**, 203113 (2006).

ACKNOWLEDGEMENT

This research was supported by National Science Council, The Republic of China, under the grant of NSC 95-2120-M-002-012 and NSC 95-2221-E-002-287, and by US Air Force Scientific Research Office under the contract AOARD-06-4052.

TuP39
14:30 – 17:30

Passivation of silicon wafer patterned by aluminum for micromachining

Ani Duan[1], Erik Poppe[2], and Xuyuan Chen[1,3]

[1]IMST, Vestfold University College, Raveien 197 Horten, 3103 Tonsberg, Norway
Tel +47 33031161, Fax +47 33031103, E-mail: Xuyuan.Chen@hive.no
[2]SINTEF ICT, Microsystems and Nanotechnology, Gaustadalleen 23 C, 0373 Oslo, Norway
[3]SAH MEMS Research Center, Xiamen University, Xiamen, China

Abstract

Low-pressure chemical vapor deposition (LPCVD) and plasma-enhanced chemical vapor deposition (PECVD) have been used for depositing silicon nitride (SiN) as passivation layer in microfabrications. SiN deposited by LPCVD and optimized PECVD can perfectly mask Si from etching attack in TMAH-water solution. After Al metallization, pinholes are always formed on SiN because of the Al crystal hillocks. Due to poor step coverage of PECVD SiN on Al, the edge etching of the Al patterns is the main reason for Si etching underneath of the Al patterns although the Al etching via the pinholes on SiN contributes. By structure designing and process tuning, we have achieved passivation techniques for micromachining after Al metallization.

Keywords: Passivation, LPCVD, PECVD, Silicon nitride, Micromachining, Wet etching

1 INTRODUCTION

After Al metallization, micromachining of MEMS/MOEMS becomes a serious problem for pattern transfer, structure generation, and formation of reliable electrical connection [1]. For developing advanced process to machine Si wafer with Al patterns, a thorough investigation of passivation method and the material is required.

2 LPCVD AND PECVD $Si_{3+x}N_{4-x}$ FOR PASIVATION

As the mask layer for Si etching in TMAH-water solution, LPCVD and PECVD $Si_{3+x}N_{4-x}$ (SiN) is often used because of its high etching selectivity. As shown in Fig. 1, SiN grown by LPCVD has high quality to resist the etching. However, anisotropic etching of Si happens under SiN grown by PECVD through the pinholes. After Al metallization, Due to

high process temperature (600°C) LPCVD can not be employed as the passivation technique. PECVD has to be used for SiN growth. Fig. 2 shows the sample of 500 nm PECVD SiN deposited on 1.2 um Al (sintered) on (100) Si substrate. The Al crystal hillock can be seen through the SiN layer. At temperature of 70°C for 3 hours in the 25% TMAH-water solution, both Al and SiN layers are etched. Etched Si surface appears (fig 2b).

Fig.2. SEM micrography of the sample surface. (a) before the etching and (b) after the etching.

3 PASSIVATION AFTER AL METALIZATION

In Fig. 3a, Si V-groove should be created after forming the Al patterns. Using poor quality PECVD SiN as passivation layer (fig.1b), both Al and SiN layers were etched away (fig. 2b). Therefore, process parameters of PECVD have been carefully tuned according the residual stress, film density, and pinhole distribution. We have succeeded in optimizing the quality of PECVD SiN to avoid the Si etching. In fig. 3c and 3d, the locations indicated by "SiN on Si" are perfectly

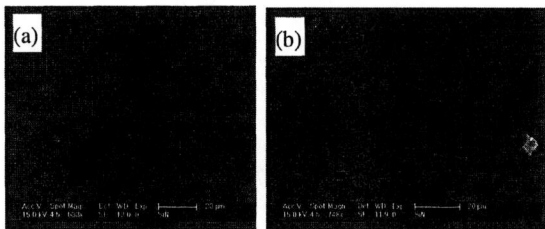

Fig. 1. SEM micrography of the etching surface in 25wt% TMAH-water solution at 80°C for 6 hours. (a) and (b) present 100 nm thick SiN grown by LPCVD and PECVD respectively on (100) Si substrate.

1-4244-0641-2/07/$20.00 ©2007 IEEE   131

passivated by PECVD SiN. There is no pinhole observed. However, the process is failure to generate the expected structure. Si underneath of the Al pattern passivated by PECVD SiN is etched (fig. 3c). Because of step coverage of SiN on Al lines, Al etching from the edge is found to be the main reason for silicon etching (fig. 3d). The quality of PECVD SiN on Al pattern can not be improved because pinholes are always created by Al crystal hillocks.

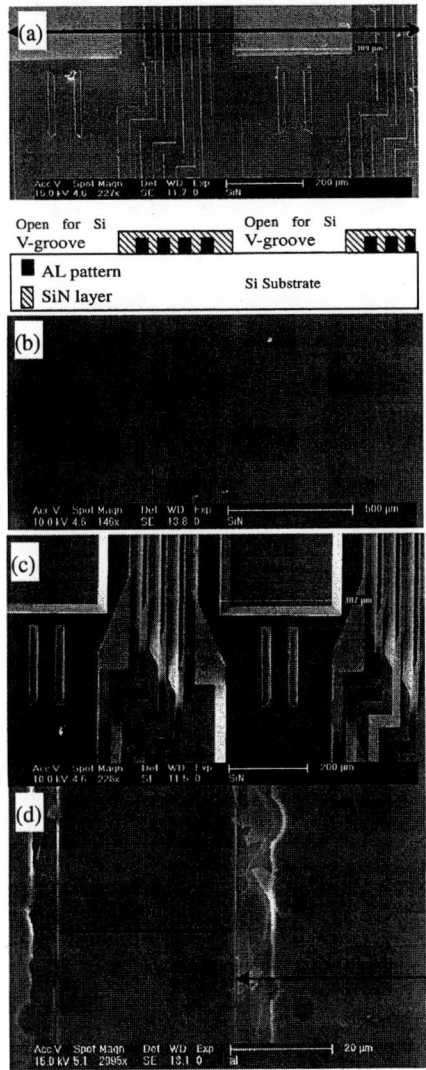

Fig. 3. The etching results in TMAH-water solution at 70°C for 30 min after Al metallization on Si substrate. (a) Top view and cross section of the sample. (b) No optimized SiN. (c) Optimized SiN, failure of structure generation (d) Low temperature etching for 30 min, Al etching starts from edge.

Fig. 4 shows the etching results for passivation with thermal SiO_2 underneath of SiN. The designed structure is created with damages observed on Al patterns (fig. 4b). Using PECVD SiN as passivation layer after Al metallization, protection of the Al during etching in TMAH-water solution must be considered. We have successfully developed the passivation technique for micromachining MEMS/MOEMS after Al metallization by optimizing PECVD process, Al etching protection, thermal SiO_2 application. A perfect result has been achieved as shown in Fig. 5.

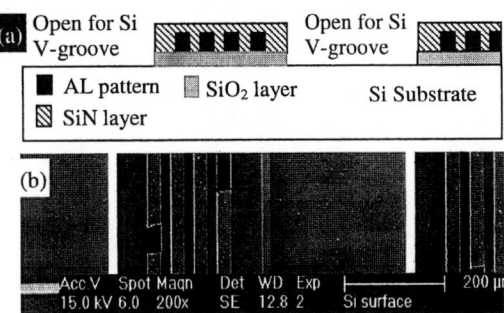

Fig. 4. The etching results in TMAH-water solution at 70°C for 30 min after Al metallization on SiO_2. (a) Cross section of the sample. (b) Optimized SiN, damaged Al lines.

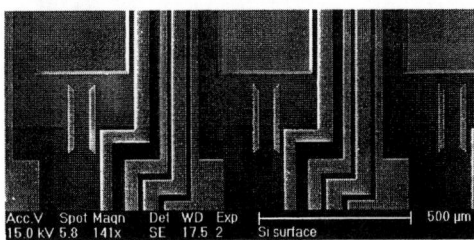

Fig. 5. Successful passivation technique after Al metallization on SiO_2.

4 CONCLUSION

SiN grown by LPCVD and optimized PECVD process can be used as passivation layer for Si etching in TMAH-water solution. After Al metallization, pinholes are formed on SiN because of Al crystal hillocks. Passivation using SiN can be done by optimizing PECVD process at 300°C, adding Si powder in etchant to avoid Al etching, and designs with application of thermal SiO_2.

REFERENCES

[1] O. Tabata, R. Asahi, H. Funabashi, K. Shimaoka, S. Sugiyama, Anisotropic etching of silicon in TMAH solutions, Sens. Actuators A34 (1) (1992) 51–57.

This page left intentionally blank

WEDNESDAY, 15 AUGUST 2007

WA   Microlenses

WB   Micro and Nano Lithography

WC   Spectroscopy

WD   Tunable Devices

WA1 (Invited)
08:30 – 09:00

MEMS-based microspectrometers for infrared sensing

C. A. Musca, J. Antoszewski, A. J. Keating, K. J. Winchester, K. K. M. B. D. Silva, T. Nguyen,
J. M. Dell, L. Faraone

Microelectronics Research Group, School of Electrical, Electronic & Computer Engineering
The University of Western Australia, Crawley, WA, 6009
Tel +618 6488 3787, Fax +618 6488 1095, E-mail charlie@ee.uwa.edu.au

Abstract

Micro-Electro-Mechanical Systems (MEMS)-based tunable optical filters, integrated with an infrared detector, select narrow wavelength bands in either the short-wavelength infrared (SWIR), or the mid-wavelength infrared (MWIR) region of the electromagnetic spectrum. The SWIR microspectrometer is based on monolithic integration of a parallel plate MEMS optical filter with a HgCdTe-based infrared detector. The fabrication process for the MEMS Fabry-Pérot filter and the integral HgCdTe detector is completed while maintaining the processing temperature less than 125°C, as the performance of HgCdTe based detectors degrades at higher temperatrues. The preliminary MWIR microspectrometer result was based on a hybrid approach, fabricating the filter separately from the HgCdTe detector, however the process temperature control were maintained during fabrication of the MWIR filter, ensuring migration of this technology into an integrated solution. A tuning range of 900 nm with linewidths of 210 nm have been achieved for the MWIR, while maintaining a relatively low tuning voltage of 17 V.

Keywords: Optical MEMS, infrared, microspectrometer, HgCdTe

1 INTRODUCTION

The combination of Micro-Electro-Mechanical Systems (MEMS) and high performance infrared (IR) detector technology offers a unique opportunity for new solutions to major challenges faced by a wide range of applications. By combining HgCdTe-based IR detectors with MEMS technologies, new systems for the detection of IR radiation, spectral data and spectral imaging will become available.

This work will describe the realisation of devices designed to select a specific wavelength band from within the short-wavelength infrared (SWIR), or the mid-wavelength infrared (MWIR) spectrum with the ability to electrically tune the selected wavelength. The fabrication process is compatible for monolithic integration of the MEMS filter on any semiconductor-based photon detector, but has been demonstrated for HgCdTe IR detectors. The test devices consist of HgCdTe-based photoconductors with either a monolithically integrated MEMS-based optical filter (SWIR), or a hybrid filter (MWIR) positioned between the detector and the radiation source. Figure 1 shows the integrated SWIR device. The tuning voltage is applied between the two mirrors, attracting the mirrors and thus altering the cavity length, d, of the Fabry-Pérot filter. It is important to note that all the process steps that have been developed for the tunable MEMS filter are compatible with device processing involving HgCdTe, although they can be readily adapted to any photon detector technology. The fabricated SWIR devices give a maximum tuning range of 440 nm with 8V tuning voltage, while the MWIR devices give a 900 nm tuning range with applied voltage of 17V.

2 DEVICE FABRICATION

The dominant technological constraint that needs to be

adhered to in the device processing, is that all aspects of the monolithic integration of the HgCdTe detector with the MEMS filter must be implemented at a low enough temperature that is compatible with the maximum allowable thermal budget for HgCdTe. The process technology for the integrated SWIR device has previously been reported [1].

For the MWIR microspectrometer a hybrid approach was employed, with two subcomponents required – the HgCdTe photoconductive detector and the MEMS tunable filter. The photoconductor was fabricated using n-type HgCdTe with a room temperature cutoff wavelength of 4.5 μm. The MEMS tunable filter was fabricated on silicon, and even though hybridization alleviates the requirement that the thermal budget be compatible with HgCdTe detectors, all process

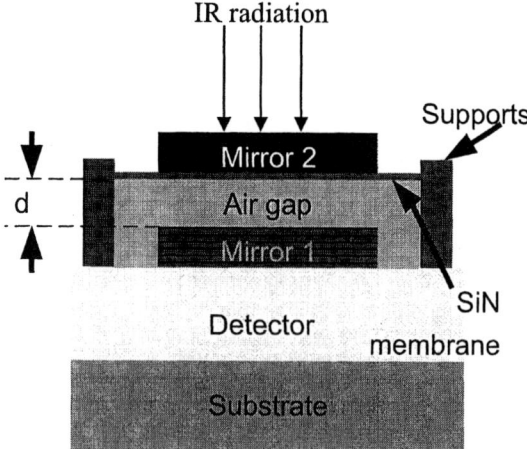

Figure 1. General concept of a Fabry-Perot filter formed on a detector

1-4244-0641-2/07/$20.00 ©2007 IEEE   137

temperatures were kept below 125°C to ensure possible future monolithic integration of the filter with the detector.

Fabrication of the MEMS filter was begun with deposition of a 500 nm layer of SiN_2 on bare un-doped silicon substrate. The lower electrode for the filter was formed by the deposition of Cr/Au layer over the SiN_2 layer, with a central $90 \times 90 \ \mu m^2$ opening as the optical aperture. The bottom distributed Bragg reflector (DBR) mirror was formed by thermal evaporation of Ge/SiO/Ge and liftoff to define features. The DBR was designed to be a quarter-wavelength at the design wavelength of 4 μm. 2.4 μm of polyimide was then used as the sacrificial layer to define the separation between the bottom and top DBR mirrors. The SiN_2 membrane was then deposited to support the top mirror. Both the top DBR mirror and the SiN_2 membrane are deposited at temperatures below 125°C. Extensive studies of low temperature SiN_2 deposition in a plasma enhanced CVD system has resulted in a set of conditions that provide structures with low internal stress allowing for significant reduction in actuation voltage. A top view of the fabricated filter structure is shown in Figure 2. Following hybridization of the MEMS filter and the HgCdTe detector the structure was assembled into a modified TO-8 package incorporating a 2-stage thermoelectric cooler.

3 CHARACTERISATION

The spectral curves across the tuning range for both the SWIR and MWIR spectrometers are shown in Figure 3. Tuning in the MWIR of 900 nm was achieved using 17V, which is almost 50% of the 3 – 5 μm MWIR region. The SWIR monolithic microspectrometer resulted in a tuning

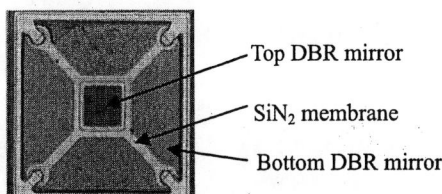

Figure 2. Optical photograph of the fabricated filter.

range of 400 nm with only 8V. Tuning in both bands was limited by voltage actuation and the maximum deflection of 1/3 of the gap in the filter allowable before snap down [3].

4 CONCLUSION

This paper presented brief details of the fabrcaition of MWIR and SWIR microspectrometers. Tuning of the filters across the MWIR and SWIR was presented.

5 ACKNOWLEDGMENTS

The authors would like to thank the Australian Research Council (ARC) and the Australian DSTO for financial assistance.

REFERENCES

[1] C. A. Musca et al., "Monolithic integration of an infrared photon detector with a MEMS=based tunable filter," IEEE EDL, vol. 26., pp.888-890, 2005.

[2] A. J. Keating, et al., "Optical characterization of Fabry-Perot MEMS filters integrated on tunable short-wave IR detectors," Photonic Lett. Technol., vol. 18, pp. 1079 – 1081, 2006

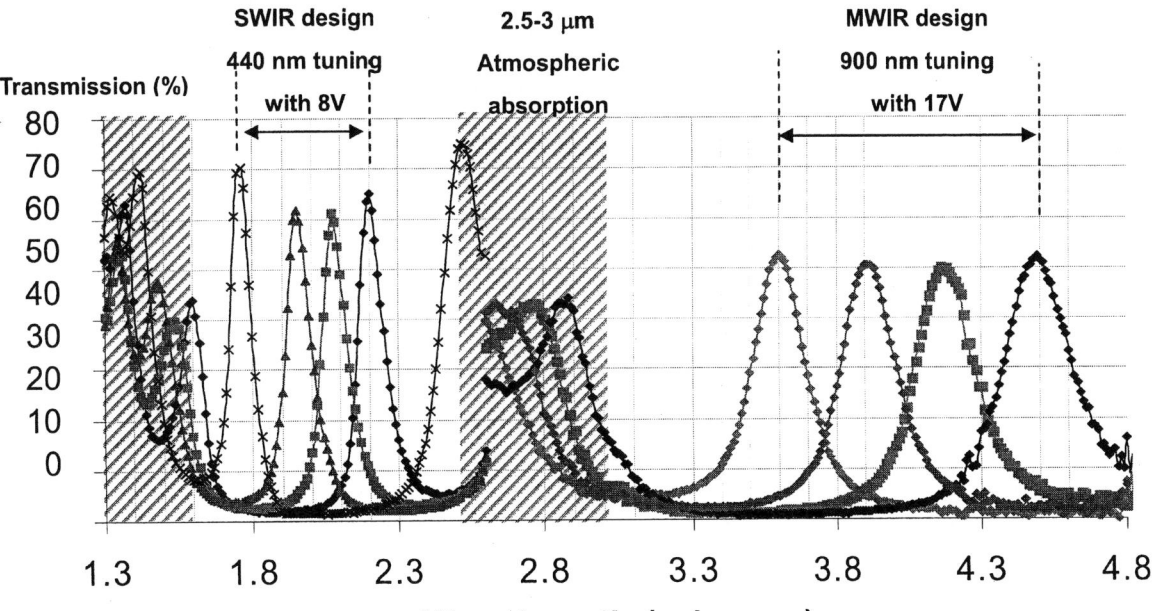

Figure 3. Spectral tuning data of microspectrometers spanning the SWIR and MWIR

WA2
09:00 – 09:15

Fabrication and Characterization of a Repositionable Liquid Micro Lens System

Riyaz P. Shaik, Leif Lasinger, Florian Krogmann, Wolfgang Mönch, and Hans Zappe

Department of Microsystems Engineering – IMTEK, University of Freiburg

Georges Koehler Allee 102, 79100, Freiburg, Germany.

Tel +49-761-203-7572, Fax +49-761-203-7562, E-mail riyaz.shaik@imtek.uni-freiburg.de

Abstract

A novel method to improve both the lateral positioning and focal length tuning accuracy of a liquid lens in a micro lens system, based on the electrowetting on dielectrics (EWOD) principle, is described. This system consists of a device, encapsulated with a packaging technique, fabricated on a transparent substrate featured with an array of electrodes and a structured surface whose periodicity is smaller than the electrode's dimension. With variable applied voltage signals, the liquid lens is actuated precisely in the lateral position and also tuned dynamically irrespective of the lens position on the substrate. The accuracy of lateral positioning is mainly dependent on the structure of the dielectric surface. Measurements of the focal length tuning range and the positioning accuracy are presented. A first design of reconfigurable micro lens system, which has diverse application as adaptive wave-front sensing, optical tweezers and etc, is demonstrated.

Keywords: EWOD, Electro-wetting on Dielectrics, Optical Microsystems, Variable focal length, Micro-lens array.

1 INTRODUCTION

Electro wetting on Dielectrics (EWOD) is an actuation mechanism which employs the change of the effective interfacial energy at the three phase contact line (TCL) of a liquid droplet on a surface as a function of the applied voltage, which results in change of the contact angle and thus incites movement of the droplet. EWOD has become a useful approach for the manipulation of liquid droplets. In micro-optical applications, EWOD may be used for focal length tuning of liquid micro-lenses. In recent years many papers have been published on tuning the focal length of single liquid lenses using this mechanism [1]. We present here a repositionable micro lens system in which the lenses can be precisely positioned laterally, and also the focal length tuned dynamically independent of the lens position on the chip [2] in a closed, packaged system.

Generally if a moving droplet system requires a high positioning accuracy, it requires a high density of electrodes. Fabrication of a large number of electrodes in a micro scale is not difficult, but controlling the potential on each electrode electrically is quite complicated. To avoid this limitation, and for achieving a high positioning accuracy, a system design has been developed which provides a control of the position of the liquid lens very precisely. The technique uses the droplet pinning at defined barriers of a grid etched into the surface, which holds the droplet at one position. By applying a voltage pulse on the electrode beyond the droplet, the droplet overcomes the barrier of the grid and comes to rest at that position. By applying a sequence of pulses, the droplet shifts to the biased electrode, which results in lateral movement towards the biased

electrode. After the droplet is shifted to the desired position, focal length can be tuned by applying the equal potential to all surrounding electrodes under the droplet.

2 THEORY

In EWOD, by applying an electric voltage between the conductive liquid droplet and the electrode under the dielectric layer, surface charges are produced at the contact line, which can induce both the reduction of contact angle and a motion of the entire droplet. This principle is mainly based on the effect of electrowetting which was described by B. Berge [3] in 1993. According to Lippmann, the contact angle of a liquid droplet as a function of applied voltage can be calculated by equation 1.

$$\cos \theta(V) = \cos\theta(0) + \frac{\varepsilon\varepsilon_0 V^2}{2\gamma d} \qquad (1)$$

Where V is applied voltage, d is thickness of dielectric, ε_0 is the permittivity of vacuum, ε is the dielectric constant, γ is the surface tension of the liquid, and $\theta(0)$ is initial contact angle. The focal length f of a spherical liquid micro-lens placed on the planar substrate can be calculated from the elementary geometry, Eq (1) and the "Lensmaker's formula" [4] and is given by

$$f = \frac{(n-1)D}{2\sin\theta(U)} \qquad (2)$$

Here, D is the diameter of the micro-lens, n is the refractive index, and the surrounding medium is air.

3 FABRICATION

A schematic view of our liquid micro-lens system is shown in Fig. 1. The system is fabricated using a Pyrex wafer (1)

1-4244-0641-2/07/$20.00 ©2007 IEEE

with buried transparent 500 μm x 500 μm Indium Tin Oxide (ITO) electrodes (2). These electrodes have zig-zag shaped edges to facilitate the movement of a liquid lens across the next electrode edge. On top of the ITO electrodes, a 800 nm PECVD SiO2 layer is deposited on the wafer (3). A grid structure with a mesh size of 50 μm x 50 μm x 300 nm is etched down into the dielectric layer (3). On top of the grid, an ITO droplet contact is sputtered (4). Finally, a hydrophobic layer is deposited by silanization on top of the dielectric grid and the ITO droplet contact. Afterwards the wafer was diced and the individual chip were glued on a PCB Board and wire bonded to obtain electrical contact which is made between the ITO droplet contact and the ITO bottom electrodes (5). For sealing the system hermetically, a housing was constructed which consisted on a sealing ring, a metal plate and a Pyrex window. Using screws the metal plate and the Pyrex window was pressed to the sealing ring which was placed on the device. By closing the system within a reservoir of surrounding liquid a hermetically closed system, filled with a surrounding liquid and the lens liquids without inclusions of air bubbles was achieved

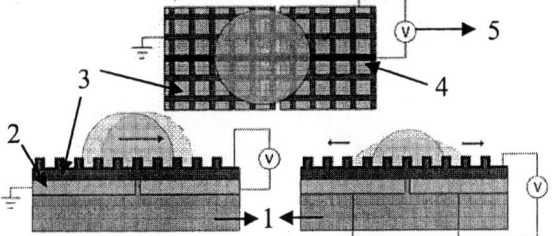

Figure 1. A liquid micro-lens system with lateral positioning and focal length tuning capabilities is illustrated. 1. Pyrex wafer, 2. ITO electrodes, 3. Dielectric layer with a grid, 4. ITO droplet contact, 5. Applied voltage

4 CHARACTERIZATION

4.1 Lateral Positioning Accuracy

For achieving the lateral positioning accuracy, a series of 70 V, 150 ms long DC voltage pulses was applied between the droplet and the electrode towards which the liquid lens is to be moved. Stills of a video sequence are shown in Fig 2. One can see that each voltage pulse moves the droplet one step more in the direction of the activated electrode. Only from step 5 to 6 a voltage of 200 V was needed, where the droplet reached the subsequent electrode which was not activated.

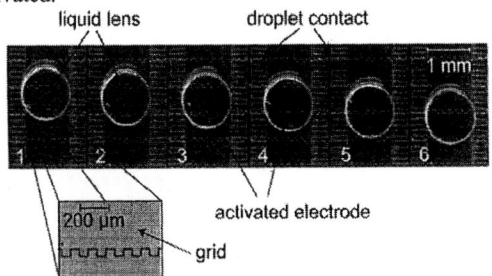

Figure 2. Lateral positioning accuracy of a liquid lens.

4.1 Focal length tuning

Tuning of the focal length is done by applying the same potential to all electrodes where the droplet is placed. With various voltage levels different focal lengths were achieved and are shown in Fig 3.

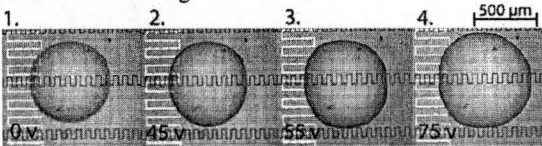

Figure 3. Back focal length tuning of a liquid lens in a closed micro-lens system.

The back focal length of a liquid lens in a closed system was measured as a function of the applied voltage from the top of the glass cover of the system to the focus produced by the lens with an optical microscope. Consequently from the diameter of the liquid lens, the contact angle was extracted. A graph which describes the experimental results and relates to the Fig 3 is shown in Fig 4. These experimental results correspond well with the Lippmann's model calculated mathematically, whereas d = 1.85 μm was considered in the calculations.

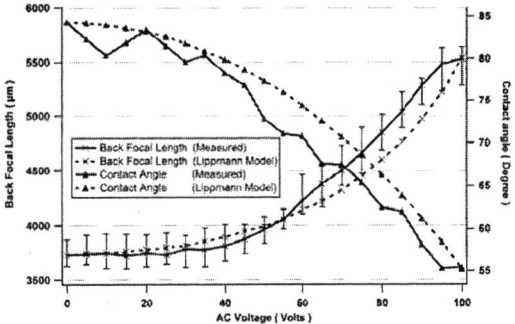

Figure 4. Experimental results of back focal length and contact angle of a closed micro-lens system compared with Lippmann model's back focal length and contact angle.

In summary, 600 μm diameter liquid lenses could be arbitrarily positioned in a range of 1000 μm with an accuracy of 70 μm on a substrate, and subsequently tuned in focal length in the range of 3.7 to 5.5 mm.

5 ACKNOWLEDGEMENT

The authors gratefully acknowledge financial support from the Landesstiftung Baden-Württemberg in the project 'ALAVARA'.

REFERENCES

[1] B Berge and J Peseux. "Variable focal lens controlled by an external voltage: An application of electro-wetting" European Physical Journal E, Volume 3, 159 -163, 2000

[2] F Krogmann, R Shaik, W Mönch, H Zappe. "Repositionable liquid lens with variable focal length", Proceedings of IEEE MEMS 2007, 21 – 25, 2007

[3] B Berge, "Electrocapillarité et mouillage de films isolants parlèau". C.R. Acad. Sci Paris, 317, 157-163, 1993.

WA3
09:15 – 09:30

A Lateral-shift-free LVD Microlens Scanner for Confocal Microscopy

L. Wu and H. Xie

Department of Electrical & Computer Engineering, University of Florida, Gainesville, FL 32611, USA

Tel: 1-352-392-1049, Fax: 1-352-846-1416, E-mail: leiwu@ufl.edu

Abstract: We report a lateral-shift-free (LSF) large-vertical-displacement (LVD) microlens scanner for tunable focusing applications such as confocal microscopy. A polymer microlens is integrated into a lens holder actuated by a LSF-LVD microactuator. The focal plane of the microlens can be vertically displaced 0.7mm at only 7.5V. The observed maximum lateral shift and tilt of the microlens during the entire vertical actuation are only about 13μm and 0.74° respectively. The resonance frequency of the vertical motion mode is 488Hz.

Keywords: Lateral shift free, Large vertical displacement, Bimorph, Tunable microlens, Confocal microscopy

1 INTRODUCTION

Microlenses with large tunable range of the focal plane are required for biomedical imaging technologies such as confocal microscopy and optical coherence microscopy, in which vertical displacements up to millimeter ranges are desired. This can be obtained either by vertically actuating a microlens or tuning its focal length [1]-[2], but the small scanning range (tens of microns) and high driving voltage limit the application in endoscopic imaging.

A large-vertical-displacement (LVD) actuation concept has been proposed [3]. The basic idea of LVD actuation is to convert the large vertical displacement at the tip of a rotating beam into a piston motion by connecting the other oppositely rotating beam to cancel the angular motion, as illustrated in Fig. 1(a). A maximum 0.7-mm vertical scan range at 23V driving voltage has been demonstrated [4]. However, there are some inherent disadvantages of the current LVD design based on two bimorph actuators. During the vertical actuation, the lens holder shifts laterally (Fig. 1(a)). For instance, for a 0.71-mm vertical scan, the lateral shift is as much as 0.42 mm [4], which is not desired for confocal imaging or many other scanning applications. This paper presents a three-bimorph based LVD microlens scanner that can generate large piston motion with almost no lateral shift.

2 LSF-LVD MICROLENS SCANNER

A new lateral-shift-free (LSF) LVD actuator design that consists of three bimorphs and two frames is proposed. As shown in Fig. 1(b), upon vertical actuation, the two frames shift laterally but in the opposite directions. Consequently, the lateral shift can be compensated by properly designing the lengths of the frames and bimorphs. The top view of the device design is shown in Fig. 1(c). The polymer lens is integrated into a hollow lens holder that is actuated by four sets of identical LSF-LVD actuators. Each actuator has three sets of Al/SiO₂ (~1μm thick each layer) bimorph beams with two frames connected in between. A Pt heater (~0.2μm thick) is embedded all along the beams and frames for uniform

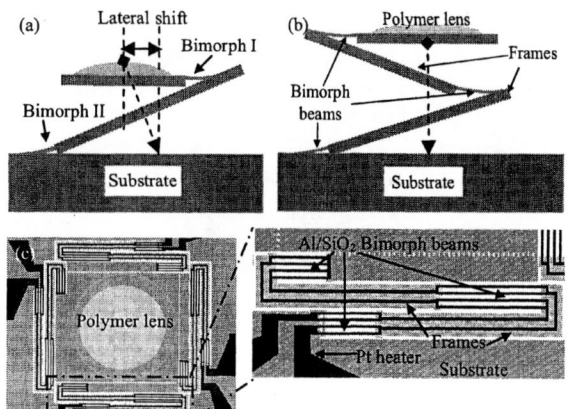

Fig. 1. (a) Two-bimorph LVD actuator with large lateral shift. (b) Three-bimorph LVD actuator with no lateral shift. (c) Top view of a three-bimorph LVD design.

heating. The frames and lens holder are flat due to the thick silicon layer underneath. The device was fabricated using a surface-bulk combined micromachining process. The fabrication process is similar to the one reported in [4]. Particularly, SOI wafers were employed. The 25μm-thick silicon device layer was used to form the hollow lens holder and the frames, while the buried SiO₂ layer (1μm thick) of SOI was kept at the opening of the lens holder. Thus a 25μm-deep and 0.6mm-diameter transparent reservoir was formed for holding the polymer lens (Fig. 2(b)). Polymer droplets were precisely dispensed into the transparent reservoir using a nanoliter-injection system and then baked in an oven to form the microlens due to the surface tension [3]. The initial elevation of the lens holder is 545μm. Fig. 2 shows some SEMs of a fabricated device before and after the polymer lens was formed.

3 DEVICE CHARACTERIZATION

An imaging experiment was performed to test the image quality of the integrated polymer lens. Fig. 3(a) shows the schematic of the test setup. The image generated by the

1-4244-0641-2/07/$20.00 ©2007 IEEE

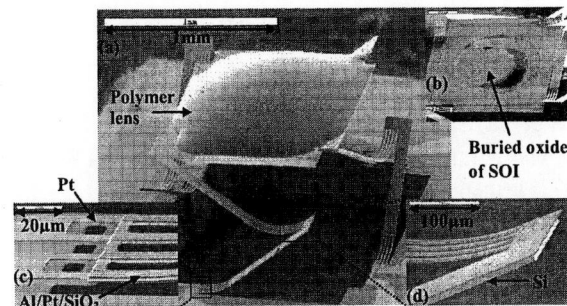

Fig. 2. SEMs of (a) a fabricated LSF-LVD microlens scanner, (b) transparent lens holder before the polymer lens being integrated, (c) Pt heater and Al/SiO2 bimorh beams, and (d) the frame with Si underneath.

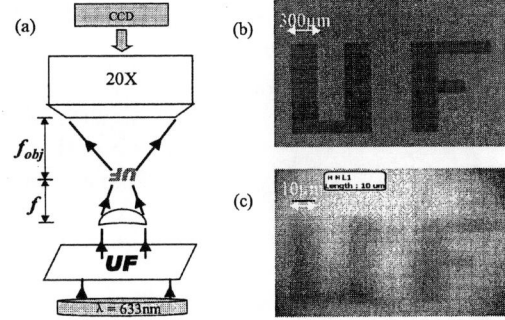

Fig.3. Imaging experiment. (a) Schematic of imaging experiment. (b) Photograph of mask. (c) Image of the test pattern generated by polymer lens

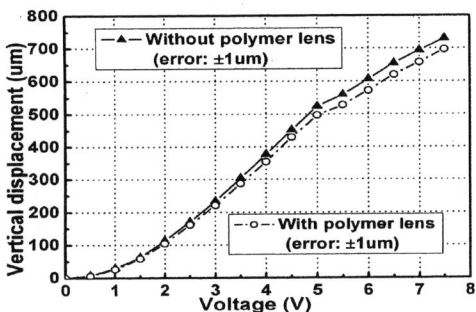

Fig.4. Vertical displacement v.s. applied voltage

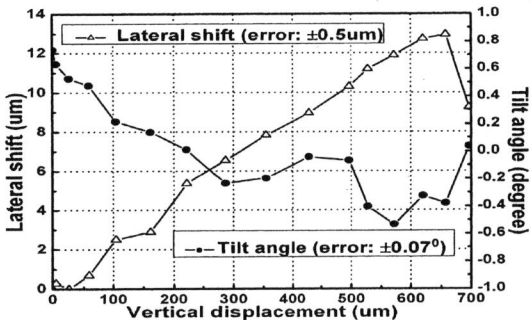

Fig.5. Lateral shift and tilt angle v.s. vertical displacement

polymer lens is shown in Fig. 3(c), where the resolution of ~5μm can be resolved without distortion. The focal length is measured to be 2mm with a depth of focus of ~0.2mm.

The LSF-LVD actuation was characterized by applying a single d.c. voltage to all the four actuators simultaneously. Large vertical displacements of 730μm and 696μm have been obtained at only 7.5V dc (323mW) from a same device before and after the polymer lens is integrated respectively, as shown in Fig. 4. Fig. 5 shows the measured lateral shift and tilt angle. The maximum lateral shift at the center of the polymer lens is only 13μm (<2% of the vertical scan range) and the maximum tilt angle is 0.74° (also the initial tilt angle) during the entire actuation range. The resonance frequency of the purely vertical motion mode is measured to be 761Hz and 488Hz before and after the lens is integrated. Fig. 6 shows a series of microscopic pictures (top view) of an LSF-LVD microlens at different actuation voltages, where no obvious lateral shift was observed.

4 CONCLUSION

A new microlens scanner based on a LSF-LVD microactuator has been presented. A 0.6mm-diameter integrated polymer lens with a 2.5mm device footprint can be vertically actuated as large as 0.7mm by only 7.5V dc (323mW) with a resonant frequency of 488Hz. The maximum lateral shift and tilt angle during the entire scanning range are only 13μm and 0.7° respectively. The LSF-LVD microlens scanner is very suitable for endoscopic confocal microscopy application. This work is supported by the National Science Foundation under award number BES-0423557 and Florida Photonics Center of Excellence.

REFERENCES

[1] S. Kwon, V. Milanovic, and L. P. Lee, "Large-displacement vertical microlens scanner with low driving voltage," *IEEE Photon. Technol. Lett.,* 14, pp. 1572-1574 (2002).

[2] T. Krupenkin, S. Yang, and P. Mach, "Tunable liquid microlens," *Appl. Phys. Lett.,* 82, pp.316-318 (2003).

[3] A. Jain, and H. Xie, "A Tunable Microlens Scanner with Large-Vertical-Displacement Actuation," IEEE MEMS'05, Miami Beach, Florida, Jan. 2005, pp. 92-95.

[4] A. Jain and H. Xie, "Microendoscopic confocal imaging probe based on an LVD microlens scanner," *IEEE J. Sel. Topics Quantum Electron.,* 13, no.2, pp.228-234 (2007).

Fig.6: Microscopic pictures of the microlens being actuated at (a) 0V, (b) 4V and (c) 6V.

WA4
09:30 – 09:45

Implementation of CMOS-MEMS Compound Lens

Chuanwei Wang[1], Sz-Yuan Lee[2], Chih-Ming Sun[2], Ming-Han Tsai[2], and Weileun Fang[1,2]
[1] Power Mechanical Engineering, [2] MEMS Institute.
National Tsing Hua University, HsinChu 300, Taiwan
Tel +886-3-574-2923, Fax +886-3-573-9372, E-mail fang@pme.nthu.edu.tw

Abstract

This study demonstrates the possibility of implementing lens system via CMOS process. The ever first CMOS based optical focusing system successfully demonstrated focusing capability, with 5um by 6um focus spot size at 614um, and capable of moving focal point 12um in axial direction.   This system is composed of a lens set and an electrothermal actuated optical bench. One PDMS dispensed planar-convex lens and CMOS made Fresnel lens forms lens system to achieve focusing power as NA equals 0.38. This kind of system has potential to monolithically integrate with optical sensing circuitry with low cost.

Keywords: CMOS, PDMS, Fresnel lens

1 INTRODUCTION

MOEMS are successfully demonstrated in various applications by means of manipulation of moving mechanical devices [1]. Consequently, various fabrication processes have been investigated to realize and to integrate micromachined components with different mechanical characteristics [2,3]. The CMOS MEMS optical devices have advantages over traditional MEMS ones due to the fine line width and monolithic integration of sensing circuitry on chip [4]. However, there are very few CMOS MEMS optical systems have been presented.

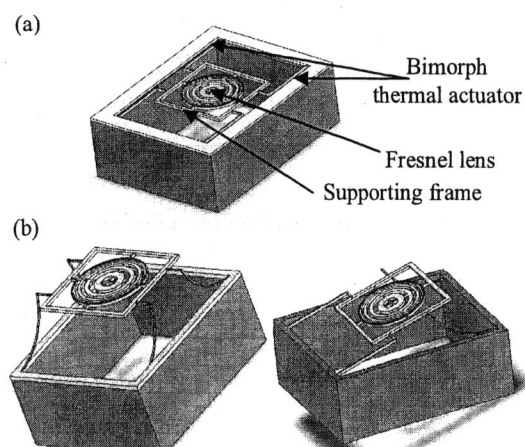

(a)

Bimorph
thermal actuator

Fresnel lens
Supporting frame

(b)

Fig. 1: The Structure of the Fresnel lens

2 FABRICATION AND RESULTS

The present optical system consists of a Fresnel lens, supporting frame, and four thermal actuators as illustrated in Fig.1a. The thermal actuator attached to the supporting frame will bend in the out-of-plane direction after Joule-heating. These actuators can be controlled individually for focusing and tilting of the Fresnel lens, as in Fig.1b. The Fresnel lens was designed according to equal phase difference and the designed effective focusing length is 1000μm. Another novel design is the tungsten-via is embedded between metal layers to prevent the incidence laser scattering.

To implement the present device, the fabrication includes the TSMC 0.35μm 2P4M CMOS process (Fig.2a), and the post-CMOS processes (Fig.2b-d). In Fig.2b, after the protective polymer was patterned, the backside of the substrate was etched by the ICP. In Fig.2c, the RIE anisotropically etching was employed to remove the dielectric materials. Finally, the XeF₂ isotropically etching was used to release the structure (Fig.2d).

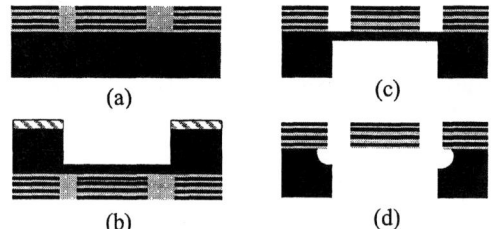

(a)   (c)

(b)   (d)

Fig 2. The post fabrication process

Fig.3a shows the SEM photo of a fabricated Fresnel lens system. The zoom-in SEM photo in Fig.3b shows the 1μm minimum line width after the post-CMOS process. Fig.4 shows the integration of Fresnel lens and polymer ball lens.

1-4244-0641-2/07/$20.00 ©2007 IEEE   143

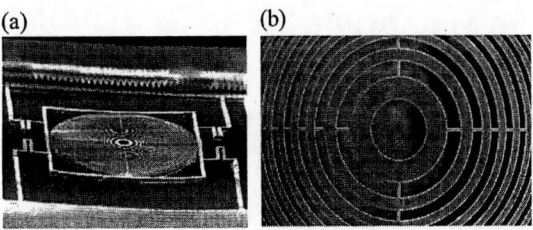

Fig. 3 The SEM photos of the Fresnel lens

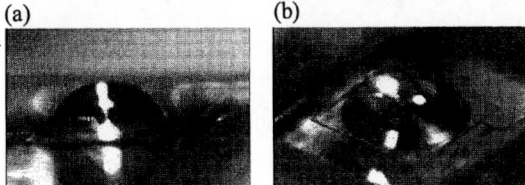

Fig. 4 The photos of the lens after curing ball lens

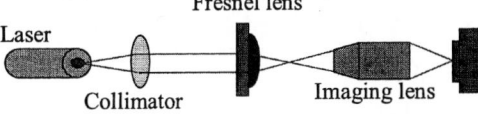

Fig. 5 The test setup

In Fig.5, He-Ne Laser (with wavelength of 632.8nm) is used as light source and a CCD beam profiler is placed in focal plane of imaging lens. Due to Gaussian Lens Formula, the focal length can be measured by moving tested device forth and back along optical axis. Simultaneously, the size of focus point can be determined via the spot size captured by CCD beam profiler and magnification of this system. The preliminary results show the Fresnel lens has focal length of 1987μm with spot size of 6.32μm by 7.69μm, as in Fig 6a. In addition, Fig.6b shows the images of focal point for Fresnel/PDMS compound lens in Fig. 4b.

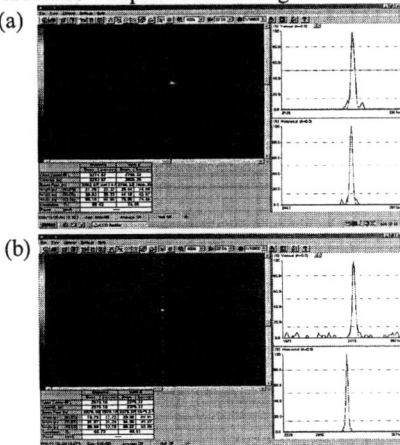

Fig. 6 The measurement results

The compound lens has focal length of 641μm, and its focal point size is of 5.05μm by 6.01μm. The flatness of Fresnel lens is measured by optical interferometer. In Fig.7, the result shows the maxima difference happened between edge and central point of Fresnel lens is 5μm. The actuating behavior was also characterized by interferometer while applying current to four thermal actuators. In Fig.8, the deflection versus temperature relation is plotted. The whole lens set departed from its original position for 12μm along optical axis direction from room temperature to 225 °C.

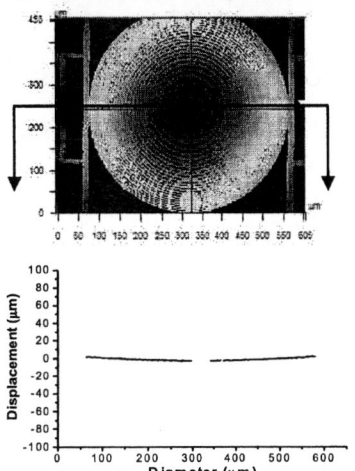

Fig. 7 The surface topology of the CMOS Fresnel lens

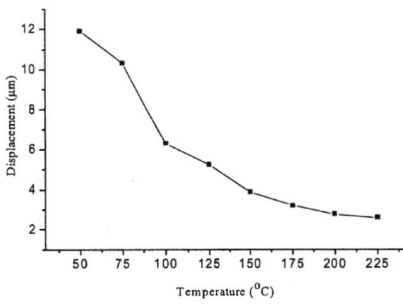

Fig. 8 The measured thermal actuator displacement

REFERENCES

[1] M.C. Wu, Proceedings of IEEE, vol. 85, pp. 1833-1856, 1997.
[2] J.M. Bustillo, R.T. Howe, and R.S. Muller, Proceedings of IEEE, vol. 86, pp. 1552-1574, 1998.
[3] V. Milanović, J. of MEMS, vol. 13, pp. 19-30, 2004.
[4] G. .K. Fedder; IEEE SENSOR, 30 Oct.-3 Nov. 2005
[5] T. Xie, H. Xie, G.K. Fedder, and Y. Pan, Electronics Letters, . Vol. 39, Issue 21, Oct. 16, pp. 1535 – 1536, 2003.

WA5
09:45 – 10:00

Low Cost Adaptive Silicone Membrane Lens

F. Schneider[+], C. Müller[*] and U. Wallrabe[+]

University of Freiburg – IMTEK, Department of Microsystems Engineering,
[+] Laboratory for Microactuators, [*] Laboratory for Process Technology
Georges-Koehler-Allee 102, 79110 Freiburg, Germany
Tel +49-761-203-7282, Fax +49-761-203-7439, E-mail schneider@imtek.uni-freiburg.de

ABSTRACT

We present the first adaptive silicone membrane lens with an integrated piezoelectric actuator. The system consists of two components, lens (5 mm diameter) and actuator (15 mm diameter). Both components are cast in a hot embossing machine to ensure a reliable and cheap fabrication. At a voltage variation of 55 V we obtain a focal length between 30 and 500 mm. Furthermore the lens shows a maximum resolution of 100 line pairs per millimeter at a focal length of 100 mm.

Keywords: adaptive lens, piezoelectric actuator, silicone

1 INTRODUCTION

Small cameras have become a key feature for many consumer devices. Currently, optical lenses in cameras are made of solid material and have fixed focal distance and lens shapes. Hence, auto-focusing and zooming rely on a change of the lens position(s) and interlens distance. Adaptive fluidic lenses eliminate expensive and sensitive mechanical moving parts in optical systems [1,2]. This paper introduces low cost adaptive membrane liquid lenses with a diameter of 5 mm including a piezoelectric actuator as a pump. We discuss the optimization of the pump and present experimental results on the lens quality and the full system behavior.

2 ASSEMBLY AND FABRICATION

The system consists of two main parts; a lens and an actuator (fig. 1). The lens chamber, which is to be filled with water or oil, consists of a silicone supporting ring and a membrane. The pump chamber consists of a bending actuator which is also embedded in silicone. The originally disc shaped piezo with a diameter of 20 mm and a thickness of 220 μm is laser cut to form a ring structure.

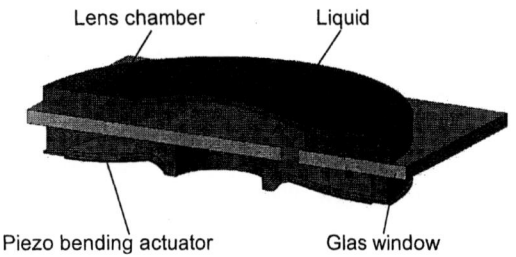

Fig. 1: Liquid lens with piezo-actuator

Both fluidic chambers are bonded to a glass plate, featuring several holes for the fluid exchange between the chambers.

Both chambers are cast in a hot embossing machine in optical quality (fig. 2). The membrane thickness can be varied easily by changing the spacer thickness.

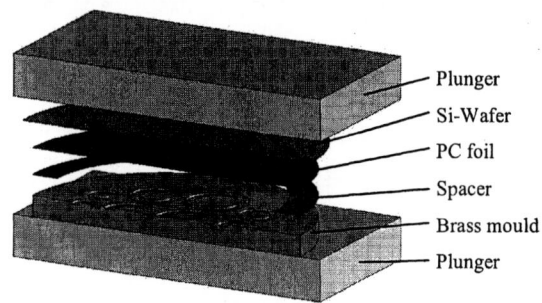

Fig. 2: Lens chamber process

3 LENS CHARACTERIZATION

The liquid lens is characterized by measuring the modulation transfer function (MTF) with the USAF 1951 test target. For this purpose, an external hydrostatic pressure from a water column is used to steer the lens. For lens optimization the membrane thickness, the aperture and the focal length are varied. The optical measurement setup is shown in figure 3. We characterize the liquid lens in parallel ray optics; thereby the liquid lens compensates the negative focal length of a plano-concave lens. For simplification we use a lens revolver, which contains eight plane-concave-lenses with focal length between -1000 mm and -30 mm. Dependent on the focal length ratio of the two achromates we obtain a 10 times amplification between test target and camera 1. Camera 2 and the beam splitter are used to measure the diameter of the iris aperture, which is important for the lens quality.

Fig. 4 shows the lens resolution as a function of the focal length for diverse membrane thicknesses and the apertures. We observe a maximum lens resolution for a membrane thickness of 150 μm at 100 mm focal length.

1-4244-0641-2/07/$20.00 ©2007 IEEE   145

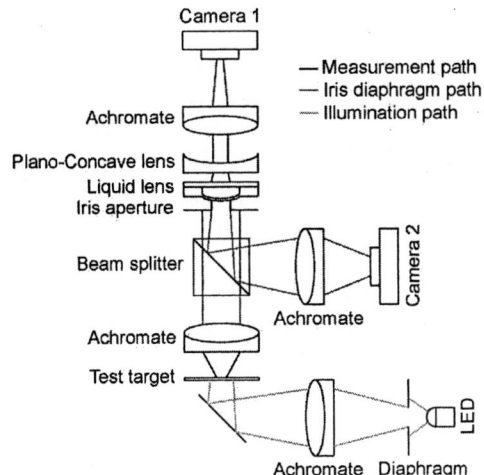

Fig. 3: Optical measurement setup

At thicker or thinner membranes the image quality decreases. The aperture is also maximal at the optimal membrane thickness. Generally, membrane lenses with a given diameter and focal range show a resolution maximum at one membrane thickness.

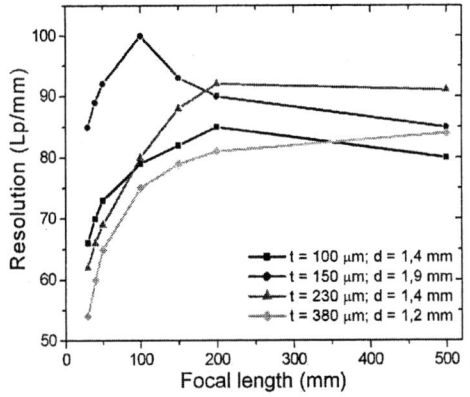

Fig. 4: Lens resolution at 50% contrast as a function of the membrane thickness t and the aperture d

4 SYSTEM CHARACTERIZATION

In order to find the optimum actuator dimensions the displaced volume is simulated, as a function of the inner ring radius while the outer radius is kept constant. Fig. 5 shows the displaced water volume as a function of the inner piezo radius r_P at 100 V. A maximum volume of 2.27 µl is displaced at radius of 1950 µm. The simulation of the whole system (piezoelectric pump and the liquid lens) is done in two steps. In a first step, the displaced volume dependent on the piezo voltage is calculated on the one hand, and the focal length of the lens as a function of the displaced volume on the other hand. In a second step, the displaced pump and lens volume are set equal to link the focal length with the piezo voltage (fig. 6).

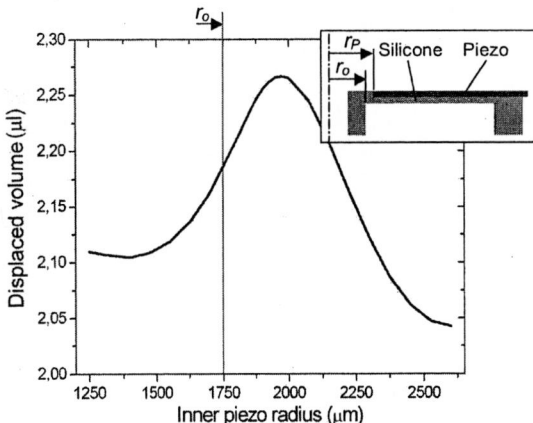

Fig. 5: Diagram of the displaced volume as a function of the inner piezo radius r_P; r_o are the outer frame radius

This is verified using the setup of fig. 3. The simulation leads to a hyperbolic behavior which is in good agreement with the measured values while the voltage increases from -9 V to +44 V. Decreasing the voltage back to -13 V shows the typical hysteresis behavior of piezo actuators.

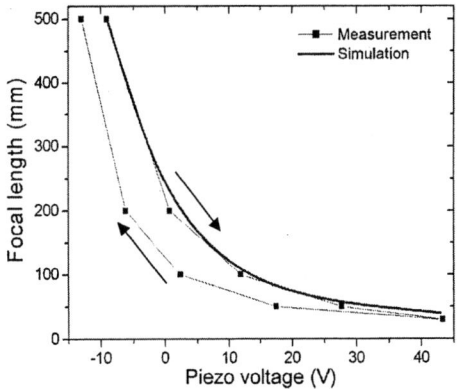

Fig. 6: Focal length dependent on piezo voltage

5 CONCLUSIONS

We have demonstrated the first membrane liquid lens with an integrated piezo bending actuator. The lens shows a maximum resolution of 100 lp/mm at a used lens diameter of 38%. Future work will focus on increasing the aperture diameter up to 50%.

6 REFERENCES

[1] De-Ying Zhang et. al., "Fluidic Zoom-Lens-on-a-Chip With Wide Field-of-View Tuning Range", Photonics Technology Letters, 16, pp.2356-2358, 2004.

[2] www.varioptic.com

WB1
10:30 – 10:45

Vortex Generation and Pixel Calibration Using a Spatial Light Modulator for Maskless Lithography

Il Woong Jung, Jen-Shiang Wang and Olav Solgaard

E. L. Ginzton Laboratory, Department of Electrical Engineering, Stanford University, Stanford, CA 94305
Tel +1-650-723-1992, Fax +1-650-725-2533, E-mail: iwjung@stanford.edu

Abstract

We present aerial images demonstrating the generation of an isolated optical vortex via and its through-focus characteristics using a SLM (Spatial Light Modulator) for optical maskless lithography. Aerial images show that a 4 phase step vortex can be formed by a group of 2 x 2 pixels. The vortex showed that there is minimal through-focus drift in the intensity profile whereas an isolated feature generated without a phase singularity showed noticeable drift of the profile. We also present experimental results in utilizing a method to improve the sensitivity in pixel calibration for an optical maskless lithography system. Using an optimized configuration of the surrounding mirrors, the calibration sensitivity of the pixel under test can be increased and even maximized at a certain defocus. The calibration sensitivity is improved by ~1.8x for our surrounding mirror configuration without defocus.

Keywords: Maskless Lithography, Spatial Light Modulator, Vortex Via, Pixel Calibration.

1 INTRODUCTION

Optical maskless lithography using a programmable SLM is seen as an economical solution to the rising costs of masks. Phase shifting SLMs have the ability to generate various patterns of interest in lithography including optical vortices [1]. Vias, or contacts, are difficult to print because they are the smallest features on the mask and inherently have a very small depth-of-focus (DOF) [2]. Vias created by optical vortices are known to have very good through-focus characteristics due to their ambiguous phase. We present aerial imaging results using the setup in Fig. 1 demonstrating the generation of an optical vortex. The numerous pixels of the SLM must be accurately and effectively calibrated due to their strong phase-shifting capability [3]. Hence, finding a way of increasing the sensitivity of calibration is important. In this paper we also present aerial imaging results demonstrating improvement in pixel calibration sensitivity.

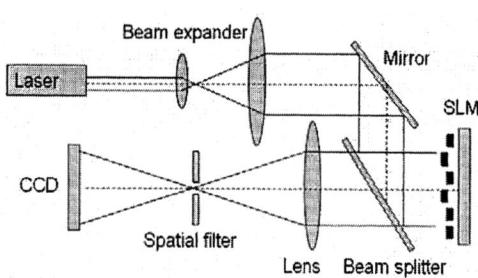

Figure 1. Diagram of an aerial imaging setup using a SLM

2 VORTEX VIA

Although through-focus asymmetries can by compensated for by

utilizing a double-exposure scheme using the positive and negative phase-shifting capability of SLMs [4], vortex masks have proven their capabilities for printing small contacts with a relatively large process window without such methods [2]. Contacts require a minimum feature size for the lateral dimensions and hence results in a very small DOF, which reduces the process window. An optical vortex forms a spot where the phase of the light is discontinuous. In the vortex the phase is not well-defined, so effects of through-focus drift caused by phase shifts in the image plane can be minimized. Our piston mirror array can be used to synthesize a vortex mask and we demonstrate the formation of an isolated vortex and its through-focus properties. An aerial image of an isolated vortex formed by 4 phase steps is shown in Fig. 2(a). Each phase step of the vortex is formed by a group of 2x2 pixels. These sub-pixels are deflected to the corresponding phase steps to form a dark vortex spot in the center. The upper left is at 0° phase, the upper right at 90° phase, the lower right at 180° phase, and the lower left at 270° phase. This clockwise configuration can be laid out in combination with a counterclockwise configuration to form a dense array of vortex contacts as shown in Fig. 2(b). However, the density of these contacts is limited by the pitch of the mirror array as they can be formed only at the corners of the pixels. They are also inherently limited in their placement as we cannot use relative phase between pixels to place them sub-grid. The observed quarter-circle pattern is a result of the interference between the 180° phase step pixels and the surrounding pixels. This quarter circle can be removed without removing the vortex by correct set-up of the surrounding pixels.

We compare the cross-section of the vortex and the edge of the quarter circle. The latter is formed by interference between the 90°, 180°, and 0° phase steps and hence is not a vortex. Results in Fig. 3 show that the profile from a non-vortex dark spot exhibits through-focus drift of the profile, whereas the vortex does not show through-focus drift. This aerial imaging experiment shows that a vortex contact exhibits superior through-focus characteristics, and therefore an improved process window.

1-4244-0641-2/07/$20.00 ©2007 IEEE   147

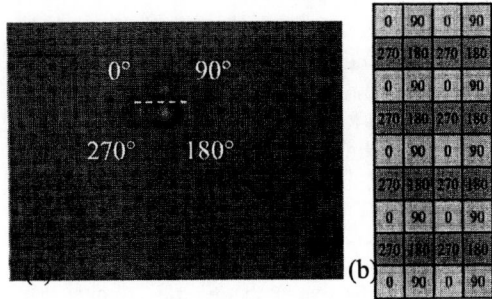

Figure 2. (a) Aerial image of a 4 phase step vortex via with 0°, 90°, 180°, 270° phase steps in a clockwise (CW) configuration (b) 4 step vortex array with CW and CCW configurations

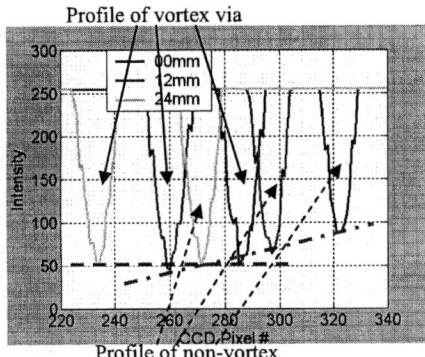

Figure 3. Through-focus intensity profile cross-sections of a vortex via and non-vortex pattern

3 PIXEL CALIBRATION

The strong phase-shifting ability of piston mirror arrays requires that the induced phase to voltage relationship be mapped out with very good resolution to acquire precise control of pattern generation on the wafer. Due to the varying characteristics from pixel to pixel resulting from non-uniformities in fabrication, an efficient and accurate method of calibrating the numerous mirror elements of the SLM is required. A method to increase the sensitivity of the calibration of the SLM has been discussed in theory and simulation [3]. It is shown that we can increase the sensitivity by finding an optimum configuration of the surrounding mirrors and optimum measuring position, i.e. the position of the CCD camera. Here, we experimentally measure and compare the sensitivity between "configuration A" with the surrounding mirrors at the initial 0° phase, and "configuration B" that has a pattern of mirrors deflected to 180° phase shift. The B configuration is expected to improve the sensitivity compared to the A configuration at the same imaging plane, i.e. the CCD at the same focus or defocus position. Fig. 4(a) shows an aerial image where

the pixel under test is deflected to form a dark spot with configuration A. In Fig. 4(b), the pixel under test is deflected with the surrounding pixels forming configuration B. The intensity profile cross-section of the center spot is plotted in Fig. 5 and the region of maximum difference measured. This measured range is where the pixel under test and the surrounding pixels have a 90° phase difference. The measurement results show that the calibration scheme using configuration B increases the sensitivity by a factor of ~1.8×.

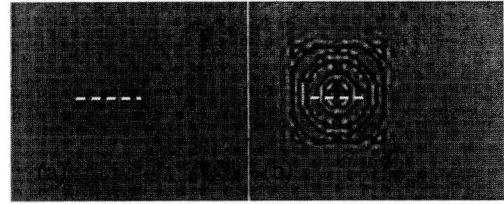

Figure 4. Aerial image of the center pixel at 180° phase and the surrounding pixels in (a) Config. A and (b) Config. B

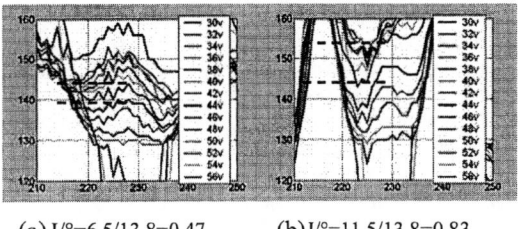

(a) I/°=6.5/13.8=0.47   (b) I/°=11.5/13.8=0.83

Figure 5. Graphs showing the intensity profile cross-section plots of the center pixel under test with (a) Config. A (b) Config. B of the surrounding pixels

4 CONCLUSION

We have demonstrated the generation of an optical vortex using a SLM in an aerial imaging experiment. The vortices show very good through-focus characteristics compared to other non-phase-singularity features. We also demonstrate increased sensitivity in pixel calibration by using a special configuration of the surrounding pixels. Using the scheme, the sensitivity is increased by ~1.8×.

REFERENCES

[1] I. W. Jung et al, "Optical Pattern Generation Using Spatial Light Modulators for Maskless Lithography," JSTQE, vol. 13, pp. 147-154, March-April 2007
[2] M. D. Levenson et al, "Optical Vortex Masks for Via Levels," JMMM, vol. 3, pp. 293-304, 2004
[3] J. S. Wang, "High Resolution Optical Maskless Lithography Based On Micromirror Arrays," PhD Thesis, pp. 58-67, 2006
[4] J. S. Wang et al, "Effects of Through-Focus Symmetry in Maskless Lithography Using Micromirror Arrays", JVST B, vol. 23, pp. 2738-2742, 2005

WB2
10:45 – 11:00

The Study on 3D Electron Beam Lithography for Sub-micrometer Diffractive Optics

Hung-Lin Yin, Joseph Y. C. Hu, Chih-Sheng Yu
Tel +886-3-5779911 ext. 335, Fax +886-3-5773947, E-mail rocks@itrc.org.tw
Instrument Technology Research Center, National Applied Research Laboratories,
20, R&D Rd. VI, Hsinchu Science Park, Hsinchu, Taiwan

Abstract
The fabrication technology of gray-scale electron beam lithography for hologram and Fresnel lens with sub-micrometer features has been reported in this paper. Mass-production of these optical elements by integrating with electroforming and hot embossing has also been demonstrated. For a more critical case, such as antireflective structure array, a novel scattering electron beam lithography has been successfully applied on this application.

Keywords: electron beam lithography, gray-scale, hologram, Fresnel lens, moth eye

1 INTRODUCTION

Various MEMS fabrication processes have been developed in recent years. However, the planar technologies might not reach the demand for several applications, such as micro lens and blaze grating. In order to generate 3D profiles of micro optical elements, various methods have been utilized such as gray-scale mask [1] and laser direct-write (LDW) [2]. A customize gray-scale mask not only makes cost a great concern, but also limits the design flexibility. On the contrary, laser direct-write technique has much flexibility but with lower resolution and poorer roughness. In this study, a 3D electron beam lithography (EBL), including gray-scale and scattering methods, has been applied to fabricate diffractive micro optics with high resolution and has better accuracy. Integrating with electroforming and hot embossing, the micro optical elements can be massively produced.

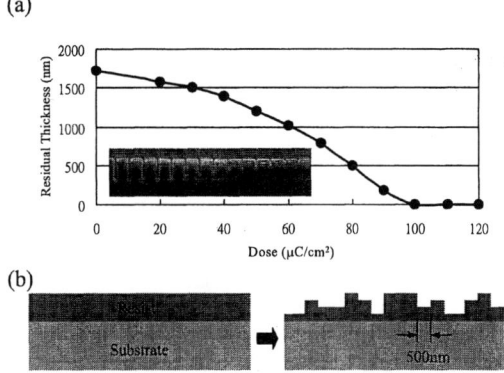

Fig. 1 (a) Contrast curve of PMMA, (b) the gray-scale EBL.

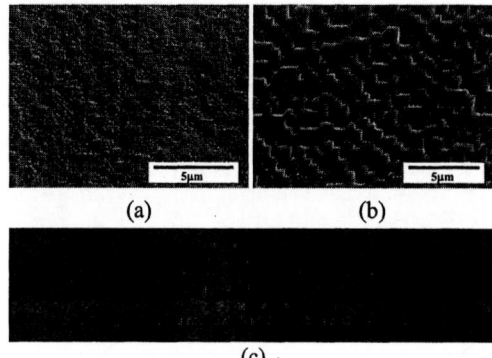

Fig. 2 The SEM picture of hologram (a) without, (b) with proximity effect correction, and (c) projection image of virtual keyboard.

2. GRAY-SCALE EBL

2.1 4-orders Hologram

The gray-scale EBL has been used for fabricating refractive optics in SU-8 resist [3]. In this study, the most commonly used resist, PMMA, has been chosen for higher resolution. The cross-section of residual thickness at various dosages and the contrast curve is shown in Fig. 1(a). For gray-scale EBL, the profile of proposed hologram has been transferred into a gray-scale pattern. Then the file has been imported to pattern generator with various-dosage distribution. After gray-scale EBL, the various-depth patterns can be transferred onto the resist, as shown in Fig. 1(b). The fabrication result of the 4-orders hologram with 500nm pixel size is shown in Fig. 2(a). The rough surface might cause by the proximity effect of EBL. After

1-4244-0641-2/07/$20.00 ©2007 IEEE

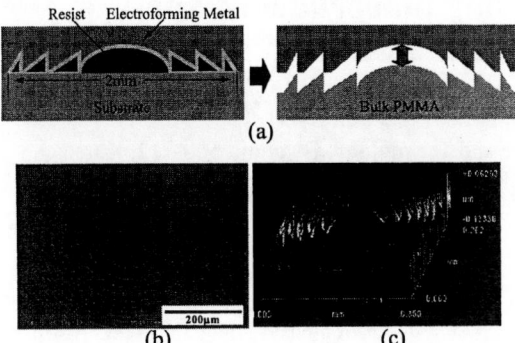

Fig. 3 (a) Fabrication process for Fresnel lens, (b) SEM picture of Fresnel lens by hot embossing, (c) measured 3D profile by an interferometer.

correcting the proximity effect, the accuracy of pattern transformation can be improved, as shown in Fig. 2(b). The 4-orders hologram has been designed for generating a virtual keyboard image with a wider viewing angle. As shown in Fig. 2(c), the far-field projection of virtual keyboard has been demonstrated.

2.2 Fresnel Lens

The process for mass–produced Fresnel lens on bulk PMMA substrate is shown in Fig. 3(a). Relying on the correction of dosage, the proposed profile of Fresnel lens can be obtained in PMMA resist with gray-scale EBL. Following with an electroforming process, an inversed Ni-Co mold has been fabricated. Finally, the Fresnel lens has been formed on bulk PMMA by hot embossing. The Fresnel lenses can be rapidly produced and are with good repeatability, as shown in Fig. 3(b). A wide variety of micro-optical components such as holograms and blaze gratings can be made by the same fabrication technique. Fig 3(c) shows the measured 3D profile within 2% depth error by an interferometer. The optical properties of the Fresnel lens have been characterized by using a 532nm green solid stated laser. The measured focal length is 5.78mm (4.6% error compared to proposed design) and the minimum spotsize is < 8 μm.

3. SCATTERING EBL

The antireflective element or so-called "moth eye" structure has attracted much interest in recent years. The requirement of fabrication technology becomes more challenge due to its half-wavelength feature and cone shape. In this paper, the scattering mechanism of electron has been utilized to generate the cone-shaped microstructure array

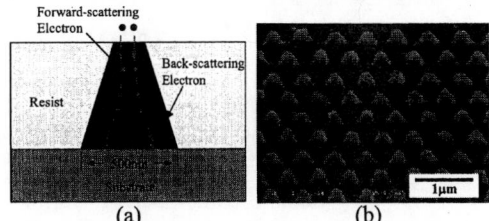

Fig. 4 (a) The mechanism of scattering EBL, (b) SU-8 antireflective structure array.

with 500nm pitch. The scattering method for 3D EBL has shown in Fig. 4(a). The SU-8 resist has been chosen due to its higher sensitivity for scattering (low energy) electrons. After single-dot exposing and developing, an approximate cone-shaped structure can be formed by the forward- and back-scattering electrons. The antireflective structure array with 500nm pitch is shown in Fig. 4(b). The height of the structure is 500nm. Combining electroforming and hot embossing, the antireflective structure array can also be transferred to a transparent polymer material.

4. CONCLUSION

In this paper, a 3D electron beam lithography for sub-micrometer diffractive optics has been demonstrated. The gray-scale EBL has been utilized to fabricate a 4-orders hologram and a Fresnel lens. High resolution and feature accuracy can be achieved by this process. Moreover, the mass-produced process combining electroforming and hot embossing for Fresnel lens has been developed. For the antireflective structure array, a novel scattering EBL has also been reported in this study. The bio-mimetic cone-shaped structure has been successfully fabricated by this method.

REFERENCES

[1] Y. Oppliger, et al., "One Step 3D Shaping Using a Gray-tone Mask for Optical and Microelectronic Applications," Microelectronic Engineering, 23, pp. 449-454, 1994.

[2] X. Wang, et al., "Rapid Fabrication of Drffractive Optical Elements by Use of Image-based Excimer Laser Ablation," Applied Optics, Vol. 36, No. 20, pp. 4660-4665, 1997.

[3] W. H. Wong, et al, "Exposure characteristics and three-dimensional profiling of SU-8C resist using electron beam lithography," *J. Vac. Sci. Technol.*, B19, pp. 732-735, 2001.

WB3
11:00 – 11:15

Fabrication of a Multi-Level Lens Using Independent-Exposure Lithography and FAB Plasma Etching

Do Kyun Woo*, Kazuhiro Hane[†], Cha Bum Lee* and Sun Kyu Lee*

*Mechatronics Department, Gwangju Institute of Science and Technology, Gwangju 500-712, Korea
Tel: +82-62-970-2430, Fax +82-62-970-2384, E-mail: wpeter@gist.ac.kr
[†]Nanomechanics Department, University of Tohoku, Sendai 980-8579, Japan

Abstract

Recently, a micro lens is need for the micro optical system requiring thin thickness. A multi-level lens can be adaptable to satisfy the requirement of this purpose. In this study, an independent-exposure of electron beam lithography and FAB plasma etching method are presented for the fabrication of a multi-level lens. The advantages of the method are the non-repetitive process and the precise fabrication due to the principle of the independent-exposure lithography.

Keywords: multi-level lens, independent-exposure lithography, proximity effect, FAB plasma etching, non-repetitive process

1 INTRODUCTION

Mobile communications has broadened into a variety of functions and devices to be applied on the entertainment field. Thus, the development of an optical micro navigator with thin thickness easily to handle a lot of data is strongly required. To satisfy the thin thickness of a micro navigator, a multi-level lens as planer optics has been interested.

To fabricate such a multi-level lens, several techniques has been developed such as photo-lithography, Electron Beam Lithography(EBL), laser fabrication and mechanical processing. Two methods using mask alignment and gray-scale mask in the photo-lithography seem to be useful to fabricate a multi-level lens[1-5].

In this study, the EBL with high resolution and Fast Atom Beam(FAB) plasma etching with 1:1 selectivity of resist and substrate are proposed to fabricate the multi-level lens. [6].

The principle of the independent-exposure lithography(IEL) is to make the resist be the profile of a multi-level lens with the appropriate exposure dose in the lithography process as shown in Figure 1. Thus, the considerable advantages of this method are the non-repetitive process and the precise fabrication for a multi-level lens.

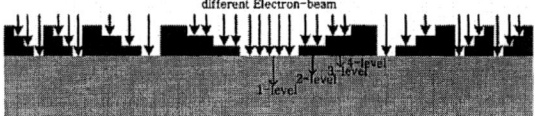

Figure 1. Independent-exposure in the EBL

2 DESIGN OF A MULTI-LEVEL LENS

Figure 2 shows the relationship between the radius R_j of j-th pattern and the focal length of the Fresnel lens and the scheme of the multi-level lens in Fresnel zone construction. The radius $R_{j,N}$ of 4-level lens can be easily calculated by geometry [7]. By using the Eq.(1), the diameter $127.4 \mu m$ of the 4-level lens, the focal length of $714.5 \mu m$ and the refractive index of 1.4981, the 4-level lens can be designed

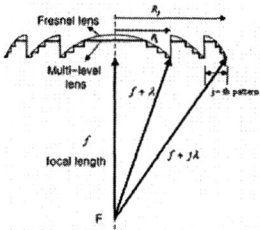

Figure 2. Fresnel lens and Multi-level lens

$$R_{j,N} = \sqrt{2\left(j - \frac{4-N}{4}\right)\lambda f + \left(\left(j - \frac{4-N}{4}\right)\lambda\right)^2} \quad (1)$$

$$d = \frac{(N-1)\lambda}{N(n_{lens}-1)}$$

where $R_{j,N}$: the radius of N-level in j-th pattern

λ : the wave length of incident light

d : the thickness of N-level lens

N : the number of level

with the maximum pattern width of $26.204 \mu m$, the minimum pattern width of $0.966 \mu m$ and the thickness of $0.723 \mu m$.

3 EXPERIMENT

3.1 E-beam dose and the developed thickness of resist

In the fabricating a multi-level lens with the IEL, it is necessary to clarify the effect of the exposure dose on the developed resist thickness. The patterns with the width of $10 \mu m$ and $80 \mu m$ for this experiment were used, and the electron beam dose was given from $20 \mu C/cm^2$ to $100 \mu C/cm^2$.

Figure 3 shows the result of the relationship between the E-beam dose and the developed thickness of resist. In this figure, it is clear that the developed thickness of resist increases with the increase of electron beam dose. Moreover, when the pattern size becomes large, the developed

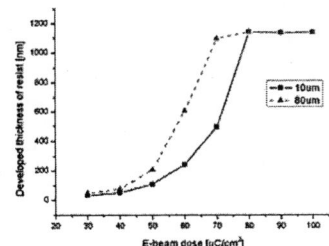

Figure 3. The developed thickness of resist

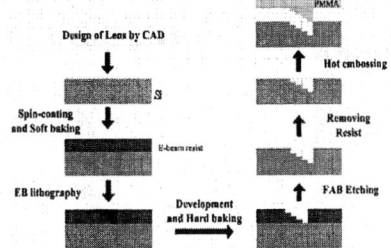

Figure 4. Fabrication procedure of 4-level lens

thickness of resist increases. Thus, in order to fabricate multi-level lens with IEL, both the pattern size and the electron beam dose should be considered. The reason why the developed thickness of resist according to the pattern size is different with the identical exposure dose can be explained by the proximity effect[6].

3.2 Fabrication procedure of multi-level lens

The procedure for the fabricating a 4-level lens with the IEL and FAB plasma etching is shown in Fig. 4. As mentioned, the prominent advantage is the ability to fabricate a precise multi-level lens with non-repetitive process.

4 RESULTS AND DISCUSSION

The 4-level lens with the focal length of $714.5\mu m$ was fabricated by the IEL and FAB etching. Figure 5 shows the fabricated mold. As shown in Figure 5, the surface of the 4-level lens fabricated by the IEL has some scattering mark; it is not flat owing to the backscattering of the electron beam on the substrate. In order to resolve this problem, electron beam resist with low sensitivity should be selected.

To evaluate the performance of the fabricated 4-level lens, MTF(Modulation Transfer Function) was measured. By using the object plane with the period of $50\mu m$ shown in Figure 6(a), the real image through the fabricated lens was obtained as shown in Figure 6(b). Comparing the object plane with image through the 4-level lens, MTF of 53.6% at the center of 4-level lens was obtained. It may be improved through alignment of the measurement.

5 CONCLUSION

In this study, to simplify the fabrication process of a multi-level lens, the independent-exposure in EBL and FAB

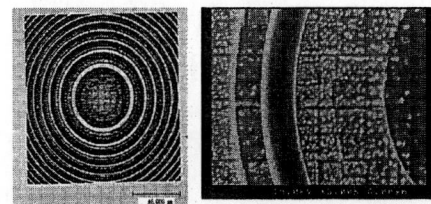

Figure 5. The fabricated mold of the 4-level lens

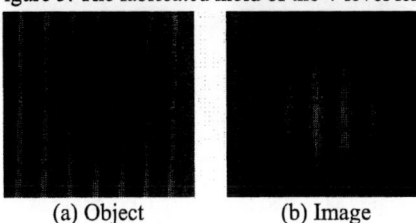

(a) Object   (b) Image

Figure 6. Image test for MTF

plasma etching is proposed. It makes possible to fabricate the precise multi-level lens with non-repetitive process of lithography and etching. Thus, it is anticipated that the time and expense for the fabrication of a multi-level lens can be reduced by the proposed method.

In the experiment, it was verified that the developed resist thickness is proportional to the pattern size as well as the exposure dose because of the proximity effect. In order to release the proximity effect, the electron beam resist with low sensitivity should be selected.

REFERENCES

[1] T. Shiono, K. Setsune, O. Yamazaki, K. Wasa, "Rectangular-apertured micro-Fresnel lens array fabricated by electron-beam lithography", appl. Opt. Vol. 26, pp 587-591 (1987).

[2] H. Hagino, C. S. Park, H. Kikuta, K. Iwata, "Multilevel computer generated hologram on an curved surface fo high power CO_2 laser beam shaping", Proc. 11th ICPE, pp 281-284 (2006)

[3] L. Kong, X. Yi, K. Lian, S. Chen, "Design and optical performance research of multi-phase diffractive micro lens array", J. Micromech. Microeng. 14(2004), pp 1135-1139

[4] J. Yan, K. Maekawa, J. Tamaki, T. Kuriyagawa, " Micro grooving on single-crystal germanium for infrared Fresnel lenses", J. Micromech. Microeng. 15(2005), pp 1925-1931

[5] H. Andersson, M. Ekberg, S. Hard, S. Jacobsson, M. Larsson, T. Nilsson, "Single photo mask, multi-level lens kinoforms in quartz and photo resist: manufacture and evaluation", appl. Opt. Vol 29, pp 4259-4267

[6] Marc J. Madou, "Fundamentals of microfabrication", CRC Press, pp 53-54

[7] J. Turunen, F. Wyrowski, "Diffractive Optics for Industrial Applications", Akademie Verlag(1997), pp 82-83

WB4
11:15 – 11:30

Self-Assembled Two-Dimensional Block Copolymers on Pre-patterned Templates with Laser Interference Lithography

C. Y. Wu [1], Y. F. Huang [1], C. M. Yang [2], N. D. Lai [1], C. H. Wu [2], C. C. Tsiang [3] and C. C. Hsu [1,2,*]

[1] Department of Physics
[2] Institute of Opto-Mechatronics
[3] Department of Chemical Engineering
National Chung Cheng University, Chiayi 621, Taiwan
*Phone: +886-5-2720411 ext. 66305, E-mail : cchsu@phy.ccu.edu.tw, d92220003@ccu.edu.tw

Abstract

We present an experimental result of the self-assembling for cylindrical nanostructures of PS-*b*-PMMA block copolymers (BCPs) on pre-patterned templates. The templates are fabricated by two-beam interference lithography at 325 nm into negative SU-8 photoresist. To align the BCPs nanostructures perpendicular the surface of templates, homopolymer polystyrene (hPS) layer is brushed on the patterns for modifying the interfacial energies. Thus, precisely control the localization and orientation of PS-*b*-PMMA nanostructures on patterned substrates is achieved.

Keywords: self-assembly, diblock copolymers, laser interference lithography

1 INTRODUCTION

Nano- and micro- phase separation (10-100 nm) via self-assembling of block copolymers (BCPs) attract large attentions for the applications in electronic and optic devices, such as photonic crystals, silicon capacitors, catalysts, etc.[1-3] Depending on the relative volume ratio, length, molecular weight and the different polymer constituents between covalence bound blocks and followed by thermal annealing, various microdomain structures and shape sizes formed as lamellae, cylinders, spheres, bicontinuous and perforated layers are available. In previous structures, there were no specific orientation and long range order, so the controllable orientation normal to the substrates and lateral order of the phase separation of BCPs for functional devices are essential. In typical topographical methods, the periodic modulation patterned structures were applied to overcome the intrinsic disorder, but the domain of BCPs self-assembled on pattering substrates were close to the scale of patterns that limit the domain size for applications.[4]

In this report, we present a new topographical method for self-assembling of PS-*b*-PMMA BCPs. Such approach emphasize with the orientation of BCPs cylindrical arrays and lateral order for large size domain. The topographic periodic substrates are pre-patterned into SU-8 photoresist by laser lithography technique and brush a PS homopolymer layer on photoresist patterns that modify the interfacial energies.

2 EXPERIMENTAL SECTION

2.1 Pre-patterned structures by laser Lithography

The SU-8 2005 negative photoresist was spin-coated on glass substrates and then use He-Cd laser at 325 nm

wavelength to interfere into SU-8 thin film. We can fabricate 1D grating, 2D hexagonal cylindrical and air hole arrays. The film thickness can be controlled by spin speed. The detailed laser interference process was described in our previous work.[5]

2.2 Self-Assembled PS Cylindrical Matrix Arrays

Asymmetric polystyrene-*b*-poly(methyl methacrylate) (PS-*b*-PMMA, Mn = 67,100, PDI = 1.09) and homopolymer polystyrene (PS, Mn = 10,000, PDI = 1.08) were used. The PMMA volume fraction of PS-*b*-PMMA was approach 31 %. The PS layer was first brushed on SU-8 structures and UV light exposed for polymerization. The PS-*b*-PMMA was then spin-coated on PS-modified surfaces and annealed. The PS cylindrical microdomains of copolymers formed after the PMMA selective etched by acetic acid and localized in precise positions between the laser interference patterns.

3 RESULTS AND DISCUSSIONS

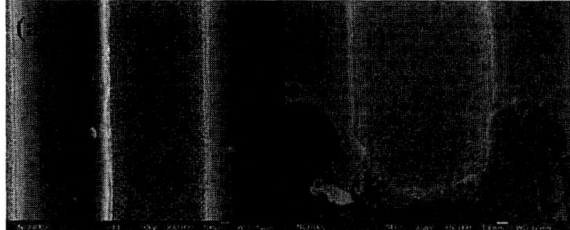

Figure 1. (a) The top view of BCPs between gratings. (b) The tiled view of BCPs between gratings.

Figure 1 shows the PS-*b*-PMMA spin-coated on PS brushed photoresist grating structures. As Figure 1a shows, after the PMMA etched, the PS cylindrical arrays appearance between the gratings. Figure 1b shows the tilted image. The grating period and depth are 1.5 μm and 1 μm. When the BCPs were spin-coated between gratings, a capillary force formed and extended into a curve by surface tension. The PS cylindrical arrays then were perpendicular the curved surface of PS brushed layer.

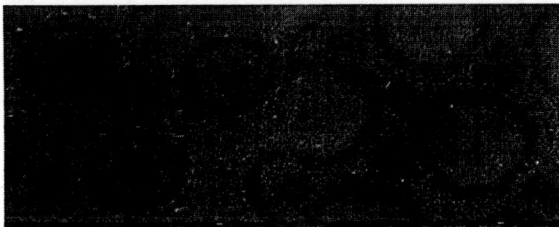

Figure 2. (a) The PS cylindrical arrays formed between the rods. (b) The PS cylindrical arrays formed inside the air holes.

Figure 2 shows the BCPs spin-coated on triangular cylinder rod and air hole structures. The periods of the cylinder rod and air hole are 1.75 μm, the diameters of rod and air hole are 1.25 μm. In Figure 2a, the PS cylindrical arrays are confined between the rods. Figure 2b shows the cylindrical arrays are formed inside the air holes.

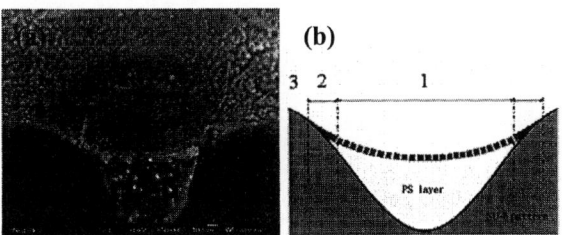

Figure 3. (a) The tiled view of PS cylindrical arrays inside the air holes. (b) The illustration from (a).

Figure 3a and 3b show the BCPs self-assembled inside the air holes, we denote the number 1, 2 and 3 of three parts, which represent the cylindrical array, flat surface and no brush modification domains. Figure 3a shows the distinct difference of three domains. The domain 1 shows the PS cylindrical array on brushed layer. Domain 2 shows flat surface, due to the thickness is thinner than monolayer. Domain 3 shows disorder fragments of PS segment and crests are not totally covered on substrates by PS brushing. Figure 3b illustrated the structure from Figure 3a. The dotted line on brushed layer represented the mono PS cylindrical arrays. The cylindrical arrays followed surface tension between the two crests. The PS cylinders in domain 1 are contact with brushing layer and formed ordered structures. Domain 2 has no cylindrical arrays due to the film thickness in thinner than monolayer. In domain 3, the brushing layer did not spin-coat over all the crests, so the BCPs formed a disordered structure.

4 CONCLUSIONS

In conclusions, we have demonstrated the topographical method of BCPs self-assembled on SU-8 photoresist structures. The laser interference lithography patterned substrates with PS brush modification layer was used to orientate the BCPs. The PS cylindrical nanoporous arrays formed and show a long range lateral order and large size domain in micro-scale. The observation of FSEM images ascertained the BCPs self-assembled ordered structures.

REFERENCES

[1] A. M. Urbas, "Bicontinuous cubic block copolymer photonic crystals," Adv. Mater., 14, pp.1850-1853, 2002.

[2] C. T. Black, "Integration of self-assembled diblock copolymers for semiconductor capacitor fabrication," Appl. Phys. Lett., 79, pp.409-411, 2001.

[3] D. A. Durkee, "Catalysts from self-assembled organometallic block copolymers," Adv. Mater., 17, pp.2003-2006, 2005.

[4] L. Rockford, "Polymers on nanoperiodic, heterogeneous surfaces," Phys. Rev. Lett., 82, pp.2602-2605, 1999.

[5] N. D. Lai, "Fabrication of two- and three-dimensional periodic structures by multi-exposure of two-beam interference technique," Opt. Express, 13, pp.9605-9611, 2005.

WB5
11:30 – 11:45

Fabrication of Large Size Photonic Crystal Templates by Holographic Lithography Technique

Ngoc Diep Lai, Jian Hung Lin, and Chia Chen Hsu
Department of Physics, National Chung Cheng University,
168 University Road, Ming Hsiung, Chiayi 621, Taiwan
Tel +886-5-242-8173, Fax +886-5-272-0587, E-mail cchsu@phy.ccu.edu.tw

Abstract

Multi-exposure two-beam interference technique was employed to fabricate large size defect-free 2D and 3D polymer photonic crystal (PhC) templates. Combining with a double-step laser scanning technique, designed defects could be precisely embedded to a 2D PhC template.

Keywords: Holography lithography; Photonic crystal, Multi-beam interference technique, Photolithography

1 INTRODUCTION

Recently, there has been considerable interest in the fabrication of two- and three-dimensional (2D and 3D) photonic crystals (PhCs), which consist of periodic dielectric structures [1]. Holography lithography (HL) technique is a very promising and inexpensive technique to fabricate large size and defect-free PhC templates. In this work we demonstrated that multi-exposure two-beam interference technique [2] is an efficient way to fabricate defect-free 2D and 3D PhC templates. In combining this technique with mask-lithography, we created arbitrary and long-range defects into uniform and large-area 2D PhC templates [3]. Moreover, adopting a double-step laser scanning technique, designed defects could be precisely embedded to a 2D PhC template [4].

2 EXPERIMENTS AND RESULTS

A He-Cd laser at 325nm was used the light source for the experiment. The laser beam was spatially cleaned and extended. A double-iris was used to select two laser beams of same profile, polarization, and intensity. These beams were overlapped and interfered. The angle between two laser beams was denoted as 2θ. A sample was fixed in a double rotation stage, which could be rotated around the z-axis (normal direction of the sample) by an angle α and around the y-axis (90^0 to the z-axis) by an angle β. We first used SU-8 photoresist (thickness = 2μm) to fabricate 2D periodic structures by choosing the angle $\beta = 0^\circ$ and changing the angle α. Figure 1(a) shows the experimental result of a 2D square structure obtained with a double-exposure at $\alpha = 0^\circ$, 90°. The structure is uniform in a very large area (6 mm × 6 mm, corresponding to the size of the iris). When

double-exposure is realized at $\alpha = 0^\circ$, 60°, 2D hexagonal structure is obtained as predicted by the theory (Fig. 1(b)). In this case, the form of "atom" is however not symmetrical, i.e. not circle or square. To obtain symmetrical form, triple-exposure was applied at $\alpha = -60^\circ$, 0°, and 60° and the hexagonal structure has a circle "atom" form as shown in Fig. 1(c).

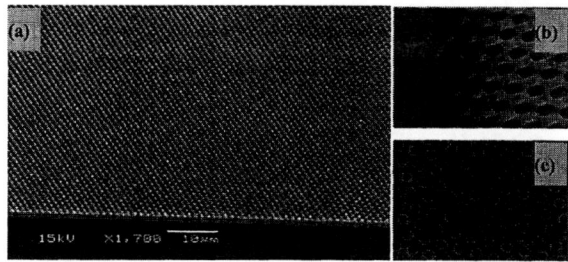

Fig.1 SEM images of 2D square and hexagonal PhC templates.

To fabricate 3D periodic structures, we varied both α and β for different exposures. A JSR photoresist with the thickness of about 6 μm is chosen for this task. Many kinds of 3D structures were successfully fabricated but only two 3D structures were shown in this paper: Hexagonal [$(\alpha,\beta) = (90^\circ, 0^\circ)$, $(0^\circ, 30^\circ)$, $(180^\circ, 30^\circ)$] and hexagonal-hexagonal [$(\alpha,\beta) = (60^\circ, 0^\circ)$, $(0^\circ, 30^\circ)$, $(180^\circ, 30^\circ)$]. The angle θ was chosen to be 24°, which resulted in the period of 400 nm. Figure 2 shows the SEM pictures of these periodic 3D structures. Structures are uniform in large area and in well agreement with the simulation results.

We combined HL and mask-lithography techniques to introduce long-range defect into above 2D PhC templates,. Following the multi-exposure of two-beam interference, we approached the mask to the sample and exposed it by a laser beam which is the same as the one used for interference. Our mask contains 3 lines defects (3 cm-length and 3 μm-width) and a Mach-Zehnder structure (1 cm-length, 3 μm-width, and 0.5 mm-separation between 2 branches). Both lines defects and Mach-Zehnder structure are well transferred from the mask to the 2D PhC templates (AZ-4620).

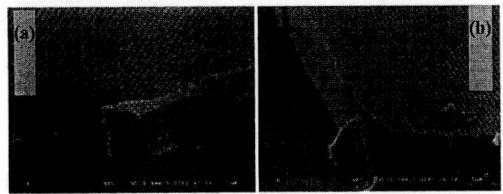

Fig. 2 SEM images of 3D hexagonal (a) and hexagonal-hexagonal (b) PhC templates.

Although any kinds of defect can be incorporated into PhC templates by the above combination technique, one question still remained is the lack of precisely control the position and orientation of defect with respect to a PhC template. Therefore, we proposed to use a double-step laser scanning technique to determine the position and orientation (mapping step) of periodic structures and embed design defects (fabrication step) into them. Figure 3 shows the experimental setup (direct laser scanning technique) used for mapping 2D PhC templates and patterning design defects. 2D periodic structures were first fabricated by interference of two laser beams into a negative photoresist (SU8). The main concept of the mapping step is based on the shrinkage effect of photoresist at positions where the material is photopolymerized by HL technique. After exposure and post-baking steps, we found that surface of periodic structure is modulated with a depth of about 30nm (AFM measurement). We then used a simple laser scanning microscopy to get a map of the periodic structure before patterning defects. After scanning the sample (no need of developing), the positions and orientations of the periodic structures were determined. By keeping the sample unmoved, the design defects were patterned in the designed positions and orientations with respect to the periodic structures. In this fabrication step, the laser power was increased to 30mW to induce multi-photon polymerization (MPP) effect. Finally, the sample was post-bake again to improve the cross-linking of polymer in the area of defects and the developing step was followed by. Figure4 shows the experimental results of 2D square structures containing several well-defined defects. Inset of each figure is the mapping image, which records positions and orientations of cylinders or holes in the 2D periodic structure. The design of defects such as point, line, and 90° bending are also drawn in the mapping images to illustrate where they are going to be embedded. The point defect was effectively introduced in an air hole as the design in Fig. 4(a). For the case of embedding the line, bending (Fig.

4(b) and (c)) or other defects, it is necessary to map the structures in a large area (more than 2 periods) in order to orient the defect in a desired direction with respect to the periodic structure. Figures 4(b) and (c) show that any kind of defect can be embedded into the periodic structure at any position and orientation with the assistance of the pre-scanning technique.

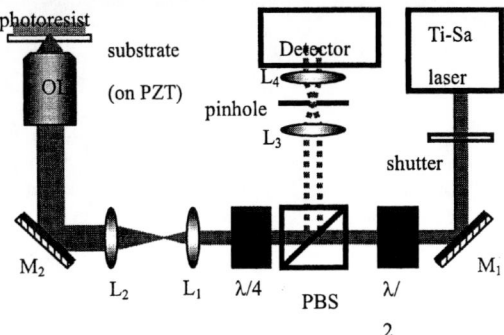

Fig. 3 Multi-photon polymerization microfabrication system used to map and pattern design defects.

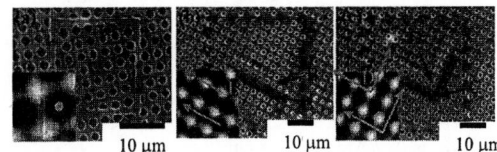

Fig. 4 SEM images of 2D periodic structures embedding with well-defined defects and their corresponding mapping images (a), (b), and (c) are point, line and 900 bending defects, respectively.

3 CONCLUSIONS

We have demonstrated different methods to fabricate 2D and 3D PhC templates with and without defects. A simple method for fabricating defect-free 2D and 3D PhC tempaltes was demonstrated. Moreover, with the assistance of a simple laser scanning microscopy technique, the defects could be fabricated with accurate position and orientation with respect to the periodic structure.

REFERENCES

[1] J. D. Joannopoulos, R. D. Meade, J. N. Winn, Photonic Crystals: Molding the Flow of Light (Princeton University Press, Princeton 1995).
[2] N. D. Lai, W. P. Liang, J. H. Lin, C. C. Hsu, C. H. Lin, Opt. Express 13, 9605-9611, 2005.
[3] N. D. Lai, W. P. Liang, J. H. Lin, C. C. Hsu, Opt. Express 13, 5331-5337, 2005.
[4] N. D. Lai, J. H. Lin, W. P. Liang, C. C. Hsu, C. H. Lin, Appl. Opt. 45, 5777-5782, 2006.

WB6
11:45 – 12:00

Extraordinary Transmission Through A Poly-SiC Membrane with Subwavelength Hole Arrays

[1]J Provine, [1]Peter B. Catrysse, [2]Christopher Roper, [2]Roya Maboudian, [1]Shanhui Fan, and [1]Roger T. Howe
[1]Department of Electrical Engineering, Stanford University
[2]Department of Chemical Engineering, University of California at Berkeley
127X Allen CIS Building, Stanford, CA 94305
Tel +1.510.717.5952, Fax +1.650.644.0464, E-mail jprovine@stanford.edu

Abstract

We report on the experimental observation of extraordinary transmission (~50% for 25% fill factor) in the infrared through a suspended polycrystalline silicon carbide (poly-SiC) membrane into which is etched a subwavelength hole array. The poly-SiC was deposited by low pressure chemical vapor deposition (LPCVD), patterned with contact photolithography, and etched by reactive ion etching (RIE). Simulation of the mid-infrared transmission is broadly consistent with measurements.

Keywords: Optical MEMS, Plasmonics, Silicon Carbide, Infra Red Spectroscopy

1 INTRODUCTION

Advances in micro- and nano-fabrication techniques have stimulated research in photonic structures and, in particular, the use of surface polariton materials for optical devices [1]. Most of this work has focused on metallic subwavelength gratings and aimed at producing surface plasmon polaritons in the visible and near-IR regimes. Recently, the phonon-polariton response of SiC has also been explored in the context of near-complete transmission and transmission suppression through subwavelength hole arrays at mid-infrared (mid-IR) wavelengths [2].

Recently, single crystal SiC gratings have been made with focused ion beam patterning [3]. This paper reports the mid-IR transmission through subwavelength hole arrays patterned in LPCVD poly-SiC using conventional optical lithography and reactive-ion etching. Film deposition via LPCVD of poly-SiC from disilabutane and dichlorosilane precursors can yield conformal films with controlled residual stress, making it an attractive structural layer for MEMS [4].

2 DESIGN & FABRICATION

Hole arrays of various permutations of pitch spacing (*a*) and hole diameter (*d*) were included into a single contact lithography mask to explore the dependency of transmission on array and hole geometry. In general the hole arrays were designed for the mid-IR regime ($a \approx \lambda \approx 10\mu$m).

Starting with a double-side polished (100) Si wafer, 1.5μm of undoped poly-SiC was deposited at 800°C. The "back-side" SiC was removed by plasma RIE etching with a mixture of Cl_2 and HBr gases [5] in a LAM TCP 9400 etcher. Next, a hard mask of low temperature oxide (LTO) was deposited by CVD at 400°C. The LTO was patterned using

contact photolithography using 1.6μm photoresist and magnetically enhanced reactive ion etched in an Applied Materials P5000 etcher using CF_3 and Ar. After stripping the photoresist the array pattern was transferred from the hard mask into the SiC film with another Cl_2/HBr RIE. The gratings were released from the Si substrate by a gas phase XeF_2 etch in a Xactix e1 Xetch tool. The release etch is extended long enough that the released hole arrays are >20μm from the handle wafer surface. The XeF2 etching leaves a rough silicon surface under the suspended hole array, so that Fabry-Perot reflection between the hole array and the substrate is insignificant. Finally, the remaining hard mask LTO and backside LTO were removed by a dip in 6:1 BOE. Figure 1 shows the fabrication process in schematic cross-section. Figure 2 shows SEMs of the completed hole arrays. The holes have a double slope profile with an overall slope of ~80°.

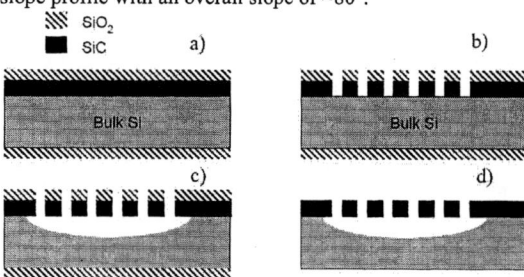

Figure 1. Schematic of fabrication process flow.

3 OPTICAL ANALYSIS

The fabricated hole arrays were spectrographically analyzed using a Nicolet 6700 FT-IR system connected to a Nicolet Continuum Infrared Microscope. Figure 3 shows the transmission spectrum for a hole array with *a*=10μm and

1-4244-0641-2/07/$20.00 ©2007 IEEE

Figure 2. (left) SEM of a full 150μm x 150μm array with a=10μm and d=5.6μm. (right) Higher magnification SEM of the same grating at a 45° angle.

thickness (*t*) of 1.5μm for different hole diameters (*d*). The transmission of a bare film is also shown to identify clearly that the hole arrays yield extraordinary transmission in the polariton gap region. Figure 4 shows simulated transmission results for a similar hole array (*d*=5.6μm, *a*=10.4μm, and *t*=4.0μm) obtained by solving Maxwell's equations using three-dimensional finite-difference time-domain (FDTD) calculations.

The experimental spectra show extraordinary transmission within the polariton gap region as high as 50% with maxima at λ≈11.6μm and 12.4μm. Furthermore, simulations indicate that strong resonance occurs only for the proper ratio of *d/a*, which is supported by the limited transmission (~10%) for a hole array with *a*=10μm and *d*=3.8μm (Figure 3). While the simulated hole arrays shows greater structure within the polariton gap, it does show two maxima at λ≈11.6μm and 12.3μm, which is similar to the experimental results in Figure 3. The differences between the experimental and simulated results may be attributed to the difference in film thickness, the surface roughness of the film, and the non-vertical hole sidewalls. Also, the simulations are for single crystal SiC, whereas the hole arrays were fabricated from poly-SiC. The various crystal orientations of SiC have similar optical properties [6] so it is possible the poly-crystalline nature may not have a dominant effect.

4 CONCLUSION

The transmission spectrum of LPCVD poly-SiC films patterned with a regular array of subwavelength circular holes has been shown to display extraordinary transmission within the polariton gap in agreement with theoretical predictions. The transmission depends on the periodicity and fill factor of the subwavelength hole array.

REFERENCES

[1] W. Barnes, A. Dereux, and T. Ebbesen, "Surface Plasmon Subwavelength Optics," *Nature*, 424, 824-830, (2003).

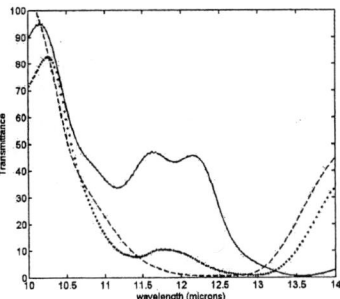

Figure 3. Experimental spectra from a 1.5μm thick poly-SiC film with no pattern (dashed), *a*=10μm and *d*=3.9μm (dotted), and *a*=10μm and *d*=5.6μm (solid).

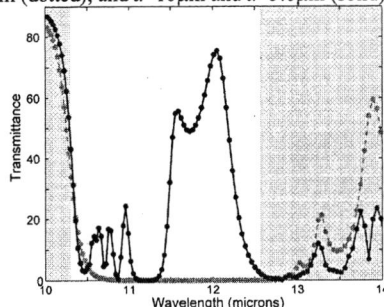

Figure 4. Simulated transmission spectra for a 4.0μm thick single crystal SiC film with no pattern (dashed) and *a*=10.4μm and *d*=5.6μm (solid). The polariton gap is represented by the clear area between the gray regions.

[2] P. B. Catrysse and S. Fan, "Near-complete Transmission Through Subwavelength Hole Arrays In Phonon-Polaritonic Thin Films," *Physical Review B*, 75, 075422, (2007).

[3] H. Hogstrom, S. Valizadeh, and C. Ribbing, "Optical Excitation of Surface Phonon Polaritons in Silicon Carbide by a Hole Array Fabricated by a Focused Ion Beam," *Optical Materials*, Article In Press (2007).

[4] C. S. Roper, R. T. Howe, and R. Maboudian, "Stress Control of Polycrystalline 3C-SiC Films in a Large-scale LPCVD Reactor using Dichlorosilane and 1,3-Disilabutane as Precursors", *Journal of Micromechanics and Microengineering*, 16 (2006) 2736-2739.

[5] D. Gao, M. Wijesundara, C. Carraro, R. Howe, and R. Maboudian, "Recent Progress Toward A Manufacturable Polycrystalline SiC Surface Micromachining Technology," IEEE Sensors Journal, Vol. 4, No. 4, 441-448, 2004.

[6] K. Narita, Y. Hijikata, H. Yaguchi, S. Yoshida, and S. Nakashima, "Characterization of Carrier Concentration and Mobility in n-type SiC Wafers Using Infrared Reflectance Spectroscopy," *Japanese Journal of Applied Physics*, 43, 8A, 5151-5156, 2004.

WC1 (Invited)
13:30 – 14:00

Compact Spectroscopic Sensor Using an Arrayed Waveguide Grating

K. Kodate, Y. Komai and K. Okamoto*
Faculty of Science, Japan Women's University
2-8-1 Mejirodai, Bunkyo-ku, Tokyo 112-8681, Japan
Tel +81-3-5981-3614, Fax +81-3-5981-3615, E-mail kkodate@fc.jwu.ac.jp
* University of California, Davis
3129 Kemper, Hall Davis, CA 95616, US

Abstract

We have previously proposed a compact spectroscopic sensor based on the arrayed waveguide grating (AWG) which makes use of wavelength multi/demultiplexer in a photonic network. We constructed a new planar sensor using a visible AWG for spectroscopic sensing of absorption characteristics with multiple dicing grooves for accommodating a liquid sample. A parabola-shaped sample injection waveguide effectively improves sensitivity because diffraction loss in the direction of its substrate is reduced. In this manuscript, discrimination between chlorophyll *a* and *b* as test samples is to be investigated using our AWG spectroscopic sensor.

Keywords: AWG spectroscopic sensor, Visible AWG, High-wavelength resolution, Environmental indicator

1. INTRODUCTION

Recently, with the emergence of the information and communication society, there has been an increasing need to develop compact sensors capable of acquiring biological and environmental sensing[1]. Optical and spectroscopic sensing is one of the promising approaches in order to meet these requirements. Although there are some compact spectroscopic sensors based on a free space optical system, it is still difficult to achieve further miniaturization and weight reduction, while keeping high wavelength resolution. In order to overcome these issues, we focused on an AWG satisfying both miniaturization of spectroscopic sensors and maintenance of high-wavelength resolution simultaneously. We proposed and fabricated a compact AWG spectroscopic sensor employing AWG with a groove for injecting a liquid sample under test[2]. Additionally, we succeeded in developing a novel compact spectroscopic sensor employing visible AWG and discriminating between chlorophyll a and b as an environmental indicator by detecting the difference between absorption characteristics[3].

In this report, we summarize the concept of our proposed AWG spectroscopic sensor, and then describe the results of differentiating chlorophylls *a* and *b*, and the sensitivity of the fabricated visible AWG spectroscopic sensor using both experimental results and numerical simulation.

2. PLANAR AWG SPECTROSCOPIC SENSOR

To produce a compact and portable spectroscopic sensor for measuring transmittance spectra of liquid samples, we have

developed a planar spectroscopic sensor using AWG as a wavelength dispersion device for the purpose of integrating a sample injection part, a light source, optical devices, and a photodetector on the planar lightwave circuit (PLC) of the AWG as shown in Fig. 1. We can design and fabricate the sample injection groove in any position of 1)-5) in Fig. 1.

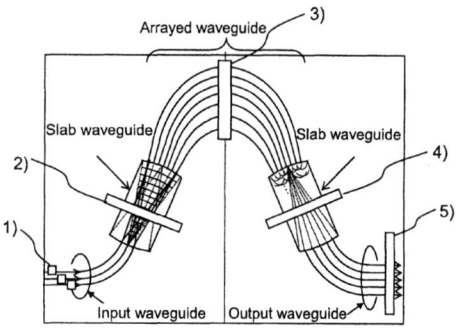

Fig. 1 Configuration of AWG spectroscopic sensor.

We fabricated a spectroscopic sensor using a near-infrared AWG in a photonic network on the basis of the original design procedure. The number of channels of the fabricated AWG was 80, and the channel spacing was 0.4nm. A liquid sample under test was poured into a groove (optical path length: 70μm, groove depth: 110μm) in the first slab region of the AWG. A preliminary experiment was carried out using a water solution of sodium acetate with the fabricated sensor. Subsequently, the concentration of the sodium acetate solute was determined with an average accuracy of ±0.25 wt%[3].

1-4244-0641-2/07/$20.00 ©2007 IEEE

3. AWG SPECTROSCOPIC SENSOR FOR VISIBLE WAVELENGTH

The next steps for the AWG spectroscopic sensor are to design a visible AWG to reduce optical losses in the waveguide device in the visible wavelength range for compact spectroscopic sensors as shown in Fig. 2. Optimizing the design parameters, which include the thickness and width of the core region, our attempt to fabricate the very first visible AWG was successful[4].

We designed and fabricated multiple dicing grooves for a sample injection in the first slab waveguide, and sought to discriminate chlorophylls *a* and *b* with various environmental indicators. Also, we employed a nonsubstrate dielectric multilayer filter as a band-pass filter for elimination of overlapped spectrum channels of multiple.

With the liquid sample inserted into the injection grooves, transmittance changes depending on the kind of liquid sample of chlorophylls, and the difference in transmittance depends on the 14 wavelength channels as shown in Fig. 3.

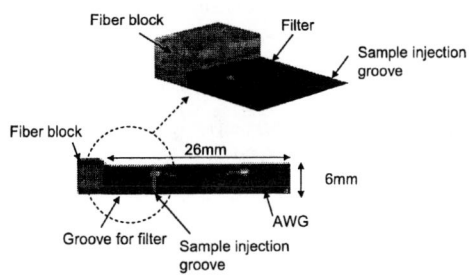

Fig. 2 Fabricated visible AWG spectroscopic sensor.

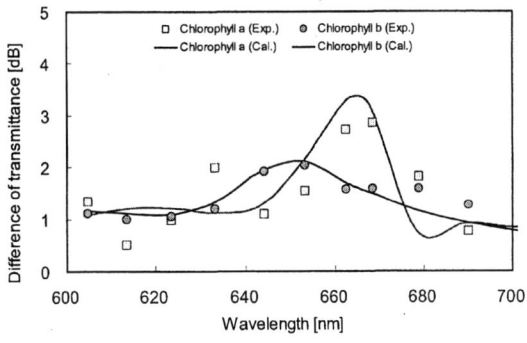

Fig. 3 Discrimination of chlorophyll solutions.

As a consequence, the experimental transmittance difference between the refractive index-matching oil and chlorophyll *a* (2.7dB) at the wavelength of 662.4nm, and that for chlorophyll *b* (1.9dB) at the wavelength of 644.0nm were obtained by absorbance spectrum sensing using fiber scanning measurement. The discrimination difference between chlorophyll *a* and *b* was 1.4dB for the absorbance of 1.0 where the optical path was set at 1 mm. The difference between the experimental value and the calculated value was 0.3 dB.

4. AWG SPECTROSCOPIC SENSOR USING PARABOLA-SHAPED SAMPLE INJECTION

We have proposed a visible AWG spectroscopic sensor using parabola-shaped sample injection waveguide and designed a parabola-shaped sample injection waveguide for discrimination of chlorophyll *a* and *b*[4]. From our simulation, the transmittance difference using the parabola-shaped sample injection waveguide increased by 2.0dB compared with that of the conventional linear sample injection grooves. The experimental results using the trial fabricated sensor will be presented in our presentation.

5. CONCLUSIONS

We have proposed a new application of an AWG for a compact planar spectroscopic sensor to meet both conditions for miniaturization and high wavelength resolution of subnanometer order. Following a preliminary experiment, we designed and fabricated a compact spectroscopic sensor using an AWG with an insertion loss of 4 dB in the visible wavelength range. To improve the sensitivity of the visible AWG spectroscopic sensor, the optimum design of multiple grooves for the liquid sample was carried out using the refractive index and transmittance of chlorophylls *a* and *b*. We succeeded in discriminating chlorophyll solutions using the AWG sensor. We also obtained a transmittance difference of 1.4dB. Relying on these theoretical and experimental investigations using a visible AWG, we can confirm that a compact AWG-based spectroscopic sensor is effective in bio-medical fields.

AKNOWLEDGMENT

The authors are much obliged to Dr. H. Takahashi, Mr. Y. Hida, and Dr. K. Suzuki for supplying us with AWG chips and for guidance on the waveguide design.

REFERENCES

[1] http://www.sci-news.co.jp/news/form.htm.
[2] Y. Komai, H. Nagano, K. Kodate, K. Okamoto, and T. Kamiya: Jpn. J. Appl. Phys., 43, 8B, 5795 (2004).
[3] Y. Komai, N. Wada, F. Moritsuka and K. Kodate: Proc. SPIE, 6025, 204 (2006).
[4] Y. Komai, H. Nagano, K. Okamoto and K. Kodate: Jap. J. Appl. Phys., 45, 6742 (2006).

WC2
14:00 – 14:15

A micro-optic-fluidic spectrometer with integrated 3D liquid-liquid waveguide

W. Z. Song[1], A. Q. Liu[1†], C. S. Lim[1], P. H. Yap[2]

[1]School of Electrical & Electronic Engineering, Nanyang Technological University
Nanyang Avenue, Singapore 639798
Tel: +65-67904336, Fax: +65-67933318, E-mail: eaqliu@ntu.edu.sg
[2] Denfence Medical & Environment Research Institute, DSO National Laboratories
27 Medical Drive #09-01 Singapore

Abstract

A micro-optic-fluidic spectrometer is demonstrated for the fluorescence detection of moving particles in a microfluidic chip. The object is excited within an integrated 3D liquid-liquid waveguide and its fluorescence spectrum is real-time recorded with resolution of 10 nm. Three different types of fluorescence beads were tested and distinguished by this spectrometer which holds a promising new concept of the multi-color fluorescence detection in microchip flow cytometry.

Keywords: Optic-fluidic, liquid-liquid waveguide, spectrometer, fluorescence, cytometry,

1 INTRODUCTION

Fluorescence activated flow cytometry is one of most powerful tools for analyzing chemicals, particles and cells in clinical diagnostic, biochemistry and biology. In the past decades, microfluidic and micro-fabrication open new opportunities for miniaturizing flow cytometry. However, nearly all of these micro-devices still rely on bulky external optical detection system which uses several filters and PMTs for multi-color fluorescence detection [1].

In this paper, we present a compact optofluidic spectrometer which use single photon-detector to obtain the full fluorescence spectrum of the particles. The moving particles are excited within a 3D liquid-liquid waveguide that is integrated in the microfluidic chip. By defining a pattern of optical encoder in the substrate underneath the microfluidic channel, the fluorescence spectrum of the particles that pass by can be obtained with high resolution of 10 nm. Three different types of fluorescence beads were tested and distinguished by this spectrometer.

2 MICRO-OPTIC-FLUIDIC SPECTROMETER

Figure 1 shows the schematic diagram of the chip. The chip consists of three parts, the microfluidic structures, liquid waveguide and optical encoder. The microfluidic structures are for cell focusing and sorting with external valves. As the core liquid has a higher refractive index than the cladding with the optically smooth liquid interface, the laminar flows also work as liquid-core liquid-cladding waveguide [2]. The pumping light being confined in the core liquid ensures the fluorescence excitation of the samples

with high efficiency and self-alignment. An opaque mask with grating pattern underneath the channel was deposited on the substrate work as the optical encoder.

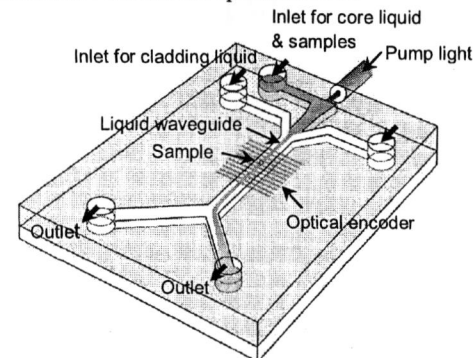

Figure1. The schematic diagram of the optic-fluidic chip.

Figure 2 shows the work principle of the optofluidic spectrometer. The sample passes through the detection area and the fluorescence light transmits into the collection lens through the optical encoder. Thus the output signal from the photon-detector (PMT) is a train of pulses for each sample and different pulses correspond to the different wavelength, namely the envelope curve of the pulse train represents the optical spectrum of the sample. As the modulated fluorescence signal can be easily extracted from the background noise, the optical encoder can also enhance the detection sensitivity.

Figure 3 shows the simulation result of a novel structure for 3-dimentional hydrodynamic focusing. By modifying the cladding flow channel, it realizes the focusing in the both

1-4244-0641-2/07/$20.00 ©2007 IEEE

x-y plane (a) and z-axis dimension (b). Hence, it forms a 3 dimensional liquid-liquid waveguide which is similar to the optical fiber.

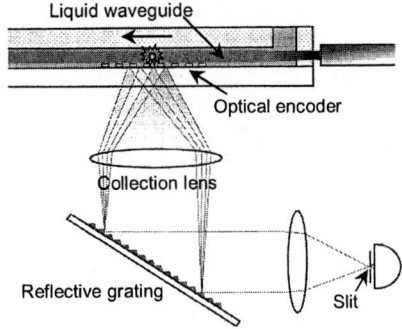

Figure2.The work principle of the optic-fluidic spectr…

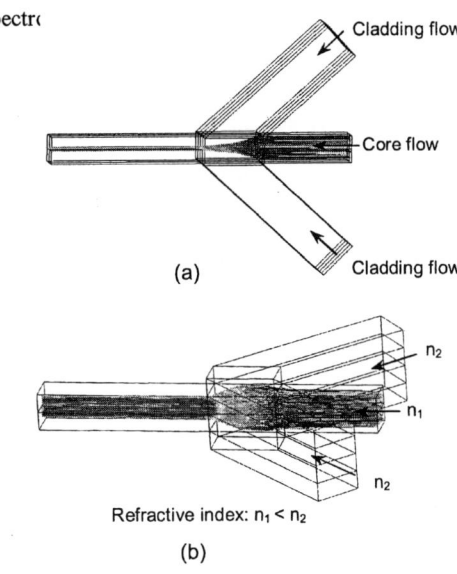

Figure3. Simulation result of 3D hydrodynamic focusing.

The chip was fabricated by replica molding with PDMS which was then bonded on the glass slide. The grating pattern on the substrate was defined by wet etching on the 100 nm thick Al layer.

3 EXPERIMENT RESULTS

Figure 5 shows the micrograph of the microchip when a fluorescence bead was passing by. A multimode optical fiber which is connected to the Argon laser ($\lambda = 488$ nm) couples the pumping light into the liquid waveguide. By adding small amount of ethylene glycol into the pure water, the core liquid has a slight high refractive index ($n = 1.35$) than the

pure water ($n = 1.33$) in the cladding.

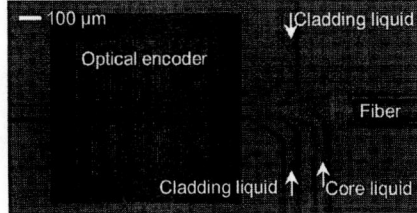

Figure 5.The micrograph of the microchip.

Figure 6 compares the measurement spectrum of three different fluorescence beads. The spectrums are all self-referenced since the pulses with the highest intensity in each spectrum correspond to the scattering light with $\lambda = 488$ nm. A resolution of $\Delta\lambda = 10$ nm can be achieved from this optofluidic spectrometer.

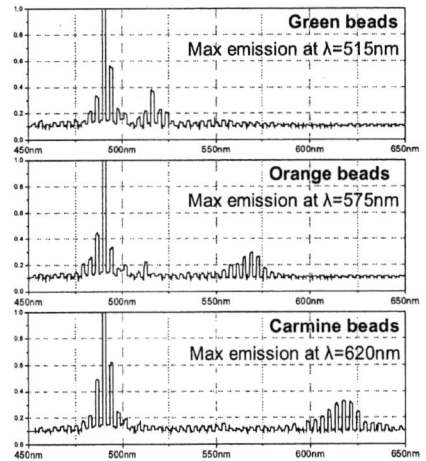

Figure 6.Spectrum of the different fluorescence beads.

4 CONCLUSIONS

An optofluidic spectrometer with integrated 3D liquid-liquid waveguide is demonstrated. Without several PMTs and filters in conventional device, only single photon-detector is needed to measure the fluorescence spectrum with high resolution of $\Delta\lambda = 10$ nm. Besides low cost and ultra compact, it also bears the advantages of self-reference, high sensitivity and accuracy. This method holds a promising new concept for multi-color fluorescent detection in microchip flow cytometry.

REFERENCES

[1] Dongeun Huh, et al., "Microfluidics for flow cytometric analysis of cells and particles," Physiol. Meas. 26(2005) R73-R98
[2] Daniel B. Wolfe, et al., "Dynamic control of liquid-core /liquid-cladding optical waveguides," Proc. Nat. Acad. Sci. USA, 2004, 101: 12434-12338

WC3
14:15 – 14:30

A Disposable Grating-integrated Multi-channel SPR Sensor Chip for Real-time Monitoring of Biomolecule Binding

Young-Hyun Jin and Young-Ho Cho

Digital Nanolocomotion Center, Korea Advanced Institute of Science and Technology
373-1 Guseong-dong, Yuseong-gu, Daejeon 305-701, Republic of Korea
Phn : +82-42-869-8691 / Fax : +82-42-869-8690 / E-mail nanosys@kaist.ac.kr

Abstract

The paper presents a multi-channel real-time monitoring of biomolecule binding using a disposable SPR (Surface Plasmon Resonance) sensor chip integrated with nano-scale pitch grating. The sensor chip has two fluidic channels for sample delivery and gratings in each fluidic channel for SPR sensing. An external mirror makes an incidence of light at different angles on each sensing channel, thus generating two SPR dip at different spectral band. The gratings on the SPR sensor chip are fabricated by micromolding process for disposable application. We monitor the biotin-streptavidin binding reaction in real-time. The binding reaction using 0.2μM streptavidin solution on the biotinylated surface is saturated in 30 min. The refractive index sensitivity of the SPR sensor chip is also characterized as 321.78nm/RI at the sensing wavelength of 607nm and 512.26nm/RI at 704nm.

Keywords: grating-integrated SPR sensor chip, multi-channel SPR sensing, real-time monitoring of biomolecule binding,

1 INTRODUCTION

The Surface Plasmon Resonance (SPR) [1] offers a fast, sensitive and label-free monitoring of biomolecule binding. Because of this feature, SPR biosensors are useful for lab-on-a-chip system. For reliable monitoring of biomolecule binding in interfering environments, SPR sensors should have multi-channel sensing configuration to distinguish specific sensor responses from background environmental perturbations. In the previous paper [2], we reported a multi-channel SPR biomolecule detection configuration using a disposable grating-integrated SPR sensor chip. In this configuration, integrated gratings and a mirror replaced bulky and expensive optical components, such as prism and CCD camera, resulting in a simple and inexpensive multi-channel SPR sensing system. We successfully demonstrated the detection of a target biomolecule. However, the target biomolecule was detected after the completion of the biomolecule binding event without any kinetics information. In this paper, we present the real-time monitoring of biomolecule binding using the grating-integrated SPR sensor chip. We also experimentally characterize the refractive index sensitivity of the SPR sensor chip using IPA solutions.

2 DESIGN AND FABRICATION PROCESS

The grating-integrated SPR sensor chip [2] has two fluidic channels for biomolecule sample delivery and two gratings in each fluidic channel for SPR coupling. An external mirror is used for sequential reflection of incident light between each grating on the SPR sensor chip. Figure 1 illustrates the working principle of the SPR sensor chip. The light is coupled into the gratings at different incident angles, thus generating SPR dips in different spectral bands (Fig.1a and b). This causes the spectrum of finally

reflected light to show two SPR dips as shown in Fig.1c. The refractive index change in each sensing channel can be monitored by observing each SPR dip. The pitch of the grating is designed as 833nm for biomolecule detection in buffer solutions. Figure 2 shows the micromolding process of the nano-scale pitch grating-integrated SPR sensor chip. The 2-hour 85°C PDMS curing process with 1.7kPa mold pressure is followed by the deposition of 40nm-thick gold layer on the PDMS grating. The fluidic channels defined in PDMS layer are covered by glass. Figure 3 compares the AFM topography of the fabricated PDMS grating (Fig.3a) and master grating (Fig.3b). The photograph of the fabricated SPR sensor chip is shown in Fig.4.

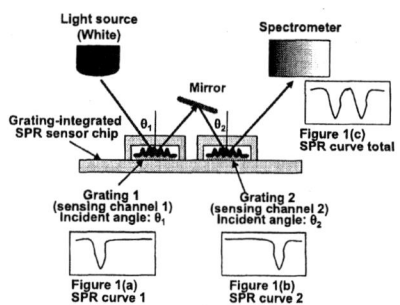

Fig. 1 Multi-channel SPR sensing configuration using a grating-integrated SPR sensor chip and an external mirror.

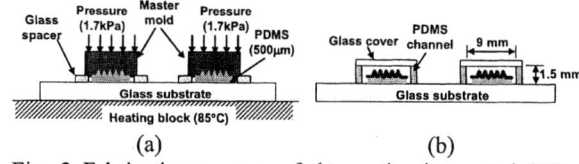

(a)   (b)

Fig. 2 Fabrication process of the grating-integrated SPR sensor chip: (a) micromolding of the PDMS nano-grating; (b) fluidic channel and cover bonding after gold sputtering.

1-4244-0641-2/07/$20.00 ©2007 IEEE   163

Table 1. Experimental refractive index sensitivity and resolution of the fabricated SPR sensor chip

	Sensing wavelength	Sensitivity	Resolution*
Sensing channel 1	607 nm	321.78 ± 8.1 nm/RI	0.00031 RI
Sensing channel 2	704 nm	514.26 ± 8.1 nmRI	0.00019 RI

* assume the resolution of the spectrometer as 0.1 nm.

3 EXPERIMENTAL CHARACTERIZATION

To verify the real-time monitoring of biomolecule binding using the fabricated sensor chip, we use biotin-streptavidin (SA) binding reaction. In both sensing channels, mixed SAM (self-assembled monolayer) is deposited on the gold layer and biotin is immobilized on the SAM. We inject 0.2μM (≈10μg/ml) SA solution in the sensing channel 2. The buffer solution is injected in the sensing channel 1 for monitoring of environmental perturbation. We measure the SPR wavelength of both sensing channels at 10-minute intervals. The results are plotted in Fig.5. In 30 minutes, SPR wavelength at sensing channel 2 increases rapidly (Fig.5a), meaning the SA is binding on the biotinylated surface. After 30 minutes, the SPR wavelength does not shifted more, that means the binding reaction is saturated. The unchanged SPR wavelength in sensing channel 1 shows no environmental perturbations during the experiments.

We also characterize the refractive index sensitivity of the grating-integrated SPR sensor chip using isoprophylalcohl (IPA) solutions with various concentrations. The results are summarized in Table 1. The sensitivity of the sensing channel 2 is about 1.6 times higher than that of sensing channel 1, which agrees with the fact that the sensitivity of SPR sensor with wavelength modulation increases with the increasing SPR sensing wavelength.

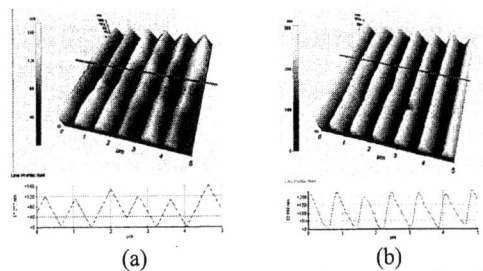

(a)   (b)

Fig. 3 AFM topography: (a) fabricated PDMS grating; (b) master mold.

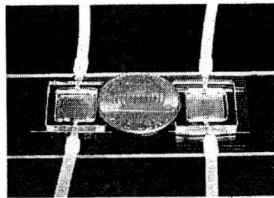

Fig.4 Photograph of the grating-integrated SPR sensor chip.

4 CONCLUSIONS

The paper presented a multi-channel real-time monitoring of biomolecule binding using a disposable nano-scale pitch grating-integrated SPR sensor chip. We observed the kinetics of the biotin-streptavidin binding reaction, as well as monitored the environmental perturbation. The sensitivity of the each sensing channel in the SPR sensor chip was also characterized. The disposable grating-integrated SPR sensor chip is promising for lab-on-a-chip application.

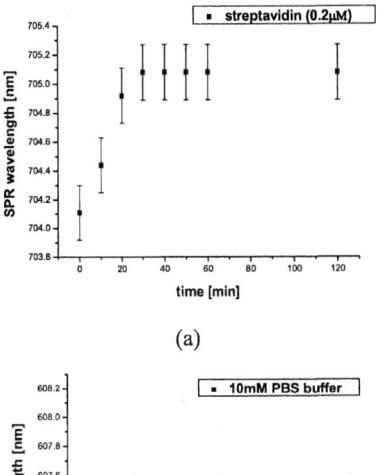

(a)

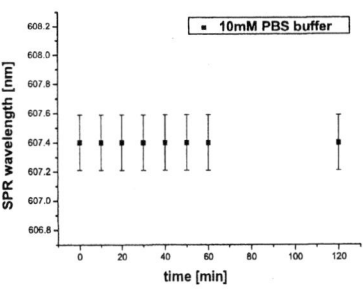

(b)

Fig. 5 Real-time monitoring of the SPR wavelength shift (a) at the sensing channel 2 (SA injected); (b) at the sensing channel 1 (buffer solution injected).

ACKNOWLEDGEMENT

This work has supported by the National Creative Research Initiative Program of the Ministry of Science and Technology (MOST) and the Korea Science and Engineering Foundation (KOSEF) under the project title of "Realization of Bio-Inspired Digital Nanoactuators."

REFERENCES

[1] J. Homola, S.S. Yee, and G. Gauglitz, "Surface Plasmon Resonance Sensors: Review," *Sensors and Actuators B*, Vol. 54 (1991) pp. 3-15.

[2] Y.-H. Jin and Y.-H. Cho, "A Multi-channel SPR Biomolecule Detection System using an Integrated PDMS Nano-scale Grating Chip," *Transducer '07*, Lyon, France (June 10-14, 2007), *in press*.

WC4
14:30 – 14:45

Tunable Resonant Cavity Enhanced Detectors using Vertical MEMS Mirrors

Niels Quack, Stefan Blunier and Jurg Dual

Center of Mechanics, ETH Zurich, Switzerland, Tannenstrasse 3, ETH Zentrum, CLA J35, CH-8092 Zürich.
Tel: +41 44 632 35 63, Fax: +41 44 632 11 45, email: niels.quack@imes.mavt.ethz.ch.

Martin Arnold, Ferdinand Felder, Christian Ebneter, Mohamed Rahim and Hans Zogg
Thin Film Physics Group, ETH Zurich, Switzerland, www.tfp.ethz.ch

Abstract

Highly sensitive photodetectors for the mid infrared have been obtained by placing a photodiode inside a Fabry Pérot cavity [1]. These Resonant Cavity Enhanced Detectors (RCED) are sensitive at the resonances only, which depend on the distance between the two mirrors of the cavity. Displacing one of these mirrors allows changing the cavity length and thus selecting the detection wavelength. The design, simulation and fabrication of a MEMS mirror and its integration with the counter mirror and the photodiode grown by molecular beam epitaxy are presented. First results with external mirrors moved by piezoactuation are described, too.

Keywords: Tunable RCED, infrared detector, vertical micromirror.

WORKING PRINCIPLE

The hybrid device consists of two parts. The complete RCED device working principle is depicted schematically in Figure 1. The lower part shows the detector containing the fixed Distributed Bragg Mirror (DBR) and the p-n photo diode. They are fabricated using lead chalcogenide narrow gap semiconductor materials. The upper part shows the movable MEMS mirror, a displacement of which changes the cavity length and therefore the detection wavelength. Simulations show a shift in detection wavelength from 4.4 µm to 5.6 µm_ for a mirror movement range of 3 µm, while quantum efficiencies are above 80%.

FABRICATION PROCESS

The movable MEMS mirrors have been realized with a process using SOI wafers as shown in Figure 2. The highly doped device layer is structured in two steps by deep reactive ion etch, which define first the electrostatic actuation gap and then the mirror and the suspension legs. The structured SOI wafer is bonded anodically to a glass wafer containing the counter electrodes, followed by removal of the handle and the buried oxide layer by dry etching.

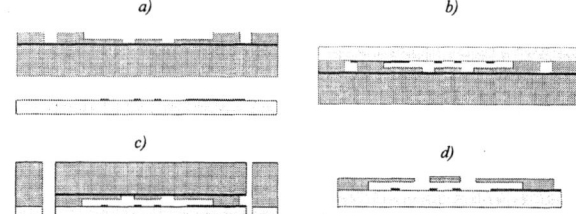

Figure 2: MEMS mirror fabrication process: a) Two step Si dry etch, glass wafer preparation with counter electrodes b) anodic bonding c) dicing d) handle layer and buried oxide removal.

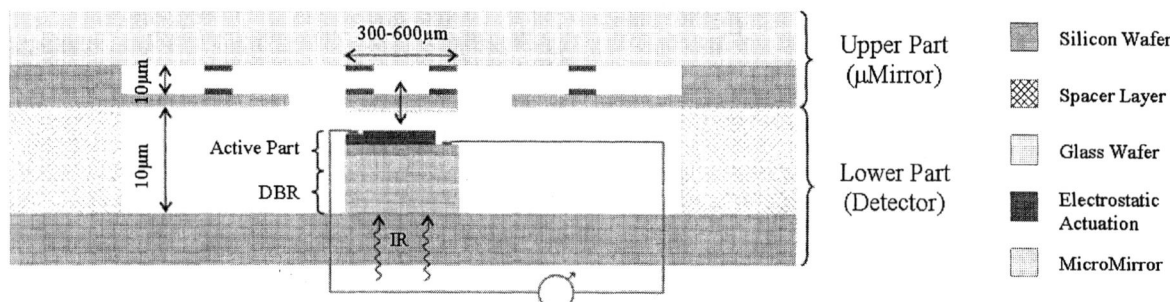

Figure 1: Cross section showing the working principle of the complete RCED Device. The upper part of the schematics shows the micromirror, the lower part comprises the detector part.

1-4244-0641-2/07/$20.00 ©2007 IEEE   165

The use of SOI wafers allowed improving mirror surface flatness and process reliability compared to the first design [2].

With the presented fabrication process, MEMS mirrors with movements of more than 3 μm were obtained (Figure 3) with actuation voltages kept below 30V. Mirror size was 400 μm x 400 μm or 500 μm x 500 μm, suspension length 1450 μm, and thickness 9 μm. Finite element simulations confirmed the obtained displacements (Figure 4).

Figure 3: White Light Interferometer image of a freestanding micromirror.

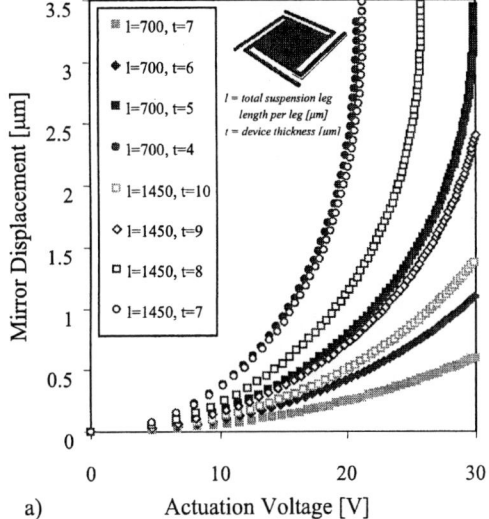

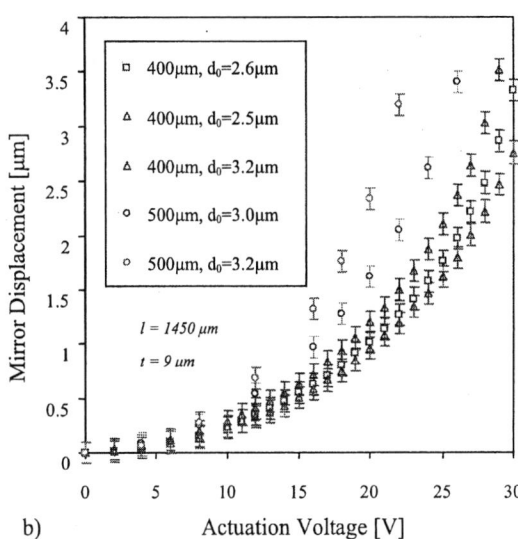

Figure 4: Finite Element simulations (a) and actual measurements (b) of the mirror displacement vs. applied voltage of the electrostatic actuators. For the measured devices, thickness was 9 μm. Measurements of different devices varying in mirror size (400 μm x 400 μm or 500 μm x 500 μm) and initial displacement d_0 are plotted.

RESULTS

First tunable detector results (Figure 5) were obtained so far with a piezoactuated movable mirror [3, 4]. At a cavity length of approximately 30 μm, the detection wavelength was shifted by about 0.7 μm with a total mirror movement of roughly 6 μm.

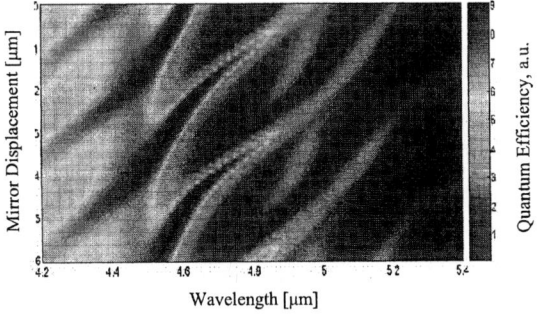

Figure 5: Measured spectral response vs. piezoactuated MEMS-mirror displacement [3, 4].

CONCLUSIONS

The presented results on the MEMS mirror and on the diode, and the demonstrated working principle of the tunable RCED device, are an essential step towards the fabrication of a truly integrated tunable narrowband mid-IR RCED employing a MEMS mirror. Such an integrated device will have a strongly reduced cavity length compared to the piezoactuated version, so that an even broader detection wavelength region can be covered with shorter mirror displacements.

REFERENCES

[1] Martin Arnold, Dmitry Zimin, and Hans Zogg, Resonant-cavity-enhanced photodetectors for the mid-infrared, Appl. Phys. Lett. 87, 141103 (2005)

[2] N. Quack, S. Blunier, J. Dual, M. Arnold, F. Felder, M. Rahim, H. Zogg, Electrostatically actuated micromirror for Resonant Cavity Enhanced Detectors, 20th IEEE Conference on Micro Electro Mechanical Systems 2007, Kobe, Japan.

[3] Ferdinand Felder, Martin Arnold, Mohamed Rahim, Christian Ebneter and Hans Zogg; Niels Quack, Stefan Blunier and Jurg Dual, Narrowband Resonant Cavity Enhanced mid-IR Photodetectors tuneable within a broad wavelength region, EUROMBE 14, 14th European Molecular Beam Epitaxy Workshop, Granada, Spain, March 5-7, 2007.

[4] Martin Arnold, Ferdinand Felder, Christian Ebneter, Mohamed Rahim, Hans Zogg, Niels Quack, Stefan Blunier and Jurg Dual, Wavelength tunable resonant cavity enhanced detectors, MIOMD VIII, Bad Ischl, Austria, May 14-17, 2007.

WC5
14:45 – 15:00

Two-state Optical Filter Based on Micromechanical Diffractive Elements

Håkon Sagberg, Thor Bakke, Ib-Rune Johansen, Matthieu Lacolle, and Sigurd T. Moe
SINTEF Department of Microsystems and Nanotechnology
Gaustadalléen 23C, Oslo, Norway
Email: Hakon.Sagberg@sintef.no

Abstract—We have designed a robust two-state filter for infrared gas measurement, where the filter transmittance alternates between a single bandpass function, and a double-band offset reference. The device consists of fixed and movable diffractive sub-elements, micromachined in the device layer of a bonded silicon on insulator (BSOI) wafer. Switching between the two states of the filter is obtained by actuation of the movable sub-elements between idle and pull-in positions, which affects the interference of reflected light. The characteristics of the filter are defined by a diffractive microrelief pattern etched on top of the sub-elements and by the position of the movable sub-elements at pull-in, the latter mechanically defined by the buried oxide layer. Thus, no accurate electrical control is needed to operate the filter. The first test components operate at $2\,\mu m$ wavelength using a displacement of $500\,nm$ and an actuation voltage of $5\,V$. No sticking or change in filter characteristics have been observed after repeated pull-in operations. The simplicity of fabrication and operation is likely to make the two-state filter an attractive component for sensors such as non-dispersive infrared gas detectors.

I. INTRODUCTION

Non-dispersive infrared (NDIR) measurements of gas concentration is a widely used technique. The principle is to measure the attenuation of light that has passed through a volume containing the gas sample. The light is band-pass filtered to match the absorption band of the desired gas. For reliable and drift-compensated operation the instrument must have a built-in reference measurement, either in the form of a reference absorption path, or a reference wavelength, or a combination of the two. The advantage of the NDIR method compared to low-cost electrochemical methods is first of all its robustness towards issues such as saturation or contamination from other gases.

It is the purpose of our work to provide a two-state optical MEMS band-pass filter for infrared gas measurements. One of the filter states defines the absorption band, and the other defines the reference wavelength band(s). Traditionally such a filter is realized with a rotating filter wheel having two or more interference filters. Compared to the filter wheel, MEMS technology has potentially the following advantages: Precise switching, smaller size, no motors, higher modulation frequency, and lower cost.

Fabry-Perot interferometers have previously been developed as MEMS devices for the above purpose [1]. Here we present a filtering device which is based on a different optical principle and requires fewer processing steps than a Fabry-Perot interferometer: The two-state controllable diffractive optical

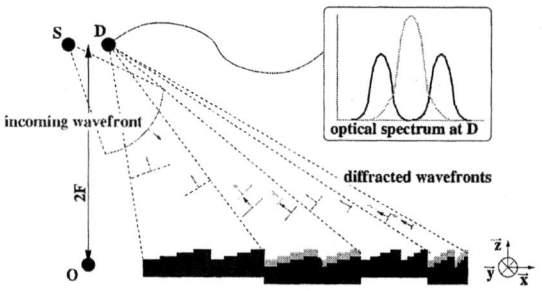

Fig. 1. Principle of the two-state controllable diffractive optical element. When a spherical wavefront propagates from the source S, it is reflected by Fresnel zone-shaped diffractive sub-elements. Here, four sub-elements are shown, each consisting of two periods (four Fresnel zones) and four height levels. The reflected wavefronts recombine in phase at the detector D only for a chosen wavelength λ_c (light gray). If we pull down every second sub-element a distance $\lambda_c/4$, the wavefronts will recombine destructively for λ_c, but constructively for the two side-bands (dark gray).

element (CDOE). The two-state CDOE has much in common with grating light valves (GLV) [2], and is a special case of a group of optical MEMS devices using phase-controlled arrays of reflectors [3].

II. DESIGN

The two-state CDOE can be viewed as grating light valves operated in a high ($m > 1$) diffraction order. Reducing the light intensity at one wavelength will increase the intensity at the neighboring wavelengths (side-bands). The principle is shown in Figure 1 with four sub-elements, and diffraction order $m = 2$. The distance between the side-bands decreases with increasing diffraction order. An external focusing lens is not needed, since focusing is achieved using a Fresnel zone-shaped microrelief pattern. For mechanical simplicity, the fabricated test component shown in Figure 2 has an array of *rectangular* diffractive sub-elements of equal size and distance. In this case the sub-element edges do not coincide with Fresnel zone borders, something that broadens and reduces the intensity of the side-bands.

The operation of the device requires only a single voltage signal applied between the suspended frames and the substrate. The optimum modulation frequency is dependent on the selected application and detector type. Since the mechanical movement is defined to sufficient accuracy by the fabrication

1-4244-0641-2/07/$20.00 ©2007 IEEE

Fig. 2. This scanning electron micrograph(SEM) shows the fabricated two-state filter. The inserts above are close-ups of the diffractive sub-element. Every second sub-element (with etch-holes) is freely suspended, and the remaining sub-elements are attached to the substrate. The suspended sub-elements are mechanically connected to a frame, which is again connected to the substrate through narrow beams that function as springs. A single filter device consists of an arbitrary number of frames. The component used for testing has 5 by 5 frames, where each frame is approximately one by one millimeter. The diffractive pattern is a quarter wavelength deep and covers the optically active area.

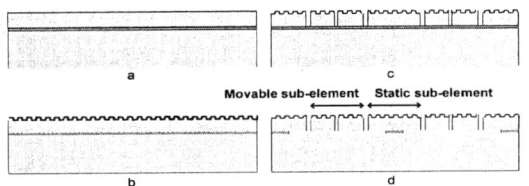

Fig. 3. Fabrication of the device. The diffractive filter is fabricated from a bonded silicon-on-insulator (BSOI) wafer using only two mask layers.

process, neither closed-loop operation nor individual calibration should be necessary.

III. FABRICATION

The fabrication process starts with the structuring of the diffractive microrelief pattern (Figure 3b). The depth of the microrelief is 500 nm, which was achieved by using a highly uniform, well tuned shallow reactive ion etching process based on a plasma that consisted of a mix of SF_6 and C_4F_8. The wafer is then coated with a 100 nm layer of aluminum by sputter deposition, followed by patterning and etching of the aluminum using wet etching. The 15 μm thick device silicon is then etched down to the buried oxide (BOX) of the SOI wafer using traditional deep reactive ion etching (Figure 3c). The final step is the release of the movable frame by partial removal of the buried oxide using vapor phase HF etching (Figure 3d). The movable parts of the filter structure contain release holes that ensure complete removal of the oxide underneath, while the absence of release holes on the static part of the filter ensure that some oxide remains, supporting the structure.

IV. MEASUREMENTS

Figure 4 shows the static deflection of the diffractive elements, measured in an optical profilometer for a range of

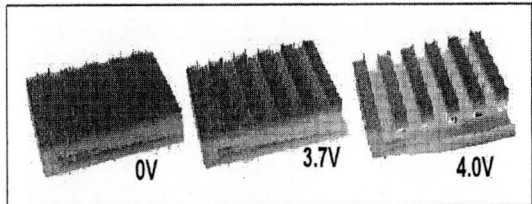

Fig. 4. Interferometric measurements of the mechanical deflection at three different voltages. The initial height difference between static and movable elements is less than 50 nm. The pull-in voltage is 4 V, creating a height difference of approximately 450 nm. At higher voltages there is little change, except that the frame becomes increasingly flat towards the edges. The measured area is 600 by 500 μm and corresponds to approximately a quarter of a complete frame.

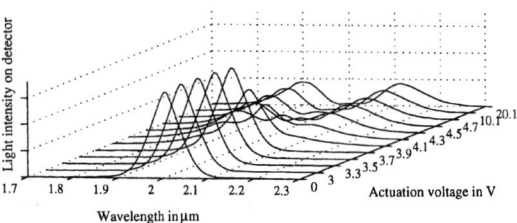

Fig. 5. Fourier-Transform Infra-Red (FTIR) measurements show the detected light intensity as a function of wavelength, for a range of actuation voltages. The two side-bands are located at $\lambda \approx \lambda_c (1 \pm 1/2m)$, where λ_c is the center wavelength and m is the order, or average number of grating periods on each sub-element. The total intensity of the reference double-band is 66% of the central band, and the band-widths are larger. Stronger and narrower reference bands can be achieved with optimal sub-element sizes and shapes.

actuation voltages. The movable frame snaps down onto the substrate at 4 Volts, creating a height modulation corresponding to the thickness of the removed oxide layer. We characterized the filtering properties by mounting the component on an optical bench with illumination and detection through fibers, using a geometry as in Figure 1 with a focal length of 41 mm. The system f-number was approximately *f/8*. The light source was a tungsten halogen lamp, and we connected the detection fiber to a Fourier Transform Infra-Red (FTIR) spectrometer. The resulting spectra are shown in Figure 5.

V. CONCLUSION

We have designed, fabricated, and characterized a controllable micromechanical diffractive element and verified that it works as a two-state optical filter for near-infrared wavelengths. The main benefits of the filter device are the simplicity of fabrication and operation, and that few external lenses or mirrors are needed in a complete sensor system.

REFERENCES

[1] Vaisala carbocap sensor: http://www.vaisala.com
[2] O. Solgaard, F. Sandejas, and D. M. Bloom, "Deformable grating optical modulator," *Optics Letters*, vol. 17, no. 9, pp. 688–690, 1992.
[3] M. Lacolle, H. Sagberg, I.-R. Johansen, O. Løvhaugen, O. Solgaard, and A. S. Sudbø, "Reconfigurable near-infrared optical filter with a micromechanical diffractive fresnel lens," *IEEE Photonics Technology Letters*, vol. 17, no. 12, pp. 2622–2624, 2005.

WD1 (Invited)
15:30 – 16:00

MEMS Tunable Filters for LWIR Spectral Imaging

J.F. DeNatale, R. Borwick, P. Stupar, W. Gunning, P. Kobrin and S. Lauxtermann
Teledyne Scientific and Imaging
1049 Camino Dos Rios, Thousand Oaks, CA USA 91360
Tel 805-373-4439, Fax 805-373-4869, E-mail jdenatale@teledyne.com

Abstract

The Adaptive Focal Plane Array (AFPA) is a new spectral imaging technology integrating an array of MEMS tunable filters with a dual-band IR focal plane array. This will provide the ability to acquire spectrally-tuned spatial information in the LWIR while preserving broadband imaging capability in the MWIR. A key component of this device is the MEMS tunable filter array, which involves consideration of multiple factors: static mechanics, dynamic response, electrical actuation, optical response, and integration compatibility. This paper presents the development and status of the MEMS element and its integration into the AFPA device.

Keywords: MEMS, imaging, tunable filter, Fabry-Perot, spectral imaging

1 INTRODUCTION

Spectral information can significantly enhance the ability to detect and identify an imaged object. This is well-accepted based on our experience with human vision, and is equally true for infrared imaging systems. Such hyperspectral / multispectral imaging systems provide enhanced contrast to identify objects in clutter, and have widespread usage in geological and botanical remote surveys. Conventional spectral imaging devices typically are characterized by significant size, weight and data processing penalties, which often limit their utility and applications. The Adaptive Focal Plane Array (AFPA) is being developed to address these limitations, enabling a compact spectrally-agile dual-band image sensor [1,2].

The AFPA device involves the integration of a spectrally-tunable filter array with a dual band (MWIR/LWIR) staring focal plane array (DB-FPA) [2]. The filters are intended to meet challenging specifications: tunability across the LWIR (8-11μm) with narrow spectral bandwidth (~100nm) while maintaining broadband transmission in the MWIR band. This must be achieved in a compact form-factor compatible with integration with the cryogenically cooled DB-FPA. The filter array must also be segmented to permit independent tuning of small areas of the imager (i.e. one filter covering a small number of imager pixels) while preserving high fill-factor.

2 APPROACH

The approach to the tunable filter is to use an array of MEMS-based Fabry-Perot filters, Fig.1. These provide transmission over a narrow bandwidth, defined by the reflector coatings and gap dimension. The mechanical motion enabled by the MEMS implementation is used to vary the gap and thus tune the wavelength of the filter peak.

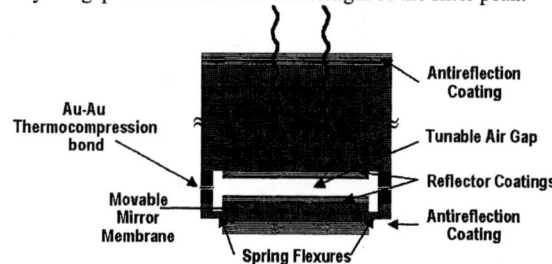

Figure 1. Schematic cross-sectional representation of an individual MEMS tunable filter element.

The general design and construction of the individual filters is shown schematically in Fig. 1. A two-layer structure is used for the individual filters, comprised of a stationary Si supporting substrate and a movable Si thin membrane. The two are attached using a Au-Au thermocompression bond process and an SOI device layer transfer process in which the Si support is bonded to an SOI wafer and the handle wafer subsequently removed. The two-wafer approach provides access to both internal surfaces of the cavity, which is where the high-reflector coatings are located which define

1-4244-0641-2/07/$20.00 ©2007 IEEE   169

the Fabry-Perot filter. Mechanical motion of the movable mirror is achieved via electrostatic actuation by applying a potential between the stationary and movable elements of the device. To maximize fill factor and maintain low actuation voltages, thin folded flexures were used for the movable element supports. A key challenge in the device is the integration with the multilayer dielectric coatings, which must be quite thick to achieve high LWIR filter transmission, narrow bandwidth, and dual-band operation. This can impose significant stresses that can deform the thin movable membrane.

3 RESULTS

The MEMS tunable filter array was successfully fabricated with filter sizes ranging from 100μm to 400μm, and the SEM micrographs of Fig. 2 illustrate the design elements: bonding supports, flexures, Si membrane, and top-surface antireflection coating to maximize filter transmission. On the underside of the element can be seen the inner surface reflector coating recessed into the membrane to minimize the standoff dimensions and thinned flexures to reduce the mechanical stiffness of the structure, permitting operation at a lower applied voltage.

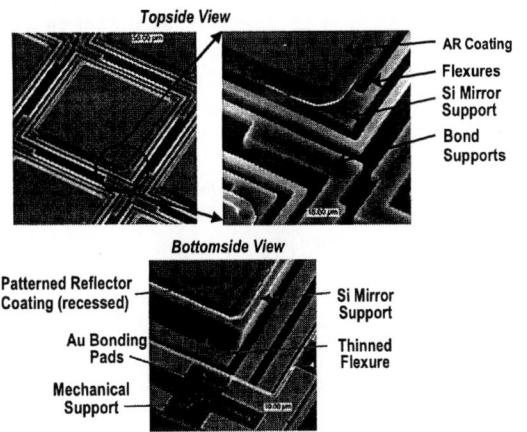

Figure 2. SEM micrographs of MEMS tunable filter array showing structural features on top- and bottom surfaces, respectively.

Transmission spectra of individual MEMS filters were taken using a Bruker Optics Hyperion 2000 IR microscope. In this instrument the sample is illuminated with a beam having a significant distribution of incident angles ranging between 12 and 25 degrees. This imposes a number of instrumental artifacts into the measured spectra, including a significant

wavelength shift, peak broadening and reduction in apparent peak transmission. The spectral transmission is shown in Fig. 3. Here, the peak wavelengths have been corrected to compensate for instrumental effects and transmission calibrated against a bare Si wafer, but filter bandwidth and peak transmission effects left uncorrected. The spectra clearly show the peak shift with applied voltage across the 8-10μm range, validating the design concept. The normal incidence transmission and bandwidth were measured by voltage scanning the filters through several lines of a tunable CO2 laser. From this we were able to determine actual filter bandwidths of 90-145nm and peak transmission of 80-90%.

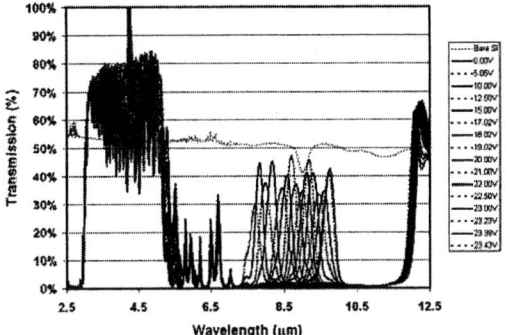

Figure 3. Measured room temperature spectral transmission of a MEMS tunable filter as a function of applied voltage. Dashed line is the transmission of an uncoated Si wafer for transmission calibration. The filter spectra are uncorrected for angle of incidence effects.

ACKNOWLEDGMENTS

This work was sponsored by DARPA MTO and under contract from the US Army RDECOM CERDEC Night Vision and Electronic Sensors Directorate. We thank Ray Balcerak, DARPA Program Manager, and Joe Pellegrino and Phil Perconti of NVESD for their strong support and guidance in this effort. The authors also acknowledge the efforts of the many Teledyne Scientific and Imaging researchers who contributed to the work described here.

REFERENCES

[1] J. Carrano, J. Brown, P. Perconti, K. Barnard, "Tuning in to Detection: Combining Tunability and High Resolution FLIR Detection Improves Target Recognition," OEmagazine, 20-22, April 2004.

[2] W. J. Gunning, J. L. Johnson, and J. F. DeNatale, "LWIR/MWIR Adaptive Focal Plane Array," SPIE Proceedings, Vol. 5612, pp. 1 – 7, (2004)

WD2
16:00 – 16:15

Tunable Erbium Doped Fiber Laser Using a Silicon Micro-Electro-Mechanical Fabry-Perot Cavity

J. Masson, S. Bergeron, A. Poulin, N. Godbout and Y.-A. Peter
Ecole Polytechnique de Montréal, Engineering Physics Department
P.O. Box 6079, Station Centre-Ville, Montréal (QC), H3C 3A7 CANADA
Tel + 1 514 340 4711 x 3100, Fax + 1 514 340 3218,
Email: {jonathan.masson, yves-alain.peter}@polymtl.ca

Abstract

We propose a novel tunable erbium doped fiber laser using a silicon micro-electro-mechanical (MEM) Fabry-Perot cavity. The cavity is made of two Bragg mirrors, one being actuated by comb drives. The MEM Fabry-Perot cavity and grooves for optical fibers are fabricated by DRIE on a 70 μm SOI wafer and integrated in a ring fiber laser configuration. The fiber laser has a tuning range of 7.7 nm in the C-band and a spectral width of 0.1 nm.

Keywords: Tunable fiber laser, Fabry-Perot cavity, silicon Bragg reflector.

1 INTRODUCTION

Tunable silicon optical filters using deformable Bragg gratings or tunable Fabry-Perot (FP) cavities were recently proposed [1] and demonstrated [2-4]. These tunable filters can be used for a variety of applications such as optical filtering in telecommunications, biochemical sensing and tunable lasers.

With the development of dense wavelength division multiplexing (DWDM) networks, numerous laser sources emitting at different wavelength are needed. The multiplication of laser sources has a large cost impact on DWDM networks. In this paper, we propose a micro-electro-mechanical system (MEMS) tunable erbium doped fiber laser, which could potentially replace several lasers at a reasonable cost. As the gain of erbium spans over a large wavelength range centered at 1550 nm, our device enables tuning over the whole C band [5]. Tunable fiber lasers using intracavity fiber Fabry-Perot filters have been previously reported [6]. In this paper, we report a novel tunable erbium doped fiber laser using an integrated silicon MEM Fabry-Perot cavity.

2 SILICON FABRY-PEROT CAVITY

The mirrors of the FP cavity are made of two silicon Bragg reflectors. One of these mirrors is fixed while the other one can be displaced by a comb drive actuator. Figure 1 shows the fabricated FP filter. When a voltage is applied, the combs get closer and thus the air gap of the FP cavity is tuned. Figure 2 is the simulated displacement of the comb versus voltage. When the combs come closer, the air gap of the FP is decreased. A smaller gap means a shorter filtered wavelength by the FP [4].

The MEM tunable FP cavity is fabricated by deep reactive ion etching (DRIE) on a silicon on insulator (SOI) wafer. A 70 μm thick silicon device layer is used in order to allow optical fiber integration in grooves etched during the same process step as the tunable FP cavity and the comb drives. The structure is released in liquid HF followed by supercritical CO_2 drying to prevent sticking of the devices. Such a FP filter was previously demonstrated with 20 nm wavelength tuning range [4].

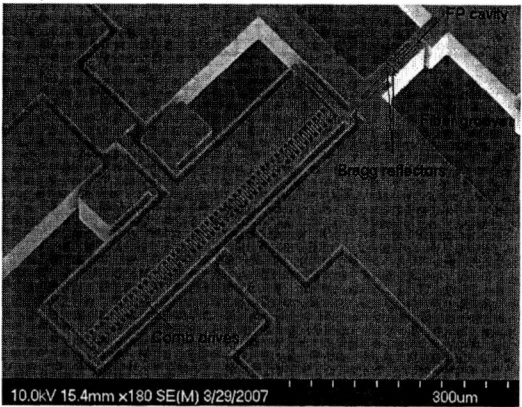

Figure 1. SEM photograph of the silicon microfabricated tunable Fabry-Perot device.

1-4244-0641-2/07/$20.00 ©2007 IEEE   171

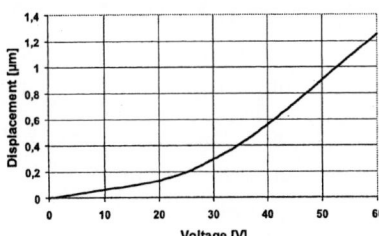

Figure 2. Simulation results of the comb drive displacement versus applied voltage.

3 TUNABLE FIBER LASER

The setup of the ring laser is shown in Fig. 3. The erbium doped fiber is pumped with a 1480 nm laser diode through a 1480nm/1550nm WDM coupler. We use an isolator to insure one way lasing direction. The MEM FP is positioned within the ring cavity to select the lasing wavelength. We use a 1 % tap as an output coupler to minimize losses in the cavity. Alignment of the optical fibers with the MEM FP is critical. The doped fiber and the output fiber are passively aligned by the silicon fiber grooves on each side of the tunable filter.

The transmission peak of the FP shifts to shorter wavelength, while increasing the applied voltage to the comb drive as shown in Figure 4. The largest tuning of the FP is reached at a voltage of 14.6 V. We measured 7.7 nm tuning range of the fiber laser from 1563.5 nm to 1555.8 nm with increasing voltage applied to the comb drive (Fig. 5). The spectral width of the fiber laser is approximately 0.1 nm (FWHM).

4 CONCLUSION

A MEMS tunable erbium doped fiber laser has been demonstrated for the first time to our knowledge. The fiber laser has a tuning range of 7.7 nm and a spectral width of 0.1 nm.

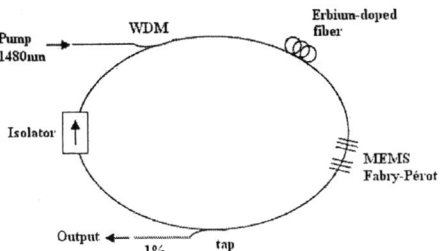

Figure 3. Optical setup of the ring fiber laser.

ACKNOWLEDGEMENTS

The authors would like to thank Maxime Rivard for the fabrication of the optical fiber couplers.

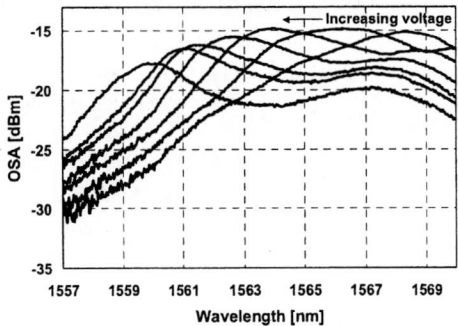

Figure 4. Transmission spectra of the tuned FP.

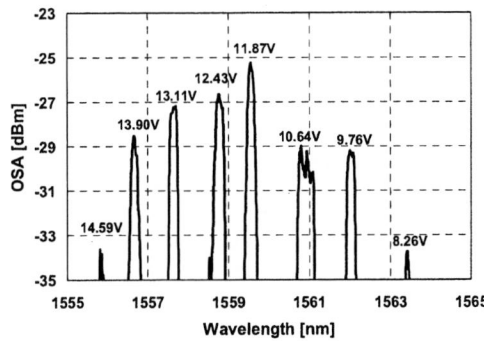

Figure 5. Measured fiber laser spectra at different voltages applied to comb drive of the Fabry-Perot cavity.

REFERENCES

[1] A. Lipson, E. M. Yeatman, "Free-space MEMS tunable optical filter on (110) silicon," IEEE/LEOS Optical MEMS, 2005, p 73-74.

[2] S.-S. Yun, K.-W. Jo, J.-H. Lee, "Crystalline Si-based in-plane tunable Fabry-Perot filter with wide tunable range," IEEE/LEOS Optical MEMS, 2003, p 77-78.

[3] M. Tormen, Y.-A. Peter, Ph. Niedermann, A. Hoogerwerf, R. Stanley, "Deformable MEMS grating for wide tunability and high operating speed," Journal of Optics A, vol. 8, no. 7, pp. S337-40, 2006.

[4] J. Masson, F. B. Koné and Y.-A. Peter, "MEMS Tunable Silicon Fabry-Perot Cavity," accepted in SPIE Optomechatronic Micro/Nano Devices and Components III, Lausanne, Switzerland, October 2007.

[5] M. J. F. Digonnet, Rare-Earth-Doped Fiber Lasers and Amplifiers, Marcel Dekker, 2001.

[6] S. Yamashita and M. Nishihara, "Widely Tunable Erbium-Doped Fiber Ring Laser Covering Both C-Band and L-Band," IEEE J. Select. Topics Quantum Electron., vol. 7, no.1, pp. 41-43, 2001.

WD3
16:15 – 16:30

Design and Fabrication of Photonic MEMS Waveguide Modulators

Akio Higo, Hiroyuki Fujita*, Yoshiaki Nakano, and Hiroshi Toshiyoshi*
Research Center for Advanced Science and Technology, The University of Tokyo
*Institute of Industrial Science, The University of Tokyo
Contact Address
Tel +81-3-5452-5155, Fax +81-3-5452-5156, E-mail higo@iis.u-tokyo.ac.jp
Research Center for Advanced Science and Technology, The University of Tokyo,
4-6-1 Komaba, Meguro-ku, Tokyo 153-8904 Japan

Abstract

We report the design and fabrication process of photonic MEMS actuators for optical attenuators integrated with a silicon photonic wire waveguide. This paper presents design and theoretical analysis of the silicon optical waveguide modulator with an electrostatic micromechanical structure actuated in the evanescent range. We observed a 100-nm MEMS displacement on a 5-um-wide and 50-um-long cantilever at a voltage of 20pp.

Keywords: Photonic waveguides, Silicon photonic MEMS, Optical MEMS,

1. INTRODUCTION

Most conventional optical MEMS use scanning micro mirrors to redirect light beams in free-space, for instance, to establish interconnection between optical fiber arrays [1]. In such systems, micro mirrors of the similar dimensions as a light beam (typically several hundreds microns in diameter) are electromechanically controlled with scan angles of a few degrees. Different from these approaches, we have recently proposed an optical MEMS device integrated with silicon photonic waveguides, where nano-to-micron scale optomechanical components are brought into the evanescent field or in full contact with lightwave for optical modulation [2,3].

2. OTPICAL DESIGN

Figure 1 shows a schematic illustration of silicon photonic MEMS modulator. We aim at the process compatibility with the silicon photonic circuits. Figure 2 shows the results of the three-dimensional FDTD (Finite Different Time Domain) simulation (Opti-FDTD^{TM} simulator), i.e., the optical intensity distribution induced by the MEMS cantilever motion. We gave a TE-mode and observed the magnetic fields. A micro anchor of ellipsoidal shape of silicon is suspended by the silicon waveguide and the optical anchors, where minimal optical loss was induced when no voltage is applied. Once the cantilever with the photonic waveguide/optical anchor[4] was brought into the evanescent field on the silicon substrate, optical power was transferred to the handle layer and hence the optical power traveling in the waveguide was calculated to decrease.

A straight silicon photonic wire waveguide made of a 500-nm wide and a 200-nm thick silicon layer was found to have minimum insertion loss when a silicon optical anchor of 1.5-um wide, 7.5-um long, and 200-nm thick (connected with the waveguide) was placed at a tip of the cantilever.

Optical modulation contrast of as large as 33 dB was estimated when the optical anchor was brought into full contact with the substrate. Having multiple MEMS modulators in series along the waveguide would give us fine control of the traveling light intensity.

3. MEMS DESIGN AND FABRICATION

In this process, we first defined the fine patterns of silicon wire waveguide by using the electron-beam pattern lithography system (F5112: ADVANTEST Corp.,) and high-aspect ratio dry etching of a 200-nm active layer of a SOI (silicon-on-insulator) wafer with Deep-RIE (reactive-ion-etching). The micromechanical structures were released in the vapor of hydrofluoric acid [5].

Figure 3 (a) and (b) show the SEM (scanning electron microscope) images of the photonic MEMS modulators. The microstructures were released in vapor HF without causing serious damage to the waveguide. Silicon photonic wire waveguide is expected to have low optical propagation loss compared with photonic crystals waveguide. We achieved 100 nm motion with an approximately 20 Vpp with the 5-um-wide and 50-um-long cantilever, measured by using the laser Doppler vibrometer .

4. CONCLUSION

We have successfully fabricated photonic MEMS modulators and observed 100 nm motion with an approximately 20 Vpp with the 5-um-wide and 50-um-long cantilever, measured by using the laser Doppler vibrometer.

Acknowledgement

We appreciate the VLSI Design and Education Center (VDEC) of the University of Tokyo for EB writing with 8-inch EB writer F5112+VD01 donated by ADVANTEST corporation.

1-4244-0641-2/07/$20.00 ©2007 IEEE

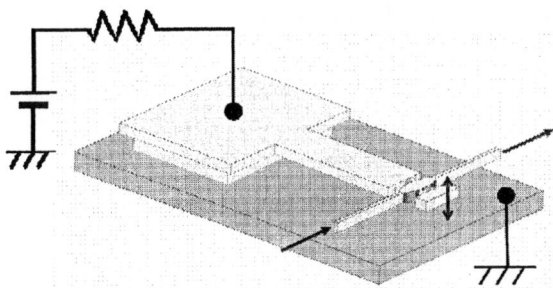

Figure 1. Schematic illustration of photonic MEMS modulator integrated with a silicon nanowire waveguide.

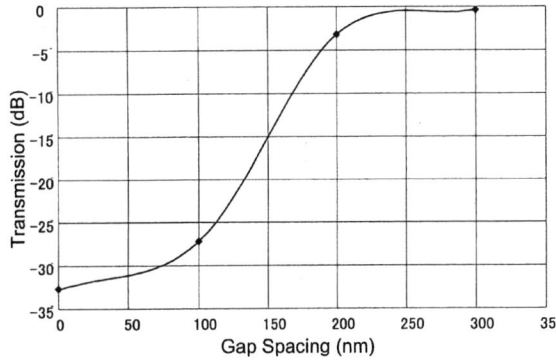

Figure 2. 3D-FDTD simulation results of transmission as a function of the gap between the anchor and the substrate.

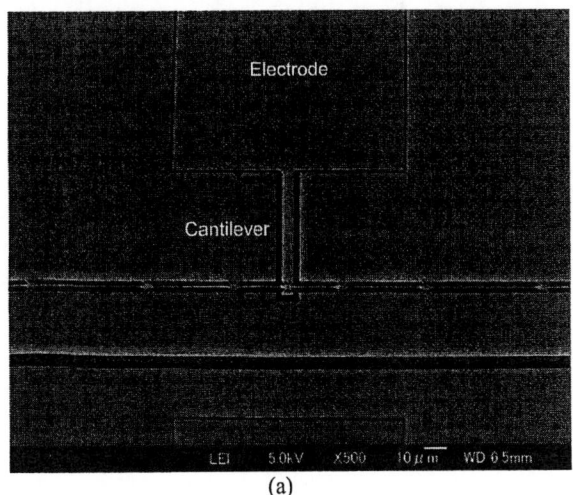

(a)

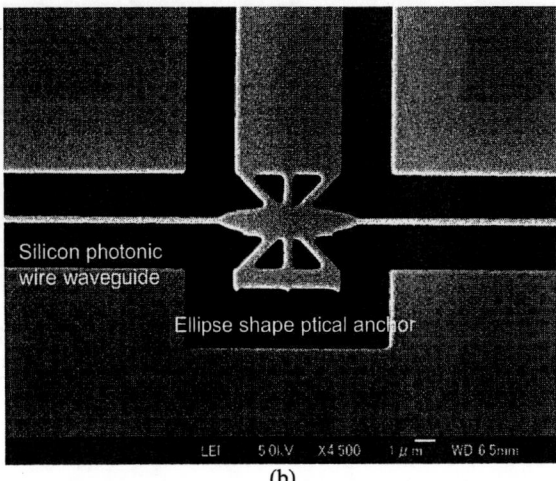

(b)

Figure 3. SEM images of fabricated device. (a) whole picture and (b) a close-up image the optical anchor at a tip of cantilever.

REFERENCES

[1] D. J. Bishop, C. R. Giles and G. P. Austion, "The Lucent Lambda Router: MEMS Technology of the Future Here Today," IEEE Communications Magazine, March, 75-79, 2002.

[2] A. Higo, S. Iwamoto, S. Ishida, Y. Arakawa, M. Tokushima, A. Gomyo, H. Yamada, H. Fujita and H. Toshiyoshi, "Development of high-yield fabrication technique for MEMS-PhC devices", IEICE Electron. Express, Vol. 3, No. 3, pp.39-43, (2006)

[3] A. Higo, H. Fujita, and H. Toshiyoshi, "Design and fabrication of optical MEMS modulator with silicon wire waveguide," Proc. IEEE/LEOS Int. Conf. on Optical MEMS and Their Applications (Optical MEMS 2006), Big Sky Resort, Big Sky, Montana, Aug. 21-24, 2006

[4] T. Fukazawa, T. Hirano, F. Ohno and T. Baba, "Low loss intersection of Si photonic wire waveguides", Jpn. J. Appl. Phys., vol. 43, no. 2, pp. 646-647, 2004.

[5] Y. Fukuta, H. Fujita and H. Toshiyoshi: "Vapor Hydrofluoric Acid Sacrificial Release Technique for Micro Electro Mechanical Systems Using Labware" Jpn. J. Appl. Phys. Vol. 42 (2003) 3690-3694

WD4
16:30 – 16:45

A Study on Optical Diffraction Characteristics
of Skewed MEMS Pitch Tunable Gratings

K.Takahashi[1,2], K.Suzuki[3], H. Funaki[3], K. Itaya[3], H. Fujita[1], and H. Toshiyoshi[1,2]

1 Institute of Industrial Science, University of Tokyo,
4-6-1 Komaba, Meguro-ku, Tokyo 153-8505, Japan
Phone: +81-3-5452-6277 / Fax: +81-3-5452-6250 / E-mail: johnny@iis.u-tokyo.ac.jp
2 Kanagawa Academy of Science & Technology, Japan, 3 R&D Center, TOSHIBA Corp., Japan

ABSTRACT

This paper presents a pitch tunable grating light valve integrated with high-voltage (40 V) driver circuits that target for image projection display devices. In our work, optical light angle is modulated by changing the period of the MEMS grating pixel by means of electrostatic actuation. In addition, we developed a MEMS grating light valve with a skewed angle for elimination of the inter-spot crosstalk. The skewed gratings were found to spread the +/- 1st order diffraction off the axis. Optical diffraction angles of 6.6 degrees (OFF-state) and 3.3 degrees (ON-state) were obtained.

Keywords: Diffraction gratings, pitch tuning, skewed gratings, projection display, on-chip electronics

1 INTRODUCTION

In this paper we present a new type of MEMS grating light valve with a skewed angle for better image quality without inter-spot crosstalk. Arrayed light valve is an indispensable device for creating 2D images of fast motion, and several types of MEMS light valves have been reported [1,2], and vertical motion of suspended micro strips has been used to control the intensity of the diffracted beam spots. Unlike these, we use lateral motion of grating elements to control the diffraction angles in an analog manner.

2 DESIGN AND OPERATION

Figure 1 schematically illustrates the optical system of projection display using a grating light valve and a Galvano scanner. A line-pattern of laser is projected on the gratings, and the diffraction angle is controlled pixel-wise. Intensity controlled pixels are vertically aligned, and the pattern is spatially swept over the screen to make a two-dimensional

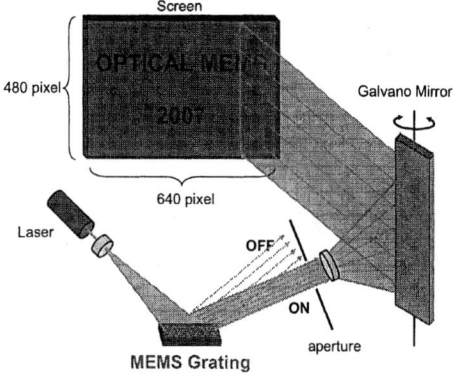

Figure 1 Principle of image projection display by using pitch tunable grating device.

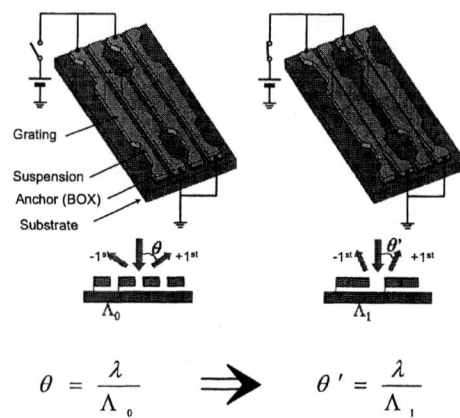

$$\theta = \frac{\lambda}{\Lambda_0} \quad \Rightarrow \quad \theta' = \frac{\lambda}{\Lambda_1}$$

Figure 2 Schematic structure of pitch tunable grating device.

image. **Figure 2** shows the schematic unit structure of the gratings. The grating slips are suspended over the substrate with both end anchored in a bridge style, and electrostatic force is used to tune the mutual distance. Hence the grating pitch is locally modulated to change the diffraction angle, resulting in the intensity change of the corresponding pixel.

In out prototype model presented at MEMS 2007 in Kobe [3], we used mechanically movable silicon slips that were arrayed in the direction perpendicular to the axis, as shown in **Figure 3** (a). A line pattern of laser beam is projected onto the gratings to make diffracted beam spots, among which only the first order diffraction pattern is used to make a 2D image by horizontal sweep. However, the ON-OFF operation of grating displaces the beam spot in the same direction as the diffraction patterns, which may superpose the existing patterns, as shown in the same figure. This may result in burred image quality when the motion of the grating

1-4244-0641-2/07/$20.00 ©2007 IEEE   175

is limited to be small. Here, we propose an alternative design of gratings by skewing the grating slips as shown in **Figure 3 (b)**. In such a case, the line pattern of laser beam makes a skewed diffraction pattern, and the ON-OFF operation of gratings make a beam spot walk off the diffraction patterns. Therefore, such walk-off beam spot can be easily stopped by the screening slit not to interfere with the image quality.

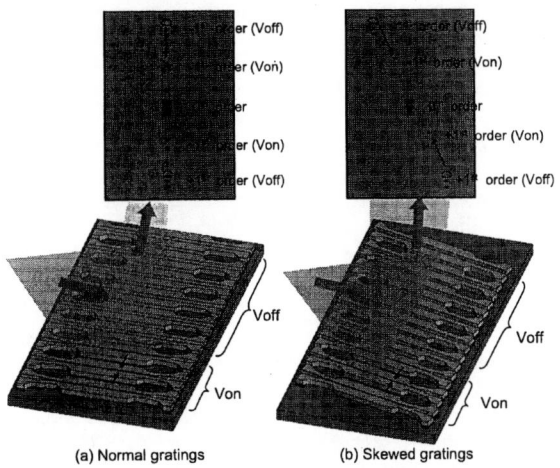

(a) Normal gratings   (b) Skewed gratings

Figure 3 Principle of light angle control by diffraction.

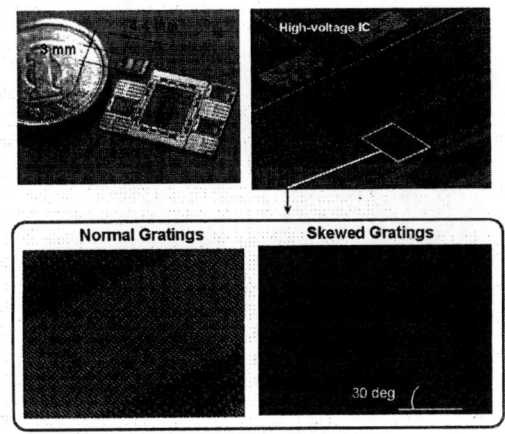

Figure 4 SEM pictures of pitch tunable light grating integrated with high-voltage driver IC.

3 FABRICATION

Electrical interconnection to the grating unit is made intra-chip by integrating the MEMS gratings with a high-voltage drive circuit by the DMOS (double-diffused metal oxide semiconductor) technology [4]. **Figure 4** shows a photograph of the chip after MEMS post-processing of implementing MEMS gratings by DRIE (deep reactive

ion etching) processing on an 8-micron-thick SOI (silicon-on-insulator) wafer of pre-fabricated DMOS driver chip. The grating slits are designed to be 4 microns wide, 8 microns thick, and 500 microns long. The skew angle was set to be 30 degrees with respect to the array direction.

4 EXPERIMENTAL RESULTS

Figure 5 shows the MEMS gratings in operation. The normal gratings that had no skew angle was ON-OFF operated to close the gap at a voltage of 40 Vdc. We used a laser diode at 632-nm wavelength of 5 mW power to observe the reflected grating patterns as shown in **Figure 5 (a)**. Due to the limited device length only two pixels were implemented this time but a clear motion of beam spot was observed in the same direction as the diffraction. On the other hand, the skewed gratings were found to spread the +/- 1st order diffraction off the axis, as shown in **Figure 5 (b)**, and one of the eight pixels were successfully found to step back and forth in the direction off the pixel arrays. Diffraction angles were found to be 6.6 degrees (OFF-state) and 3.3 degrees (ON-state) for the both gratings.

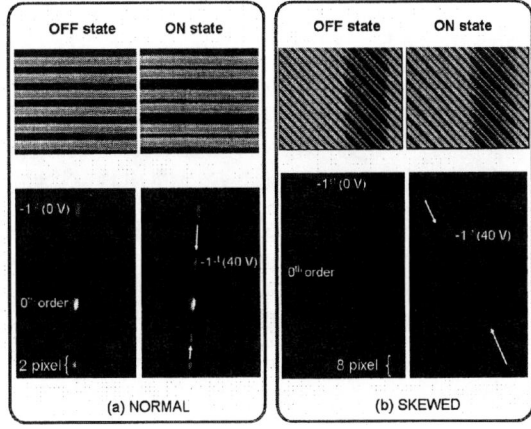

Figure 5 ON-OFF demonstration of the pitch tunable grating.

Acknowledgment

The photomasks used in this work are made (Direct writing is done) by using the University of Tokyo VLSI Design and Education Center (VDEC)'s 8-inch EB writer F5112+VD01 donated by ADVANTEST Corporation.

References

[1] R. B. Apte, et al., Proc. Solid-State Sensors and Actuators Workshop '94, pp. 1-6.

[2] M. W. Kowarz, et al., SID '06, pp. 1908-1911.

[3] K. Takahashi, et al., IEEE MEMS '07, pp. 147-150.

[4] H. Funaki, et al., IEEE IEDM'95, pp. 967-970.

WD5
16:45 – 17:00

Tunable MEMS Actuated Microring Resonators

Ming-Chun Tien[1], Sagi Mathai[2], Jin Yao[1], and Ming C. Wu[1]

[1]*Berkeley Sensor & Actuator Center (BSAC) and Dept. Electrical Engineering and Computer Sciences*

University of California, Berkeley, California 94720, USA, Tel +1-510-642-1023, Fax +1-510-643-6637, E-mail mctien@eecs.berkeley.edu

[2]*Quantum Science Research, Hewlett-Packard Laboratories, Palo Alto, California 94304, USA*

Abstract

A simplified process has been developed to fabricate MEMS tunable microring resonators on six-inch silicon-on-insulator wafers. Deep UV lithography is used to create 220-nm-wide waveguides and microrings. The process is CMOS compatible. The transmission spectra change from a double resonance dip (under-coupling) to a broader single resonance dip (over-coupling) when the waveguide is moved closer to the microring. This is explained by coupled mode theory that includes the effect of backscattering in the microring.

Keywords: optical resonator, MEMS, optical filter, doublet

1. INTRODUCTION

Silicon microresonators, including microdisks, microtoroids, and microrings, are versatile components to realize high density photonic integrated circuits (PIC). A parameter widely used to evaluate the performance of a resonator is its quality factor (Q). Ultra-high Q silica microtoroidal resonators and silicon microdisk resonators evanescently coupled to an off-chip tapered fiber have been reported [1, 2]. Waveguide integrated microresonators, patterned by e-beam lithography, with fixed power coupling ratio were also demonstrated [3]. Previously, our group reported tunable Si microdisk and microtoroidal resonators with integrated Microelectromechanical-system (MEMS) actuators on two-layer silicon-on-insulators (SOI) using wafer bonding process [4, 5].

In this paper, we report on a simplified process for MEMS tunable *microring* resonator integrated with laterally coupled waveguides and electrostatic actuators for coupling control on the SOI platform. Combining thermal oxidation and deep ultraviolet (DUV) photolithography, 220-nm-wide waveguides and microring resonators were fabricated. No wafer bonding is needed. The process is compatible with complementary metal-oxide-semiconductor (CMOS). Thermal oxidation provides an additional advantage of smoothing out surface imperfections [6], which results in a Q ~ 80,000.

2. CMOS COMPATIBLE FABRICATION PROCESS

The integrated device is fabricated from a 150 mm SOI wafer with 0.8 μm device layer and 3 μm buried oxide (BOX) layer. The process starts with thermal oxide hardmask growth followed by two DUV lithographies and silicon dry etch steps to form the microring, submicron waveguides, and MEMS actuators. Thermal oxidation is then utilized to reduce the surface roughness. Finally, the MEMS actuated waveguides are released in buffered oxide

etchant (BOE) followed by super critical drying. The remaining silicon in the center of the microring prevents the microring from being released. All the processes are CMOS compatible and e-beam lithography is not needed. The SEM picture of a fabricated microring resonator is shown in Figure 1. The waveguide dimensions are 220 nm wide by 500 nm high, which satisfies the single mode condition for TE and TM polarizations. By controlling the gap between the waveguide and the microring using electrostatic force, different coupling regimes can be achieved. At 80 V, the travel of the waveguide is ~ 1 μm.

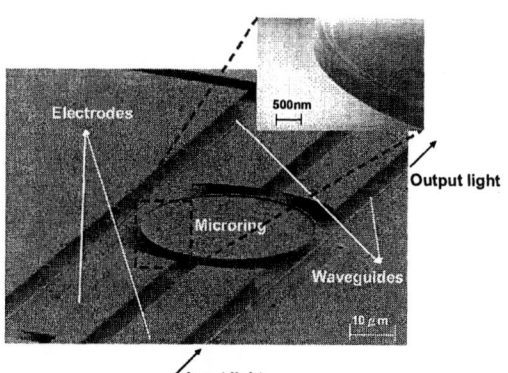

Figure 1: SEM of a released tunable MEMS actuated microring resonator. The upper right image zooms in on the microring edge.

3. OPTICAL CHARACTERIZATION

The tunable microring resonator is tested using an Agilent 81680A tunable laser. Light is coupled into the waveguide through a polarization maintaining (PM) lensed fiber, and another single mode lensed fiber collects the transmitted light. The measured transmittance is shown in Figure 2. Only one waveguide is actuated while the other parallel waveguide is decoupled from the microring resonator. In this case, the device behaves as an optical notch filter.

1-4244-0641-2/07/$20.00 ©2007 IEEE   177

By applying different voltages, the resonator can be operated in the under-coupled or over-coupled regimes. In the under-coupled regime, a double resonance dip (doublet) caused by backscattering inside the resonator is observed. The clockwise (CW) and counterclockwise (CCW) propagating modes couple to each other via backscattering, which lifts the degeneracy and splits the resonance dip. However, in the over-coupled regime, the doublet is not observable, as shown in Figure 2 at a bias of 70.2 V.

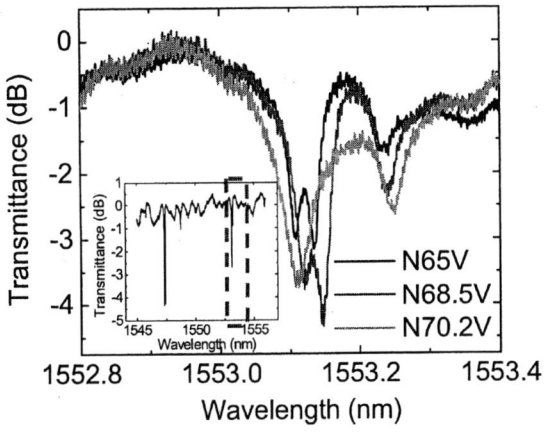

Figure 2: Measured transmission spectra near the resonance at 1553.1 nm under different bias voltages. (Inset) Measured transmittance over the 1550 nm regime.

4. COUPLED MODE MODEL FOR DOUBLETS

In order to investigate the doublet behavior caused by coupling between the CW and CCW modes in the microring resonator, a modified coupled mode theory is used [7, 8]. The coupled mode equations are written as follows.

$$\frac{da_{cw}}{dt} + \left[\frac{1}{2}\left(\frac{\gamma}{T} + \frac{\kappa}{T}\right) + i\Delta\omega\right]a_{cw} = i\frac{R}{2T}a_{ccw} + i\frac{\sqrt{\kappa}}{T}A_{in},$$

$$\frac{da_{ccw}}{dt} + \left[\frac{1}{2}\left(\frac{\gamma}{T} + \frac{\kappa}{T}\right) + i\Delta\omega\right]a_{ccw} = i\frac{R}{2T}a_{cw}$$

where a_{cw} and a_{ccw} are the amplitudes of the CW and CCW modes. T is the round-trip propagation time inside the resonator. κ is the power coupling ratio between the waveguide and resonator while γ is the round-trip intrinsic loss. The excitation frequency is detuned by $\Delta\omega$ with respect to the resonance frequency. R is the backscattering power ratio, and A_{in} is the input amplitude.

The measured transmission spectra for the under-coupled and over-coupled regimes were fitted to the coupled mode model as shown in Figure 3 (a) and (b), respectively. In the under-coupled regime, an obvious doublet was observed due

to backscattering in the resonator. The asymmetric doublet is caused by different intrinsic round-trip loss experienced by each eigenmode, represented by γ_1 and γ_2 in Figure 3. In the over-coupled regime, the doublet is no longer observable because the loaded Q broadens the linewidth and hides the doublet. The measured asymmetry in the maximum transmission is due to the overlap between the adjacent resonant modes as shown in Figure 2. Due to backscattering in the resonator, the condition for critical coupling is shifted and obtained when $\kappa^2 = \gamma^2 + R^2$ for a high-Q resonator [7, 8]. The extracted intrinsic Q is around 80,000 and the backscattering power ratio is around 3.5%.

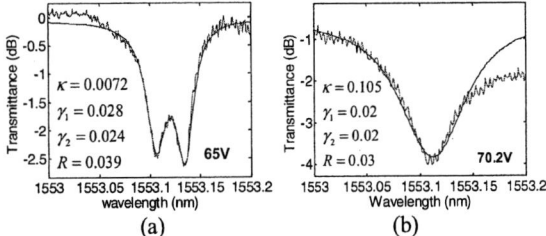

Figure 3 Transmittance curves fitting in the (a) under-coupling (b) over-coupling regimes. Blue lines are theoretical fits with the coupled mode theory while red lines are measured spectra.

5. SUMMARY

We have successfully demonstrated a CMOS compatible tunable microring resonator based on MEMS actuators using thermal oxidation and DUV photolithography techniques. An obvious doublet was observed in the transmission spectrum when the resonator was operated in the under-coupled regime. With a modified coupled mode model, the behavior of the doublet was analyzed, and the measured Q of the resonator was around 80,000.

REFERENCES

[1] D. V. Armani, et al., "Ultra-high-Q toroid microcavity on a chip," *Nature*, vol. 421, pp. 925-8, 2003.

[2] M. Borselli, et al., "Beyond the Rayleigh scattering limit in high-Q silicon microdisks: theory and experiment," *Optics Express, vol.13, no.5, 7 March 2005,*, 2005.

[3] L. Martinez and M. Lipson, "High confinement suspended micro-ring resonators in silicon-on-insulator," *Optics Express, vol.14, no.13, June 2006,*, 2006.

[4] M. C. M. Lee and M. C. Wu, "Tunable coupling regimes of silicon microdisk resonators using MEMS actuators," *Optics Express*, vol. 14, pp. 4703-4712, May 2006.

[5] J. Yao, et al., "Silicon microtoroidal resonators with integrated MEMS tunable coupler," *IEEE Journal of Selected Topics in Quantum Electronics*, vol. 13, pp. 202-8, 2007.

[6] K. K. Lee, et al., "Fabrication of ultralow-loss Si/SiO2 waveguides by roughness reduction," *Optics Letters*, vol. 26, pp. 1888-1890, Dec 2001.

[7] M. L. Gorodetsky, et al., "Rayleigh scattering in high-Q microspheres," *Journal of the Optical Society of America B-Optical Physics*, vol. 17, pp. 1051-1057, Jun 2000.

[8] T. J. Kippenberg, et al., "Modal coupling in traveling-wave resonators," *Optics Letters*, vol. 27, pp. 1669-1671, Oct 2002.

THURSDAY, 16 AUGUST 2007

ThA   Nanofabrication

ThB   Micromirrors

ThA1 (Invited)
08:30 – 09:00

One-way waveguide and strong photon-photon interaction in nanophotonic structures

Shanhui Fan, Jung-Tsung Shen, Zongfu Yu, Georgios Veronis, and Zheng Wang
Ginzton Laboratory, Stanford University, Stanford, CA 94305 USA

Abstract – New opportunities of manipulating light propagation as well as controlling light-matter interaction arise with the use of nanophotonic structures. Here we discuss two recent examples of such opportunities: a one-way electromagnetic waveguide that fundamentally suppresses the effects of disorder-induced backscattering, as well as the possibility of creating strong photon-photon interactions at a few photon level with the use of photonic crystals.

I. INTRODUCTION

We consider some of the new optical effects that arise in nanophotonic structures. In particular, we show that by embedding a two-level atom into a photonic crystal waveguide, one can create strong photon-photon interaction and correlations at few photon-level [1][2]. Such a capability is potentially important for quantum information processing. We also show that a one-way waveguide can form at the interface between a dielectric photonic crystal, and a plasmonic free-electron metal subject to a static DC magnetic field [3]. In such a waveguide, disorder-induced backscattering is completely suppressed. Such a one-way waveguide represents a new regime of photon transport in nanostructures, and can have important implications for on-chip integrated photonic circuits.

II. STRONGLY-CORRELATED TWO-PHOTON TRANSPORT IN PHOTONIC CRYSTAL WAVEGUIDES

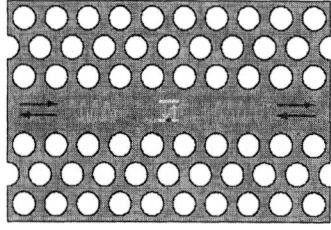

Fig. 1. A two-level system coupled to a photonic crystal waveguide in which the photons, shown as wiggly waves, propagate in each direction.

Creating a strong photon-photon interaction at the few-photon level is of great interest for quantum information sciences. In atomic gases, such an interaction can be accomplished either with systems exhibiting electromagnetically induced transparency (EIT) [3,4], or by reaching the strong-coupling regime of a two-level atom in a high-Q cavity [5]. However, in an on-chip, solid-state environment, which is crucial for practical applications, there have been significant challenges in implementing these concepts. For example, it is difficult to create the long-lifetime dark state, which is required for EIT effects, in most practical solid-state environments [6]. While the strong-coupling regime has been reached by placing a quantum dot in a high-Q photonic crystal microcavity [7,8], doing so requires very accurate tuning of both the electronic and optical resonances to ensure simultaneous spectral and spatial overlaps [9].

Here we propose and analyze in detail an alternative scheme to create strong photon-photon interaction [1]. Our approach exploits a unique one-dimensional feature for photon states in many nano-photonic structures. In a photonic crystal with a complete photonic band gap, for example, a line-defect waveguide forms a true one-dimensional continuum for photons, since there are no other states within the gap. Here we show that by coupling a two-level system to such a continuum, strong photon-photon interactions can be created (Fig. 1). In this system, the strong interaction arises from the fact that, in a one-dimensional system, the re-emitted and scattered waves from the atom inevitably interfere with the incident waves [2]. Moreover, since the atom, intuitively speaking, can at most absorb only one photon at a time, the transport properties of multiphotons are strongly correlated.

Compared with previous solid-state approaches, our scheme does not require the presence of a long-lifetime dark state. Neither does this scheme necessitate detailed spectral tuning or spatial control of the two-level system, since it operates in the weak-coupling regime, and thus the one-dimensional continuum can be broadband. Moreover, the Hamiltonian of the system actually describes an exact photonic analogue of the Kondo effect, which is important for processing electronic quantum bits [10]. Our approach may

1-4244-0641-2/07/$20.00 ©2007 IEEE

therefore open a new avenue toward practical photon-based quantum information processing on-chip.

III. ONE-WAY ELECTROMAGNETIC WAVEGUIDE

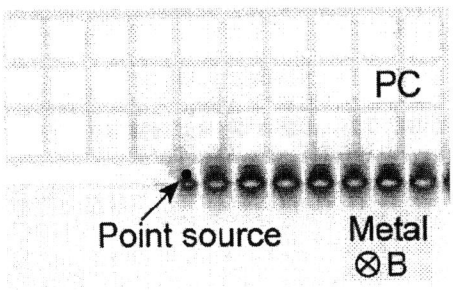

Fig. 2. One-way waveguide formed at the interface between a dielectric photonic crystal and a free-electron plasmonic subject to a static out-of-plane magnetic field. In such a waveguide, a point source radiates only to one direction but not the other.

Understanding and controlling the effects of disorders on wave propagation are both of fundamental interest, and are becoming increasingly important in practical nano-device applications. In general, the effects of disorders are drastically influenced by time-reversal symmetry properties of a system. In reciprocal nanophotonic systems, which include most photonic crystals and plasmonic waveguides, at any given frequency for each forward propagating mode, there is a corresponding backward propagating mode with identical mode shapes. In the presence of disorders, these two modes, having maximal modal overlap, always scatter into each other, resulting in back-reflection. Such back-reflection can be particularly worrisome for slow light systems, which are of current interest for optical signal processing applications. As the group velocity v_g is reduced, the back reflection increases as $1/v_g^2$, and in slow light systems may dominate over all other loss mechanisms [12, 13].

In systems with broken time-reversal symmetry, the effect of disorders can be suppressed with the use of a one-way waveguide [14]. Such a waveguide supports a single forward propagating mode in a given frequency range, while having neither radiation nor backward propagation modes in the same range. For electronic waves, an example of a one-way waveguide is the edge states for a two-dimensional electron gas in the quantum hall regime [15]. Here we introduce a mechanism for creating a one-way electromagnetic waveguide [3]. The waveguide is formed at the interface between a dielectric photonic crystal that is assumed to be reciprocal, and a free-electron plasmonic

metal under a static magnetic field along the direction perpendicular to the plane of propagation. (Fig. 2) We show that such a waveguide supports a one-way waveguide mode, provided that the surface plasmon frequency of the metal surface lies within the band gap of the photonic crystal. In such a system, the presence of the band gap is intrinsic to the photonic crystal and the metal itself, independent of time-reversal symmetry breaking. Hence the size of band gap can be made very large and can survive significant disorders [16]. As a proof of concept, we present direct numerical evidence demonstrating suppression of disorder effects. We also show that the nature of waveguide-cavity interaction, which is at the heart of many integrated devices, can be fundamentally altered. Such a waveguide therefore represents a complete new regime of light transport with important implications for on-chip integrated photonic circuits.

IV. REFERENCES

1. J. T. Shen and S. Fan, Phys. Rev. Lett. 98, 153003 (2007).
2. J. T. Shen and S. Fan, Opt. Lett. 30, 2001 (2005).
3. Z. Yu et al (submitted).
4. S. E. Harris et al., Phys. Rev. Lett. 81, 3611 (1998).
5. A. Imamoglu et al., Phys. Rev. Lett. 79, 1467 (1997).
6. K. M. Birnbaum et al., Nature 436, 87 (2005).
7. A.V. Turukhin et al., Phys. Rev. Lett. 88, 023602 (2001).
8. J. P. Reithmaier et al., Nature 432, 197 (2004).
9. T. Yoshie et al., Nature 432, 200 (2004).
10. A. Badolato et al., Science 308, 1158 (2005).
11. D. M. Zumbuhl et al., Phys. Rev. Lett. 93, 256801 (2004).
12. S. Hughes et al, Phys. Rev. Lett. 94, 033903 (2005).
13. S. G. Johnson et al, Appl. Phys. B 81, 283 (2005).
14. F. D. M. Haldane and S. Raghu, cond-mat/0503588.
15. S. Datta, Electronic Transport in Mesoscopic Systems (Cambridge University Press, Cambridge, UK, 1995).
16. S. Fan et al, J. Appl. Phys. 78, 1415 (1995).

ThA2
09:00 – 09:15

Flower-structured InGaN/GaN quantum-well nanodisk crystals on micromachined Si pillars

R. Ito, F. R. Hu, K. Ochi, Y. Zhao and K. Hane

Tel +81-22-795-6965, Fax +81-22-795-6963, E-mail hane@hane.mech.tohoku.ac.jp

Department of Nanomechanics, Tohoku University, Sendai 980-8579, Japan

Abstract

Periodic Si pillars were fabricated by Si-micromachining and unique flower-structured InGaN/GaN quantum-well nanodick crystals were deposited on the Si pillars. Split photoluminescence peak positions indicated stronger emission from the high In included quantum dots, as compared with the emission from the quantum-well matrix layers. Optical microscopy images indicate clear vertical and horizontal photoluminescence distributions of the flower structure on the pillared Si substrate.

Keywords: Photoluminescence, InGaN/GaN, Quantum well, Nanodisk

1 INTRODUCTION

High efficient light-emitting GaN-based semiconductor crystal on Si substrate attracts much attention since Si is the most widely used semiconductor and GaN has a wide application prospect in light-emitting diode (LEDs) and laser diode (LDs). Here, we use periodic Si pillars as substrate and flower-structured InGaN/GaN quantum-well (QW) nanodisk crystals were deposited on the Si pillars. The details of the deposited structure was examined.

2 EXPERIMENTAL PROCEDURE

The pillared Si substrate was fabricated by Si-micromachining process. As shown in figure 1(a), the diameter and period Λ of the Si pillars is about 0.3 μm and 1.0 μm, respectively. Because of the nonuniformity of etching during the micromaching process in some region, the length of the Si pillars is between 3.0-7.0 μm. The top surface of the Si pillars is Si (111) and the side-wall surface is not a specific plane of the Si crystal. The InGaN/GaN QW nanodisk crystals were deposited on pillared Si with Riber 32 molecular beam epitaxy system. Detailed deposition process was the same as described in Ref. 1. Crystal microstructure was characterized by scanning electron microscopy (SEM) images. Photoluminescence (PL) measurement was carried out using a Hg-Cd laser (60 mW, 325 nm) as light source. Optical microscopy images of the PL distribution of the QW crystals were obtained with a Hg lamp used for the excitation light source. A 405 nm HgI ($^3S_1$-$^3P_0$) line filter was used to filter the Hg lamp light source. As for the collection of the detection, a high-pass filter of wavelengths longer than 450 nm was used to detect the emission. Laser scanning confocal microscopy (LSCM) was used to investigate the fluorescent distribution along the vertical direction in the flower structure. The incident laser source was 487 nm blue light and the detection was carried out in three different wavelength ranges, namely 487-552 nm green region, 552-652 nm yellow region and 634-734 nm red region.

3 RESULTS AND DISSCUSION

Figure 1(b) shows the deposited InGaN/GaN QW nanodisk

crystals on the Si pillars. Clear image of the upper part of the crystals is shown in figure 1(c). The QW nanodisk crystals grow around the top and side-wall surface of the Si pillars, forming unique flower structure. The tip part of the crystals is the InGaN/GaN QW and cap layers. The lower part is GaN underlayer. Clear image of the flower structure is shown in figure 1(d) and figure 1(e). It is demonstrated the nanodisk crystals also grow in the bottom of the pillared Si. Figure 1(f) shows two single nanodisk crystals. The GaN underlayer diameter is about 180 nm and 280 nm. The diameter of the tip of the nanodisk is about 230 nm.

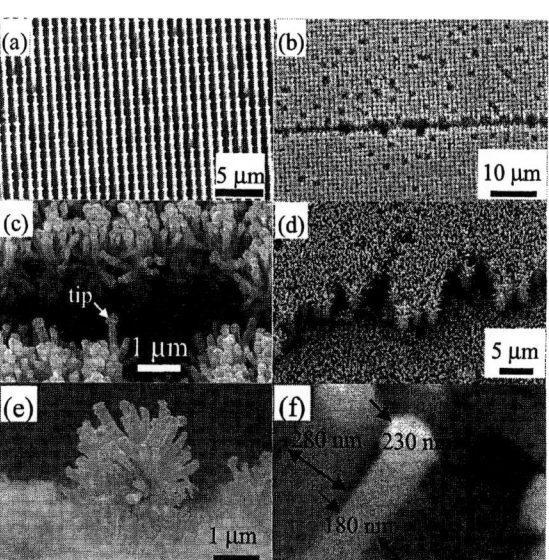

Figure 1. SEM images of the micromachined Si pillars and the InGaN/GaN QW disk crystals

Figure 2 shows the PL measurement results of the QW nanodisk crystals on the pillared Si substrate. At room temperature, wide range wavelength of PL from 370 nm to 650 nm with two clear peak positions of 497 nm and 544 nm are demonstrated in the inset of figure 2. As the

1-4244-0641-2/07/$20.00 ©2007 IEEE   183

measurement temperature decreases to about 230 K, another low wavelength peak of about 412 nm appears. At low temperature of 15 K, the low wavelength PL peak position shifts to 391 nm and 399 nm and their intensity is much stronger than that of the high wavelength peak of 492 nm. This is very similar to the PL properties of the InGaN layer in Ref. 2, in which the low wavelength emission of 432 nm is dedicated to the InGaN matrix and the high wavelength emission of 511 nm is from the quantum dots (QDs) in the InGaN matrix. In our case, the wavelength of 391nm and 399nm is from the InGaN matrix and the wavelength of 492 nm is from the quantum dots. Because of the very strong localized confinement and a comparatively low electron-phonon scattering effect in the QDs, PL from the QDs is strong at room temperature and the peak intensity only increases by about one order of magnitude with the measurement temperature decreasing to 15 K. For the PL from the QW matrix, however, it is much weak at room temperature and is overlapped by PL from the QDs. This may be because of the suppression of PL by the scattering effects of the phonon-electron at room temperature. When temperature decreases, the scattering effect decreases. PL from the QW matrix starts to increase fast and appears obviously at 220 K, as shown in figure 2.

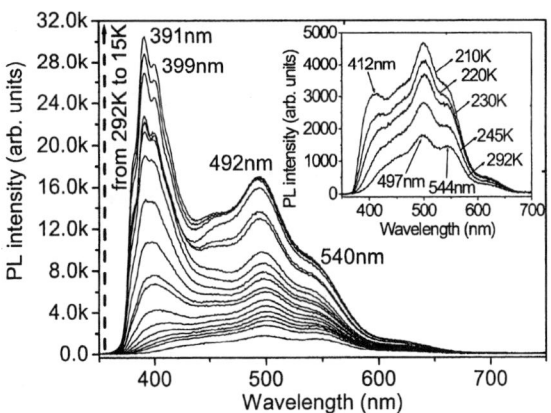

Figure 2.   temperature-dependent photoluminescence of the flower structure on the Si pillars

Figure 3 shows the optical microscopy images of the PL distribution of the QW crystals. Different PL is emitted from different flower structure in figure 5. The flower structure on the long Si pillars (A, 7 μm) mainly emits red emission. Flower structure on the middle length Si (B, 5 μm) and short Si pillars (C, 3 μm) mainly emit yellow and blue-green emission, respectively. This maybe because of the different growth temperature of the flower structure. There existed graded growth temperature from the top to the bottom part of the Si pillars during the deposition of the QW layer, which resulted in the In concentration in the QW crystals decrease from the top part to the bottom part of the flower structure.

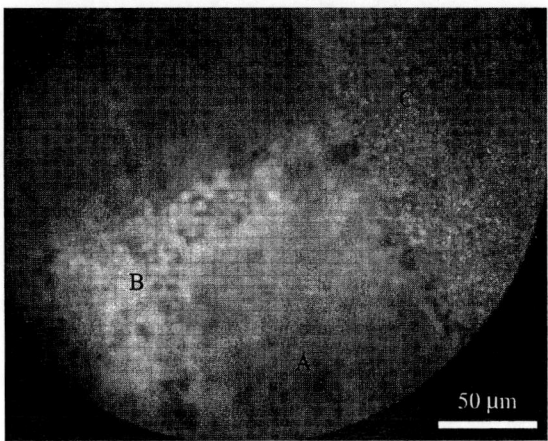

Figure 3.   Optical microscopy images of the PL distribution of the QW crystals on the pillared Si

LSCM was carried out with confocal plane scanned from the bottom to the top of the flower structure. As for the 487-587 nm detection region, green PL gradually weakens from the bottom to the top of the flower structure on the short Si pillars. While for the flower structure on the long Si pillars, green PL gradually increases from bottom to the middle part and then gradually decreases from the middle to the top part of the structure. As discussed above, the growth temperature of the QW crystals decreases from the bottom to the top part of the Si pillars, so the In concentration and corresponding PL peak wavelength increases from the bottom to the top part of the flower structure. However, because the intensity of the incident exciting laser light decreases from the top to the bottom part of the flower structure, the green PL intensity on long Si pillars does not increases monotonically from the bottom to the top. It reaches the highest in the middle part of the flower structure on the long Si pillars. As for the 552-652 nm detection region, the yellow PL gradually increases from the bottom to the middle and then decreases from the middle to the top part of the structure on the short Si pillars. A similar trend is demonstrated that the yellow PL gradually increases from the bottom to the middle and then decreases from the middle to the top part of the structure on the long Si pillars. As for the 634-734 nm detection region, the red PL gradually increases from the bottom to the middle and then decreases from the middle to the top part of the structure on the short Si pillars. While for the flower structure on the long Si pillars, the red PL gradually increases from the bottom to the top part of the structure.

REFERENCES

[1] F. R. Hu, K. Ochi, Y. Zhao and K. Hane, Appl. Phys. Lett., Vol 89, 171903, 2006.
[2] I. K. Park, M. K. Kwon, S. H. Baek, Y. W. Ok, T. Y. Seong, S. J. Park, Y. S. Kim, Y. T. Moon and D. J. Kim, Appl. Phys. Lett. Vol. 87 061906, 2005.

ThA3
09:15 – 09:30

Fabrication of Wafer-level Antireflective Structures in Optoelectronic Applications

C. Max Hsieh[1*], J. Y. Chyan[2], Wen-Ching Hsu[2] and J. Andrew Yeh[1,2]
1 Institute of MicroElectroMechanical Systems
2 Institute of Electronics Engineering
National Tsing Hua University, Hsinchu, Taiwan
*Phone: 886-3-5715131 ext. 33730, E-mail: g945007@oz.nthu.edu.tw

Abstract

Moth-eye structure (MES), biologically antireflective structure, can be regarded as an impedance-matching layer in optics. MES possesses remarkable antireflective effects and plays a big role in optoelectronic applications. The wafer-level MES achieved using chemical etching with metal catalyst has been demonstrated. The method proposed in this paper provided a rapid and simple fabrication and could be integrated with current fabrication of solar cells and light-emitted devices (LED).

Keywords: moth-eye structure (MES), antireflective, wafer-level

1 INTRODUCTION

MES discovered in a nocturnal moth has been extensively studied since 1967 [1]. MES refers a grade-index structure and provides impedance-matching in optics. It bears tiny optical reflectance even though incident light is oblique with a large angle. A scheme of MES is illustrated in Figure 1. Many methods for manufacture of MES patterns were demonstrated, including interference, holography, e-beam lithography, anodic aluminum oxide (AAO) mask, and plasma nanofabrication [2-6]. Majority of the studies in MES require advanced lithography and/or plasma etching, which are time-consuming. Thus, the method demonstrated in this paper was used to replace advanced lithography and time-consuming plasma etching.

Approaches to manufacture of nanostructures using chemical etching with noble metal catalysts have been demonstrated [7-8]. In this article, Ag nanoparticles (NPs) within tens of nanometers are reduced on the silicon (Si) surface and function as catalysts for anisotropic etching. The sizes of catalysts determine the dimension of holes caused by the following etching. The metal catalysts replace the lithography and facilitate the anisotropic etching. Consequently, wafer-level fabrication of MES can be achieved using several simple chemical reactions.

2 PRINCIPLE AND FABRICATION

A rapid and simple method for wafer-level manufacture of MES is depicted in Figure 2. The wafer is dipped into the 0.01M silver nitrate solution for 5 minutes. Silver ions (Ag^+) in the solution will reduce on the Si surface because electronegativity of Ag is higher than Si. Then the wafer rinses in DI water followed by blowing the wafer with N_2 and heating it by hotplate for dehydration. After formation of metallic catalyst layer, the wafer is immersed in the etchant comprising HF (49%wt) and hydrogen peroxide (H_2O_2, 30%wt) with the mixture ratio of 3:1 (v/v). At the beginning, HF mainly reacts with the silicon dioxide (SiO_2) which is produced during the reduction of Ag. The next step, H_2O_2 oxidizes Ag to Ag^+ again and Ag^+ reduces on the Ag side for the second reduction according to the activity difference between Si and Ag. Therefore, anisotropic etching is achieved and etching time can be shortened effectively by addition of H_2O_2.

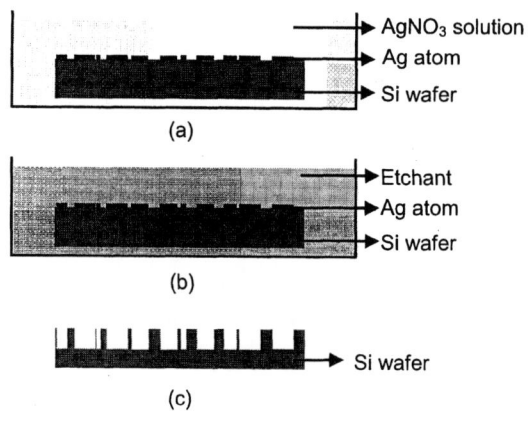

Figure 2: Process flow for wafer-level manufacture of MES (a) Reduction of Ag catalyst (b) Etched in HF+H_2O_2 solution (c) Formation of MES after removing Ag.

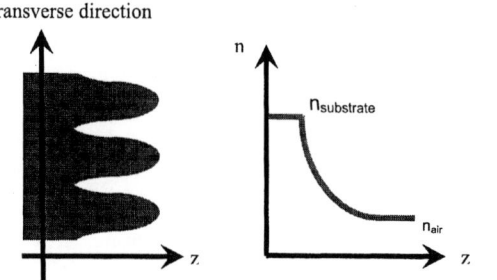

Figure 1: A scheme for MES concept. MES can be viewed as a graded index structure briefly for the function of optical impedance matching.

1-4244-0641-2/07/$20.00 ©2007 IEEE   185

Ag reduces on the Si surface randomly, so the catalytic layer produces porous characteristics and fluorine ions (F⁻) can reach the bottom of the metal through the crack inside the catalytic layer and react with SiO_2. Total reaction time does not exceed 10mins. Finally, Ag was removed by dipping in the H_2O_2 solution.

3 RESULTS

Performances of wafer-level fabrication of MES are shown in Figure 3(a). Both samples 1 and 2 are 4" wafer. Sample 1 is the result after the etching assisted by Ag catalyst and sample 2 is a chemically etched as-cut wafer for reference. Sample 1 shows dark-brown color with a high uniformity all over the wafer. SEM picture reveals that there are tapered pillars on the sample 1 surface which is consistent with MES. The size of tapered pillar Figure 3(b) is about 75nm in width and 600nm in depth, which corresponds to an aspect ratio of 8. The reflective spectrum of the sample is shown in Figure 4. Reflectivity of the sample is measured by the system being inclusive of a light source, an integrating sphere and a spectrometer. The results of reflectivity measurement exhibit a broadband antireflective characteristic. The average reflectivity within visible light is about 5%.

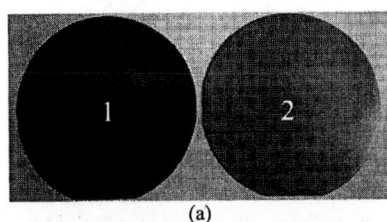

(a)

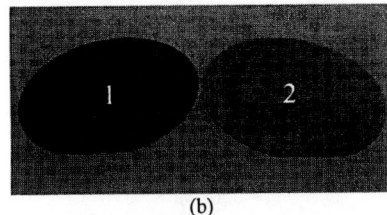

(b)

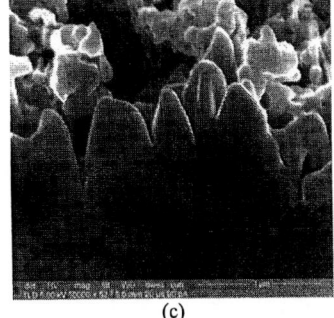

(c)

Figure 3: Wafer-level fabrication of MES: (a) and (b) are top and oblique view of two 4" Si wafers before and after etching assisted by Ag (c)Cross section view of the MES on the Si wafer

This paper demonstrates a chemical etching method assisted by Ag catalysts. The rapid and simple processes have potential to replace the plasma etching in the manufacture of MES. Another advantage of the method in this article is the tens of nanometers feature can be achieved by Ag catalysts without the advanced lithography. The proposed process can be simply integrated into any fabrication of optoelectronic devices like antireflective layer of the solar cells and LED.

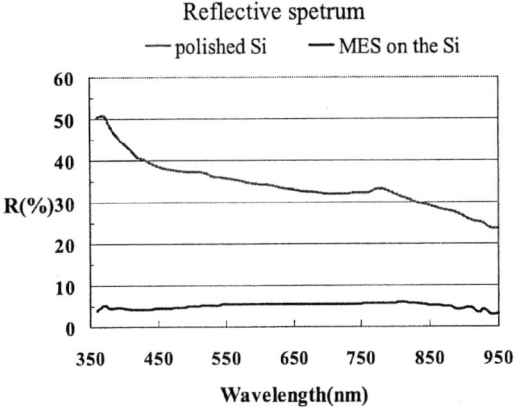

Figure 4: Reflective spectrum of MES

4 ACKNOWLEDGEMENTS

The authors are indebted to Sino-American Silicon Product in Taiwan for wafer support and technical assistance.

5 REFERENCES

[1] C. G. Bernhard, 1967, Endeavour, 26, 79

[2] S. J. Wilson and M. C. Hutley, Optica Acta, Vol. 29, No. 7, 993-1009, 1982

[3] A. Gombert et al, Optical Engineering, Vol. 43, No. 11, pp2525-2533, November, 2004

[4] Y. Kanamori, M. Sasaki, and K. Hane, Optics Letters, Vol. 24, No. 20, October 15, 1999

[5] H. Sai et al, Applied Physics Letters, Vol. 88, 201116, 2006

[6] J Shieh, C. H. Lin and M. C. Yang, J. Phys. D: Appl. Phys. 40 2242–2246, 2007

[7] X. Li and P. W. Bohn, Applied Physics Letters, Vol. 77, 2572, 2000

[8] K. Peng et al, Angew. Chem. Int. Ed., Vol. 44, 2737-2742, 2005

ThA4
09:30 – 09:45

Structural and Optical Properties of III-V Nanowires and Nanowire Heterostructures Grown by Metalorganic Chemical Vapour Deposition

H. J. Joyce[1,*], Q. Gao[1], Y. Kim[2], H. H. Tan[1], C. Jagadish[1]

[1]Department of Electronic Materials Engineering, Research School of Physical Sciences and Engineering, The Australian National University, Canberra ACT 0200, Australia
[2]Department of Physics, College of Natural Sciences, Dong-A University, Hadan 840, Sahagu, Busan 604-714, Korea
*Tel +61 2 6125 1595, Fax +61 2 6125 0511, E-mail hjj109@rsphysse.anu.edu.au

Abstract

We have investigated the structural and optical properties of III-V nanowires grown by metalorganic chemical vapour deposition. Binary GaAs, InAs and InP nanowires, and ternary InGaAs and AlGaAs nanowires, have been fabricated and characterised. A variety of axial and radial heterostructures have also been fabricated, including GaAs/AlGaAs core-multishell and GaAs/InGaAs superlattice nanowires. GaAs/AlGaAs core-shell nanowires exhibit strong photoluminescence as the AlGaAs shell passivates the GaAs nanowire surface reducing the surface nonradiative recombination.

Keywords: nanowire, MOCVD, heterostructure

1 INTRODUCTION

Semiconductor nanowires have recently become a topic of intensive research, owing to their great potential as nano-building blocks for future optoelectronic devices [1]. Nanowires of III-V materials are particularly promising owing to the direct bandgap and high carrier mobility of such materials. These nanowires are successfully fabricated by metalorganic chemical vapour deposition (MOCVD). Gold nanoparticles are used in the growth process: these behave as sinks for Group III reaction species and seed epitaxial nanowire growth.

Nanowires offer great flexibility in materials systems. High quality binary nanowires, such as GaAs, InAs and InP have been demonstrated. We have developed a two-temperature procedure for the growth of GaAs nanowires with optimised structural, crystallographic and optical properties [2]. In addition to binary nanowires, InGaAs [3] and AlGaAs nanowires have been grown and characterised.

A wide variety of axial and radial heterostructure nanowires, incorporating quantum dots, axial quantum wells and radial quantum wells, can be achieved. These will play a major role in future nanowire-based devices. We report the fabrication and properties of GaAs/InGaAs axial heterostructure nanowires, and GaAs/AlGaAs core-shell and core multi-shell nanowires.

2 EXPERIMENTS

Nanowires were grown on semi-insulating GaAs (111)B substrates, with the exception of InP nanowires which were grown on InP (111)B substrates. Substrates were functionalized by immersion in poly-L-lysine (PLL) solution, and treated with gold colloid solution (10, 20, 30 or 50 nm diameter Au particles). The Au nanoparticles are attracted to, and immobilized on, the positively charged PLL layer. Nanowires, catalysed by these nanoparticles, were grown by horizontal flow MOCVD at a pressure of 100 mbar. Trimethylgallium, trimethylindium, trimethylaluminium, AsH_3 and PH_3 precursors were used. Prior to nanowire growth initiation, the substrate was annealed under arsine (or phosphine for InP nanowires) pressure at 600°C to desorb surface contaminants. Nanowires were grown at temperatures between 350 and 510 °C for times of approximately 30 minutes. For nanowire heterostructures, nanowire core growth was carried out at 450 °C. Shells of GaAs and AlGaAs were grown at 650 °C.

Nanowires were characterised by field emission scanning electron microscopy (FESEM), low temperature (10 K) photoluminescence (PL), room temperature micro-photoluminescence (micro-PL) and transmission electron microscopy (TEM).

1-4244-0641-2/07/$20.00 ©2007 IEEE   187

3 RESULTS

2.1 Binary Nanowires

GaAs (Figure 1), InAs and InP nanowires grew in the vertical [111]B direction on (111)B substrates. We observe a material-dependent minimum growth temperature, above which nanowires are straight and vertically aligned, and below which nanowire growth becomes kinked and irregular.

For GaAs nanowires, this minimum temperature is approximately 410 °C. However, using a novel two-temperature growth procedure we obtained vertically aligned GaAs nanowires at growth temperatures as low as 350 °C, as illustrated in Figure 1b [2]. These low-temperature grown GaAs nanowires (Fig 1b) exhibited minimal tapering compared to nanowires grown at higher temperatures (Fig 1a). Low-temperature grown nanowires were free of the planar crystallographic defects found in high-temperature grown nanowires [2, 4], and exhibited enhanced photoluminescence compared to high-temperature grown nanowires [2].

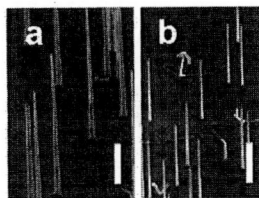

Figure 1. FESEM images of GaAs nanowires grown at (a) 450 °C and (b) 350 °C. Scale bar is 1 μm. Sample is tilted at 40°.

2.2 Ternary Nanowires

The morphology of InGaAs nanowires was strongly dependent on nanowire density. Low density regions exhibit greater nanowire height and tapering. This density dependence arises because Group III adatoms are adsorbed on the substrate, diffuse and contribute to nanowire growth, and nearby nanowires compete for these adatoms. The diffusion length of In species is larger than that of Ga species. This results in a greater In composition in low density nanowires, which is in turn associated with a PL peak redshift from high density to low density regions [3].

2.3 Axial and Radial Heterostructures

GaAs/InGaAs superlattice nanowires were grown. These nanowires were straight and vertically oriented. PL and EDS measurments confirm the presence of InGaAs segments.

GaAs/AlGaAs core-shell nanowires are illustrated in Figure 2 together with a room temperature micro-PL image and 10 K PL spectrum. The AlGaAs shell passivates the GaAs core surface. This leads to strongly enhanced GaAs-related PL compared to bare GaAs nanowires. The peak occurs at 1.515 eV corresponding closely to excitonic emission in bulk GaAs [5].

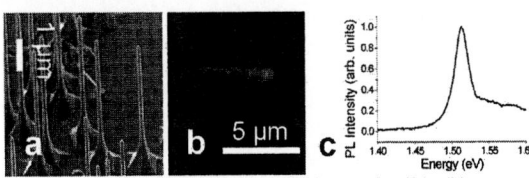

Figure 2. (a) FESEM image (40° sample tilt), (b) room temperature micro-PL image and (c) 10 K PL spectrum of GaAs/AlGaAs core-shell nanowires.

ACKNOWLEDGMENTS

The Australian Research Council is gratefully acknowledged for financial support. Thanks are due to our collaborators at the University of Queensland (J. Zou, M. Paladugu, H. Wang, G. J. Auchterlonie, Y. Guo and X. Zhang), the University of Cincinnati (T. B. Hoang, L. V. Titova, H. E. Jackson and L. M. Smith), Miami University (J. M. Yarrison-Rice) and Ohio University (A. O. Govorov).

REFERENCES

[1] Y. Huang, X. Duan, and C. M. Lieber, "Nanowires for Integrated Multicolor Nanophotonics," *Small*, vol. 1, pp. 142-147, 2005.

[2] H. J. Joyce, Q. Gao, H. H. Tan, C. Jagadish, Y. Kim, X. Zhang, Y. Guo, and J. Zou, "Twin-Free Uniform Epitaxial GaAs Nanowires Grown by a Two-Temperature Process," *Nano Lett.*, vol. 7, pp. 921-926, 2007.

[3] Y. Kim, H. J. Joyce, Q. Gao, H. H. Tan, C. Jagadish, M. Paladugu, J. Zou, and A. A. Suvorova, "Influence of Nanowire Density on the Shape and Optical Properties of Ternary InGaAs Nanowires," *Nano Lett.*, vol. 6, pp. 599-604, 2006.

[4] J. Zou, M. Paladugu, H. Wang, G. J. Auchterlonie, Y. Guo, Y. Kim, Q. Gao, H. J. Joyce, H. H. Tan, and C. Jagadish, "Growth Mechanism of Truncated Triangular III-V Nanowires," *Small*, vol. 3, pp. 389-393, 2007.

[5] L. V. Titova, T. B. Hoang, H. E. Jackson, L. M. Smith, J. M. Yarrison-Rice, Y. Kim, H. J. Joyce, H. H. Tan, and C. Jagadish, "Temperature dependence of photoluminescence from single core-shell GaAs--AlGaAs nanowires," *Appl. Phys. Lett.*, vol. 89, pp. 173126, 2006.

ThA5
09:45 – 10:00

Magnetic Alignment of Carbon Nanotube Interconnects

H. J. In, A. J. Nichol and G. Barbastathis
Dept. of Mech. Eng., Massachusetts Institute of Technology,
77 Massachusetts Ave. Room 3-466, Cambridge, MA, USA
Tel +1-617-258-0649, Fax +1-617-258-9346, E-mail hji@mit.edu

Abstract

In recent years, much research effort has focused on advancing the fundamental understanding, growth, and utilization of carbon nanotubes (CNTs). In particular, the precise handling, placement and anchoring of individual nanotubes remain a challenge to realizing practical systems that exploit CNTs' superior electrical and mechanical properties. In this paper, we investigate the use of a novel magnetic alignment technique for establishing CNT interconnections in a 3-D device. Our proposed configuration is compatible with a number of nanoelectronic, subwavelength photonic, and electrochemical energy storage devices and systems.

Keywords: carbon nanotubes, nanomanufacturing, integrated microsystems, MEMS

1 INTRODUCTION

Carbon nanotubes, due to their unique mechanical, electrical, and optical properties, have been utilized in a wide range of research as well as commercial applications in the fields of electronic and optoelectronic sensors and computing systems. However, controlled directional placement of CNTs or secure anchoring of both ends of the nanotube remain a serious challenge. In this paper, we present a novel magnetic alignment technique that can be used to create CNT interconnects in a 3-D device. The 3-D device is assembled through the Nanostructured Origami™ method[1], which enables the creation of complex, nanostructured devices in 3-D space through the self-aligned folding of 2-D membranes patterned through conventional planar micro- and nanofabrication tools and processes.

2 MAGNETIC ALIGNMENT

In recent work, Nichol *et al.* demonstrated that arrays of nanomagnets can be used to align folding membranes to each other to the order of 200 nm[2]. Using a similar technique, we propose to align and attach the free ends of plasma enhanced chemical vapor deposition (PECVD) grown multiwalled carbon nanotubes (MWCNTs) onto corresponding magnetic receptor sites as well as to each other. Figures 1 and 2 show the proposed alignment and attachment schemes.

2.1 Alignment Theory

In PECVD grown MWCNTs, the magnetic catalyst material remains at the top of the nanotubes as a teardrop-shaped cap. Essentially, these materials can be regarded as batch-produced nanomagnets. It has been shown that magnetic properties of catalyst materials can cause CNTs to assemble onto ferromagnetic contacts[3] as well as force the nanotubes to cluster together[4].

The magnetic attraction force between a CNT tip and a magnetic attachment site located approximately 100 nm from the tip can be estimated in a similar fashion as for nanomagnet arrays [2] from

$$ F_m \approx \frac{(M_s A)^2}{4\pi\mu_0 d^2}, \tag{1} $$

where A is the cross-sectional area of the magnets and d is the initial misalignment between the nanotube tip and the receptor site (see Figure 1(b)). For our purpose, the magnetic tip can be approximated as a cylinder with diameter of 100 nm and length of 200 nm, while the nanotube is modeled as a solid cylinder with diameter of 100 nm, length of 10 μm, and Young's modulus of 1 TPa. Standard beam theory can be used to estimate the maximum magnetically-induced lateral tip deflection of each nanotube to be approximately 100 nm. Accordingly, this tip movement should be sufficient to

1-4244-0641-2/07/$20.00 ©2007 IEEE   189

overcome the initial misalignment of 100 nm between the nanotube tip and the receptor site.

2.2 Alignment Schemes

Figure 1 shows our proposed schemes for 3-D magnetic alignment and attachment of carbon nanotubes. In Figures 1(a) through 1(b), two membranes are folded to face each other through well-established membrane-folding techniques of the origami process[1] and remain coarsely aligned. In the final magnetic alignment step taking place between Figures 1(b) and 1(c), an external magnetic field is applied perpendicular to the substrate and subsequently causes the nanotube tips to become magnetized to saturation. Each tip then behaves as a magnetic dipole and completes the final folding, alignment, and attachment process. The opposing membrane can also be covered with corresponding nanotubes instead of the magnetic receptor sites as shown in Figure 2.

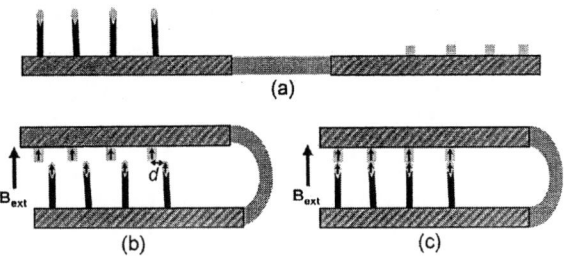

(a)

(b)   (c)

Figure 1. Illustration of the origami fabrication of CNT-covered membranes a) before folding, b) after folding and coarse alignment, and c) after magnetic alignment and attachment

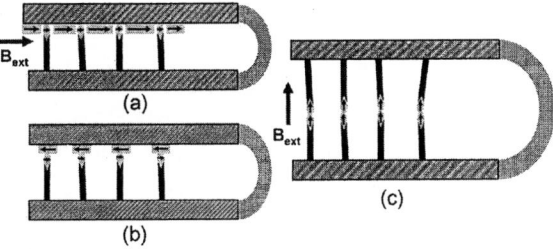

(a)

(b)   (c)

Figure 2. Other possible schemes for CNT alignment: a) horizontal magnetic field is applied; b) magnets are magnetized prior to folding; and c) CNTs on opposing membranes are aligned and attached to other

2.3 Initial Fabrication Results

We have demonstrated *in-situ* growth of MWCNTs on origami membranes for nickel, iron, and cobalt catalysts.

Furthermore, dense arrays of MWCNTs have been grown on conducting TiN membranes to ensure electrical conductivity throughout the structure. Figure 3 shows a dense array of MWCNTs grown on a folding TiN membrane. The rectangular frame around the nanotubes is part of the wiring that provides electrical connection to the membrane and the nanotubes. Furthermore, initial magnetic testing of cobalt MWCNTs has shown some nanotube clumping, indicating that magnetic forces arising from the tips can overcome the mechanical stiffness of the nanotubes. Site-selective growth of nanotubes can be accomplished by pre-patterning the catalyst layer through e-beam lithography.

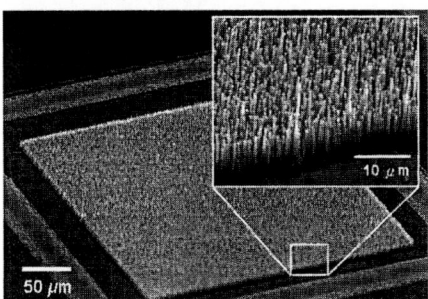

Figure 3. PECVD grown MWCNTs on a folding titanium nitride membrane

3 CONCLUSION

We have shown that, by using the Nanostructured Origami™ process along with nanomagnetic tips that remain on the nanotubes from the PECVD-based growth process, we can create devices with 3-D interconnectivity provided by carbon nanotubes with defined endpoints. MWCNTs have been successfully grown on folding TiN membranes, and initial calculations and experiments show that the attractive force between magnetically saturated nanotube tips should be sufficient to provide massively parallel alignment and attachment of nanotubes onto corresponding bonding sites on the opposing membrane.

REFERENCES

[1] H. J. In, W. J. Arora, T. Buchner, S. M. Jurga, H. I. Smith and G. Barbastathis, Proc. of the 4th IEEE Conference on Nanotechnology, Munich, Germany, August 16-19, 2004, pp.358-360.

[2] A. J. Nichol, W. J. Arora and G. Barbastathis, J. Vac. Sci. Technol. B, vol.24 no.6, pp.3128-3132, 2006.

[3] S. Niyogi, C. Hangarter, R. M. Thamankar, Y.-F. Chiang, R. Kawakami, N. V. Myung and R. C. Haddon, J. Phys. Chem. B., vol.108 no.51, pp.19818-19823, 2004.

[4] Y. Wu, P. Qiao, J. Qiu, T. Chong and T.-S. Low, Nano Letters, vol.2 no.2, pp.161-164, 2002.

ThB1 (Invited)
10:30 – 11:00

Optical MEMS for Future Instruments in Astronomy

Frederic Zamkotsian[1], Arnaud Liotard[1], Patrick Lanzoni[1],
Severin Waldis[2], Wilfried Noell[2], Nico de Rooij[2], Veronique Conedera[3], Norbert Fabre[3]

[1] Laboratoire d'Astrophysique de Marseille, 2 place Leverrier, 13248 Marseille Cedex 4, France
[2] IMT, University of Neuchatel, Rue Jaquet-Droz 1, CH-2002 Neuchatel, Switzerland
[3] LAAS-CNRS, 7 av. Colonel Roche, 31077 Toulouse cedex, France

Corresponding author: F. Zamkotsian, Tel +33 4 95 04 4151, e-mail: frederic.zamkotsian@oamp.fr

Abstract

MOEMS devices are under study in order to be integrated in next-generation astronomical instruments for ground-based and space telescopes. Their main advantages are their compactness, scalability, specific task customization using elementary building blocks, and remote control. Several laboratories, including Laboratoire d'Astrophysique de Marseille in close collaboration with microtechnology institutes, are engaged since several years in the design, realization and characterization of programmable slit masks for multi-object spectroscopy and micro-deformable mirrors for wavefront correction. First prototypes have been developed and show results matching with the requirements.

Keywords: Multi-Object Spectroscopy, wavefront correction, Adaptive Optics, programmable slit masks, micro-deformable mirror.

1 INTRODUCTION

The NASA's Origin Program and equivalent programs in Europe bring into fashion what astronomy always wanted to do, explaining where we are coming from by studying the formation of the galaxies and their evolution, as well as the formation and evolution of the planets around nearby stars.

The following gives two applications of MOEMS in observational astronomy:

1) Programmable Multi-Object Spectroscopy masks
Thanks to its multiplexing capabilities, Multi-Object Spectroscopy (MOS) is becoming the central method to study large numbers of objects. However, it is impossible to use traditional ground-based MOS in space. New methods need to be defined and technologies developed. For one of the most central astronomical program, deep spectroscopic survey of galaxies, the density of objects is low and it is necessary to probe wide fields of view. MOEMS provides a unique and powerful way of selecting the objects of interest (whatever the criteria distance, color, magnitude, etc.) within deep spectroscopic surveys. This saves time and therefore increases the scientific efficiency of observations.

2) Wave front correcting deformable mirrors
Telescopes and instruments are designed to obtain the best scientific performances. To reach the faintest objects, we must get the best Point Spread Function (PSF) with the

minimum of energy scattered within the outer areas of the PSF. Also sharp PSFs will allow to reach the best spatial resolution (limit of diffraction) and therefore to potentially resolve objects such as remote interacting building blocks in their way to become giant galaxies, star-forming regions within nearby galaxies or disks around forming planetary systems. The wave front perturbations are either residual optical aberrations in the design of the optical train of the instrument or dynamic deformation of the instrument PSF due to the atmospherical perturbations. MOEMS devices should enable the wavefront correction in next generation of extremely large telescopes on ground (Adaptive Optics systems) as well as big telescopes in space.

MOEMS devices are under study in several laboratories around the world, providing key advantages as compactness, scalability, specific task customization using elementary building blocks, and remote control. At Laboratoire d'Astrophysique de Marseille, we are engaged since several years, in close collaboration with microtechnology institutes, in the design, realization and characterization of programmable slit masks for multi-object spectroscopy and micro-deformable mirrors (MDM) for wavefront correction.

MOEMS have not yet been used in space astronomy, but this technology will provide the key to small, low-cost, light, and scientifically efficient instruments, and allow impressive breakthroughs in tomorrow's observational astronomy.

1-4244-0641-2/07/$20.00 ©2007 IEEE

2 PROGRAMMABLE SLITS

Programmable multi-slit masks are required for next generation Multi-Object Spectrograph (MOS) for space and ground-based instruments, including NIRSpec for JWST. A promising solution is the use of MOEMS devices such as micromirror arrays (MMA) or micro-shutter arrays (MSA), which both allow the remote control of the multi-slit configuration in real time. MSA has been selected to be the multi-slit device for NIRSpec and is under development at the NASA's Goddard Space Flight Center. They use a combination of magnetic effect for shutter opening, and electrostatic effect for shutter latching in the open position [1]. Within the framework of the JRA Smart Focal Planes (part of the European FP6 Opticon program), we have engaged a collaboration with the Institut de Micro-Technologies (IMT) of University of Neuchatel (Switzerland) in order to get a first demonstrator of a European MOEMS-based slit mask. We develop and microfabricate a novel micro mirror array suited for this application. The requirements are: high contrast, optically flat mirrors in operation, high fill factor, uniform tilt angle over the whole array and low actuation voltage. In order to fulfill these requirements we use a combination of bulk and surface micromachining in silicon, compatible for cryogenic operation. The mirrors are actuated electrostatically, and a system of multiple landing beams has been developed, which passively locks the mirror at a well defined tilt angle when actuated. The mechanical tilt angle obtained on 100 x 200 μm^2 micromirrors (prototypes of 5x5 mirrors array, Fig. 1) is 20° at a pull-in voltage of 90V. The tilt angle of the actuated and locked mirror is stable with a precision of one arc minute over the whole array. The surface quality of the mirrors in actuated state is better than 10nm peak-to-valley and the local roughness is around 1nm RMS [2]. For future MOS, MMA will be mostly designed for infra-red applications, leading to instruments needing vacuum environment and cryogenic temperatures. Large arrays are under development and cryo operation will be tested soon.

Fig. 1: 5x5 micro-mirror array (100x200 μm^2 mirrors).

3 MICRO-DEFORMABLE MIRRORS

Next generation space telescopes as well as future giant telescopes rely on the availability of highly performing wavefront correction systems. These systems require deformable mirrors with very challenging parameters, including number of actuators up to 250 000 and inter-actuator spacing around 500μm. MOEMS-based devices permit the development of a complete generation of new deformable mirrors. Three main MDM architectures are under study in different laboratories: the bulk micro-machined continuous membrane deformable mirror, the segmented deformable mirror (piston-only or piston-tip-tilt moving surfaces), the surface micro-machined continuous-membrane deformable mirror. The third concept is certainly the most promising architecture [3]. All these devices are based on silicon or polysilicon materials. We are currently developing a MDM based on an array of electrostatic actuators with attachment posts to a continuous mirror on top, in collaboration with a microtechnology laboratory, LAAS (Toulouse, France). The originality of our approach lies in the elaboration of layers made of polymer materials, in order to reach high strokes for low driving voltages. Mirrors with very efficient planarization and active actuators have already been demonstrated; the first polymer piston-motion actuator (Fig. 2) exhibits a stroke of 2μm for 30V and a resonance frequency of 6.5kHz, measured on our dedicated characterization bench using phase-shifting and time-averaged interferometry. The electrostatic force provides a non-linear actuation, while wavefront correction systems are based on linear matrices operations. We have successfully developed a dedicated 14-bit electronics in order to "linearize" the actuation [4]. Comparison with FEM models shows very good agreement, and realization of a complete polymer-based MDM is under way.

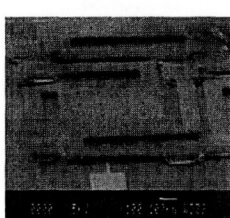

Fig. 2: Polymer-based actuator(500μm)
for micro-deformable mirrors.

REFERENCES

[1] S. H. Moseley, R. Arendt, R. A. Boucarut, M. Jhabvala, T. King, G. Kletetschka, A. S. Kutyrev, M. Li, S. Meyer, D. Rapchun, "Microshutter arrays for the JWST near-infrared spectrometer", *Proc. SPIE* **5487**, pp. 645-652, 2002.

[2] S. Waldis, P.-A. Clerc, F. Zamkotsian, M. Zickar, W. Noell, N. F. de Rooij, "Uniform tilt-angle micromirror array for multi-object spectroscopy", *Proc. SPIE* **6466**, 2007.

[3] Y. Zhou, T. Bifano, "Characterization of contour shapes achievable with a MEMS deformable mirror", *Proc. SPIE* **6113**, 2006.

[4] F. Zamkotsian, V. Conedera, H. Granier, A. Liotard, P. Lanzoni, L. Salvagnac, N. Fabre, "Electrostatic polymer-based micro deformable mirror for adaptive optics", *Proc. SPIE* **6467**, 2007.

ThB2
11:00 – 11:15

Passivated Piezoresistive Rotation Angle Sensor Integrated in Micromirror

Minoru Sasaki[1], Motoki Tabata[2], and Kazuhiro Hane[2]

[1]Dept. of Advanced Science and Technology, Toyota Technological Institute
Hisakata 2-12-1, Tenpaku-ku, Nagoya 468-8511, Japan
Phone: +81-52-809-1840, E-mail: mnr-sasaki@toyota-ti.ac.jp

[2]Dept.of Nanomechanics Eng., Tohoku University, Aramaki 6-6-01 Aoba-ku, Sendai, 980-8579, Japan

Abstract

A piezoresistive rotation angle sensor integrated in a micromirror device is improved with the passivation film. The sensor detects the shear stress inside the torsion bar generated by the mirror rotation under the electrostatic driving. The rotation angle is measured during the actuation. The passivation film reduces the leak current. The sensor signal shows the better performance showing the smaller hysteresis.

Keywords: Passivation film, Rotation Angle Sensor, Micromirror

1 INTRODUCTION

In many applications, the optical beam reflected by mirror must be controlled accurately in its position and direction. The bulk device of the galvanometric mirror is combined with the rotation angle sensor (e.g., encoder). Integrating the sensor inside the micromirror is ideal. We have reported a piezoresistive rotation angle sensor integrated in the micromirror [1]. The previous sensor has bare Si structure and the mirror is driven by the outside actuator.

In this study, the sensor is improved with the passivation films and the sensor and the actuator work at the same time.

2 DESIGNS

Figure 1 shows the vertical design of the sensor. The device Si layer has *pn* junction. The sensor is in *p*-type region at the top. When the mirror rotates, the shear stress is generated inside the torsion bar. The stress is proportional to the mirror rotation angle. The stress magnitude will be maximum at the sensor region. The inset shows three cross-sectional designs. In the previous design, Si etching crosses *pn* junction. The reverse bias is applied for confining the current in the *p*-type region. The leak current will decrease the sensitivity or increase the drift in the signal. In the design 1, the *p*-type region is 3 μm inside from the sidewall. In the design 2, the top and side surfaces are covered by SiN and SiO$_2$ passivation films, respectively. The resistance in design 2 decreased by ~1/3 at the same time.

Figure 2(a) shows the whole view of the device. Figure 2(c) is the vertical comb drive actuator. Figure 1(b) shows the sensor included in the torsion bar. The bias current flows along the torsion bars and the mirror. The transverse voltage V_{out} is measured through the meandering springs.

3 RESULTS

Figure 3 shows I-V curves in the reverse bias region applying the voltage between *p*- and *n*-type regions.

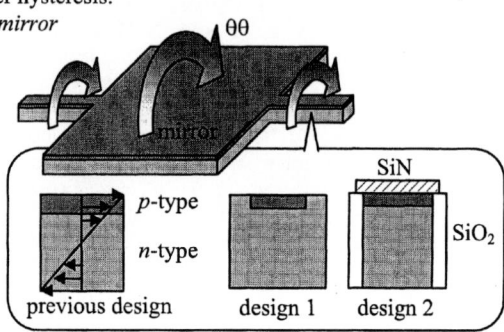

Fig. 1 Vertical design of the micromirror and the distribution of the shear stress inside the torsion bar.

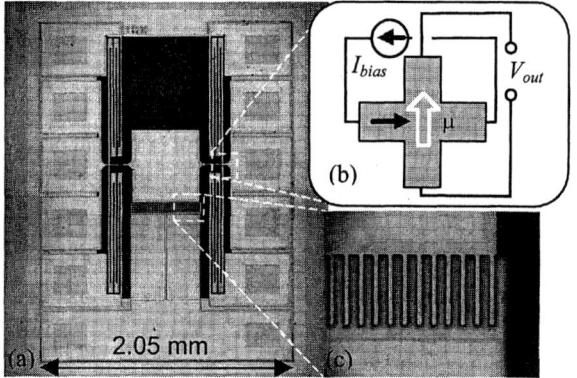

Fig. 2 Fabricated micromirror. (a) Whole view. (b) Schematic drawing of the sensor. (c) Comb drive part.

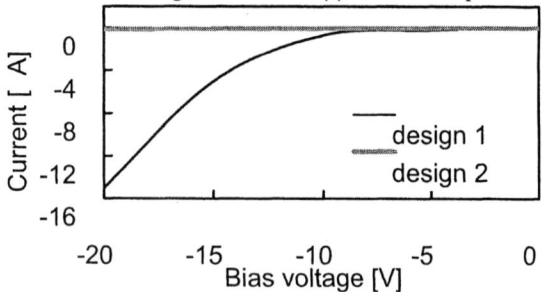

Fig. 3 I-V curves from devices in designs 1 and 2.

1-4244-0641-2/07/$20.00 ©2007 IEEE

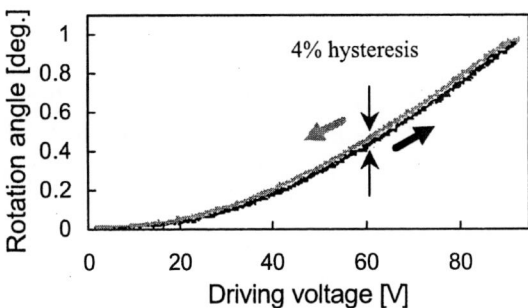

Fig. 4 A round-trip of the mirror rotation angle as the function of the driving voltage.

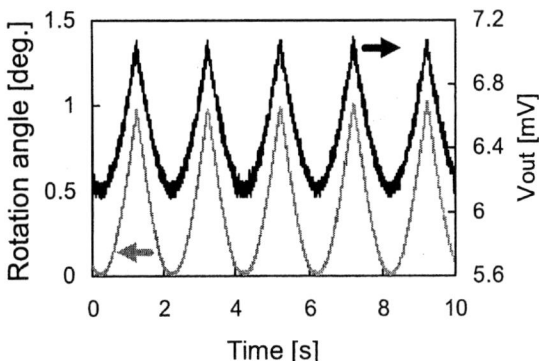

Fig. 5 Time responses of the mirror rotation angle and the sensor signal.

Compared to the design 1 showing the large leak current (15 µA at -20 V), the device design 2 with the passivation film reduces the leak current (0.12 µA at -20 V).

Figure 4 shows a single cycle of the mirror rotation angle as the function of the driving voltage. The mirror rotation angle is measured using the optical lever method. The mirror has ~4% hysteresis. The hysteresis value is almost same for the device designs 1 and 2. The hysteresis is the obstacle for controlling the angle accurately.

Figure 5 shows the time responses. The triangular driving voltage is applied. The black and gray curves are the sensor signal and the mirror rotation angle, respectively. The bias current is from −1 to 0 mA modulated at 5 kHz. The time constant of the lock-in amplifier is 0.1 ms. Figure 6 plots the sensor signal for 4 round-trips as a function of the mirror rotation angle. There is a linear relation.

Table 1 lists the fitted lines to the experimental data obtained from the device design 1. The first cycle of the sensor signal shows 1.4% hysteresis. Figure 4 with 4% hysteresis corresponds to the first cycle. The hysteresis evaluated from 4 cycle data is 2.9%. Table 2 lists the data obtained from the device design 2. The hysteresis evaluated

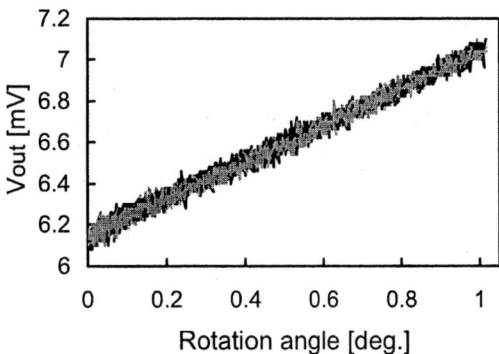

Fig. 6 The sensor signal for 4 round-trips as the function of the mirror rotation angle.

Table 1 Fitted lines of V_{out} [mV] data against the mirror rotation angle θ [deg.] in design 1.

cycle	forward[mV]		backward[mV]		hysteresis[%]
1	0.911	+6.145	0.895	+6.148	1.39
2	0.896	+6.146	0.915	+6.144	1.89
3	0.898	+6.156	0.887	+6.150	1.91
4	0.891	+6.142	0.890	+6.143	0.10
all					2.94

Table 2 Fitted lines of V_{out} [mV] data against the mirror rotation angle θ [deg.] in design 2.

cycle	forward[mV]		backward[mV]		hysteresis[%]
1	0.132	+0.693	0.132	+0.696	0.36
2	0.134	+0.691	0.132	+0.694	0.73
3	0.136	+0.689	0.132	+0.692	0.27
4	0.136	+0.687	0.132	+0.692	0.42
all					1.17

from 4 cycle data is improved to 1.2%. The bias current is modulated at 25 kHz, which is realized with the lower resistance design. In spite of the lower sensitivity shown in the coefficient of θ, the lock-in amplifier works at the higher gain with almost same SN ratio. In addition to the smaller leak current, the possible reason for improving the hysteresis will relate to the increase of the modulation frequency over the resonant frequency (14 kHz) of the micromirror.

This research was supported by a grant-in-aid for scientific research on priority areas (no. 17040003). The facilities used for this research include the micro/nano-machining research and education center, at Tohoku University.

REFERENCE

[1] M. Sasaki, M. Tabata, T. Haga, K. Hane, *Jap. J. Appl. Phys.* vol. 45, no. 4B, pp. 3789-3793, 2006.

ThB3
11:15 – 11:30

Integrated Piezo-resistive Positionssensor for Microscanning Mirrors

Thilo Sandner, Holger Conrad, Thomas Klose, and Harald Schenk
Fraunhofer Institute for Photonic Microsystems (IPMS), Grenzstr. 28, D-01109 Dresden
Phone: +49-351-8823-152, Fax: +49-351-8823-266, E-mail: thilo.sandner@ipms.fraunhofer.de

Abstract

Microscanning mirrors with integrated piezoresistive positionsensors are presented. The novel sensor approach is based on intrinsic piezoresistivity of SOI material. It is fully compatible to microscanner technology and requires no additional technological efforts, enabling a cost efficient fabrication process. Integrated 2D position sensors with amplitude sensitivity of $S_f = 2.0\text{mV/V} @ 6°$, similar to metallic strain gauges, as well as a good linearity of $\leq 0.5\%$ error of linearity has been realized.

Keywords: Optical SOI-MEMS, micro scanning mirror, integrated piezo-resistive position sensor, angular sensor.

1 INTRODUCTION

Micro scanning mirrors are essential components in miniaturized and highly accurate scanning systems for various applications like portable laser projection displays or scanning grating spectrometers [1]. For precise control of amplitude, phase and timing of the scanning system a mirrors position feedback is required. The commonly used optical position feedback sensors are limited regarding size, miniaturization and cost due to their hybrid assembly. Hence, an integrated position sensor is required for mass fabrication. In this paper we present a novel integrated piezo-resistive position sensor, which is fully compatible to the current SOI technology, enabling a cost efficient fabrication [2].

3 PIEZOELECTRIC POSITION SENSORS

The piezo-resistive effect [3], i.e. the strain induced change of valence band structures of semiconductors (e.g. Si, Ge) observed as changes of resistivity, is frequently used for mechanical sensors of pressure or force. A piezo-resistive sensor consists in general of two parts a) the mechanical stressed mechano-elastic transformer, which transforms the mechanical input value (i.e. fource, pressure or angular tilt position) in a mechanical strain field, and b) the mechano-electric transformer, transforming the strain field in a changed field of electrical conductivity. In conventional piezo-resistive sensors the piezo-resistors are integrated only in small, surface near parts of the stressed mechano-elastic transformers, because of the symmetric strain fields to prevent a compensation of the electrical field domain within the piezo-resistors. For this the piezo-resistors are implanted in regions of max. strain and electrically separated from the common SOI bulk material. Integrated piezo-resistive sensors for measuring the angular tilt position of micro mirrors have been presented so far in [4] [5]. In [4] the shear

stresses are measured using a transverse voltage gauge, whereas in [5] the normal stresses have been used. In booth chases the piezo-resistive sensor elements have been locally implanted within the mechanical deformed and electrically separated spring structures. Hence, additional technological effort is required for these piezo-resistive sensors increasing the complexity of the fabrication process. The objective of our novel sensor approach was to develop an integrated piezo-resistive angular position sensor, which is fully compatible to our current SOI scanner technology without the need of any technological changes [2].

2.1 Sensor Principle and Design

In our novel sensor approach we use the SOI bulk material of the monolithic mechano-elastic transformer itself as the piezo-resistive mechano-elastic transformer (see fig. 1).

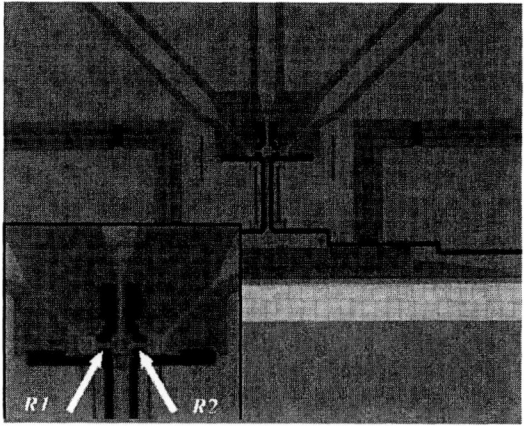

Fig. 1: Microscopic photograph of the integrated piezo-resistive position sensor (*R1*, *R2* piezo-resistors of a half bridge gauge)

1-4244-0641-2/07/$20.00 ©2007 IEEE
195

The mechano-elastic transformers are designed as auxiliary parallel bending beams attached to the anchors of the torsional springs (see fig. 1). Depending on the actual scan angle the torsional springs and auxiliary bending structures are deformed. The induced bending stresses of the mechano-elastic transformers results in a symmetric normal strain field shown in figure 2. By means of the lateral arrangement of the electrical surface contacts an asymmetric electrical field is generated within the monolithic sensor structure defining the electric active part of the piezo-resistive sensor (see fig. 3). Hence, a change of resistivity can be externally measured. It should point out that the mechano-elastic transformers and the piezo-resistors are made of the same monolithic highly doped SOI material, whereas no local changes of doping concentration exist.

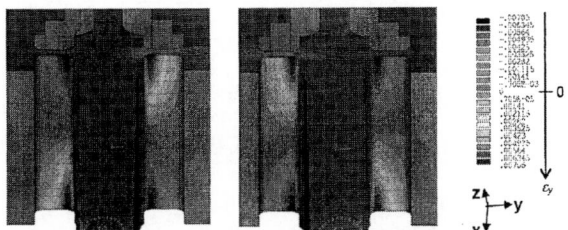

Fig. 2: Symmetric strain field in yz at maximal scan angle

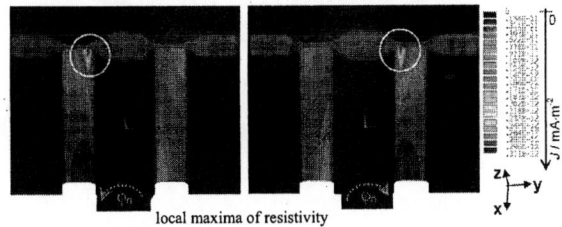

local maxima of resistivity

Fig. 3: Asymmetric electrical current field of piezo-resistive sensor

2.2 Experimental results

Several variants of micro scanning mirrors with integrated piezo-resistive 1D/2D position sensors have been fabricated using [100] oriented and highly p-doped ($N_A = 10^{18}$) 30μm thick SOI. In the chase of 1D mirrors Wheatstone full bridge gauges as well as half bridge gauges for 2D mirrors have been implemented, whereas for the 2D chase the SOI springs itself are used as electrical interconnects of the inner mirrors sensor. The experimental results of a 2D mirror with integrated piezo-resistive angular 2D position sensor are summarized in fig. 4 and fig. 5. Shown are the amplitude and phase signals of the piezo-resistive sensors vs. mechanical scan angle (MSA) for the movable frame (see fig. 4) and inner mirror (see fig. 5). An amplitude sensitivity $S_{f/m}$ of $S_f = 2003$ μV/V @ 6° (fig. 4) and $S_m = 2657$ μV/V @ 15°

(fig. 5), similar to metallic strain gauges, as well as a good linearity with $\leq 0.5\%$ error of linearity has been measured.

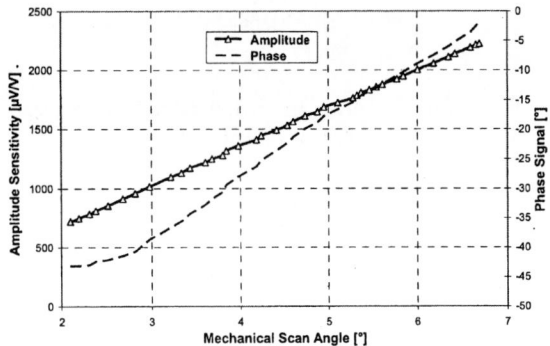

Fig. 4: Piezo-resistive 2D-position signals of movable frame

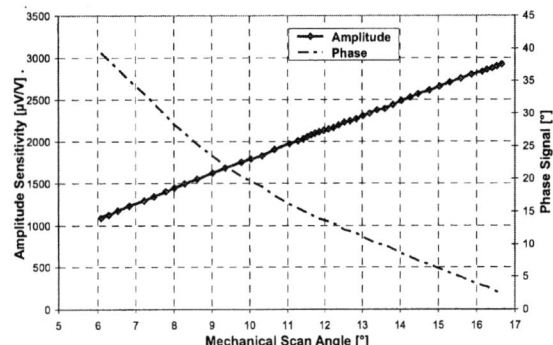

Fig 5: 2D-position sensor signals of inner mirror

2 CONCLUSION

Integrated piezo-resistive position sensors for feedback control of micro scanning mirrors have been presented. Several 1D/2D sensor variants have been fabricated. A sensitivity of ≥ 2.0 mV/V at 6° MSA as well as a good linearity with $\leq 0.5\%$ error of linearity has been achieved. The novel sensor approach is fully compatible to the current SOI technology, enabling a cost efficient fabrication.

REFERENCES

[1] A. Wolter et al.: "Applications and requirements for MEMS scanner mirrors", Proc. SPIE vol. 5719, pp. 64-75 (2005).
[2] H. Schenk: PhD thesis, University Duisburg, 2001.
[3] C. S. Smith: "Piezoresistance effect in germaninum and silicon". Physical Review 94, 1954.
[4] M. Sasaki, T. Haga, K. Hane: "Piezoresistive Rotation Angle Sensor Integrated in Micromirror", Jpn. J. Appl. Phys. 45 (4B), pp. 3789-3793.
[5] K. Kehr, PhD thesis, Technical University Chemnitz, 2000.

ThB4
11:30 – 11:45

Ultra flat high resolution microscanners

Shu-Ting Hsu, Thomas Klose, Christian Drabe, Alexander Wolter, and Harald Schenk
Fraunhofer Institute for Photonic Microsystems (IPMS), Maria-Reiche Str. 2 01109 Dresden, Germany
Phone: +49-351-8823-241, Fax: +49-351-8823-266, E-mail: shu-ting.hsu@ipms.fraunhofer.de

Abstract

We present a high frequency (30 kHz) micro-scanner with 27 nm dynamic deformation at $\pm 10°$ mechanical scan angle. To achieve that, a dry-wet combination process is utilized to fabricate reinforcement frames for scanner flatness improvement.

Keywords: micromirror, dynamic deformation, display scanner, backside reinforcement

1 INTRODUCTION

The micromirror described in this work is designed to perform the fast-axis scanning in a SVGA (800 pixels × 600 pixels) quality laser projector. The basic requirements for a high resolution scanner are a large amplitude-diameter ratio θD, and a high frequency f. However, increasing the above parameters also dramatically reduces scanner dynamic flatness δ, which, in return, degrades the optical resolution. (1) describes the relationship between maximum surface error δ_{max} and the design parameters [1]. The equation is valid for a rectangular mirror plate with two torsion hinges, but it is reasonable to assume that the the relationship is valid also for root-mean-square deformation δ_{rms} in a round mirror with multiple springs.

$$\delta_{max} \propto \frac{D^5 f^2 \theta}{t^2} \qquad (1)$$

where t is the thickness of the mirror plate.

Methods to reduce dynamic deformation have been reported in [2]–[5]. Here, we present a flat scanner design, which combines the multiple-spring layout [5] with backside reinforcement frame to further reduce the δ_{rms} to meet the optical quality requirement of $\delta_{rms} < \lambda/10$ (40 nm for violet). The geometry of reinforcement frame is optimized with finite element analysis to increase the rigidity of the mirror plate with minimal additional mass. A dry-wet combination etch process is utilized [6] to fabricate the backside frame with reproducible results.

2 REINFORCEMENT FRAME FABRICATION

The fabrication process takes advantage of the loading effect in a DRIE process to create two etch depths in one process step. A TMAH wet-etch step then proceeds to remove the sidewalls, leaving a 86 μm high frame within the cleared backside cavity. This process produced a very uniform etch depth within the five 6-inch wafers fabricated. The process flow is illustrated in Fig. 1.

The scanners are fabricated on SOI wafers with a 30 μm B-doped device layer, a 1 μm buried oxide, and a 390 μm Si handle wafer. The backside process begins after all the front-side patterning is completed. Only one etch step is left to be performed afterwards to etch the comb-drive actuators

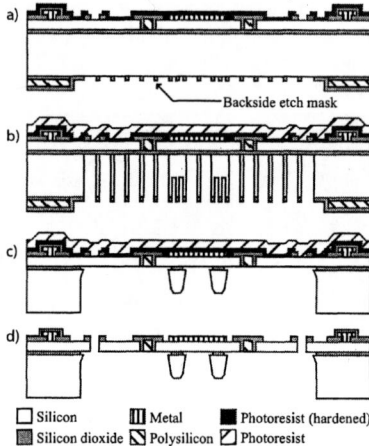

Fig. 1. Fabrication process flow, a) Backside oxide mask and last frontside mask patterned, b) Frontside protected with PR during backside DRIE etch, c) Back-side TMAH etch and oxide mask removal, d) Frontside trench etch.

and torsion hinges. During the backside process, a 6 μm photoresist (PR) is deposited on the frontside to protect the UV-hardened PR mask and other frontside structures. The protective PR is later removed in a developer.

The backside oxide mask consists of two trench sizes, the 30 μm × 30 μm squares populate the area where reinforcement frame is located; the rest of the backside cavity is filled with trenches larger than 110 μm × 110 μm. During DRIE backside etch, the small trenches experience slower etch rate due to reduced etchant mass transportation. The depth of 30 μm wide trenches is therefore 80-90 μm shallower than their larger counterparts. After DRIE etch, the wafer is placed in a capsule to protect its front-side before being dipped into a 75°C TMAH bath to remove the sidewalls. At the end of the process, the backside cavity is cleared with only reinforcement structures standing on the mirror plate (Fig. 1-d).

3 RESULT

3.1 Fabricated device

The scanners have a 1 mm mirror plate and a ~30.4 kHz resonant frequency. The average height of the backside frames

1-4244-0641-2/07/$20.00 ©2007 IEEE   197

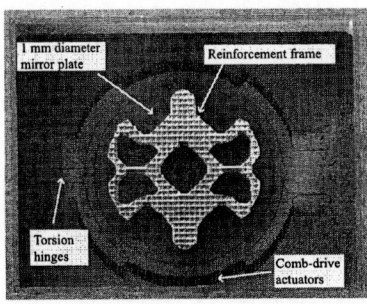

Fig. 2. Backside SEM micrograph of a fabricated device. The cavity is 390 μm deep and the reinforcement structure height is 83 μm.

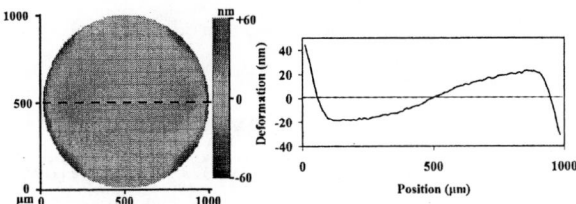

Fig. 3. Dynamic deformation measurement by a stroboscopic interferometer.

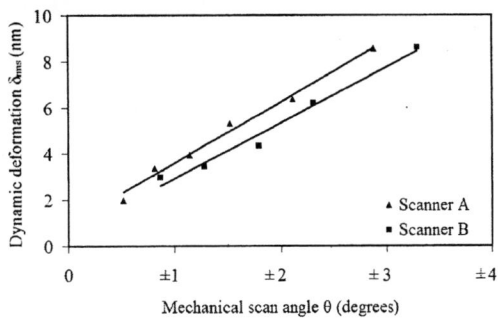

Fig. 4. Dynamic deformation of scanner A and B at various angles. The static deformation is removed.

TABLE I
RMS DEFORMATIONS AND NORMALIZED DEFORMATION FACTORS.

Scanner	Frequency	θ, δ_{rms}	Deformation factor
A	30.84 kHz	10°, 28 nm	0.52
B	30.11 kHz	10°, 25 nm	0.49
C (no frame)	17.02 kHz	4.4°, 7 nm	1

is 86 μm. Fig. 2 shows the backside structure of a fabricated scanner. The grid pattern is the remanent sidewalls, which does not affect the performance of the device and can be removed with longer TMAH wet-etch.

3.2 Dynamic deformation

Dynamic deformation is measured with a stroboscopic interferometer [7]. During scanner actuation, the largest mirror deformation occurs at peak mechanical scan angle. To measure the dynamic deformation, we pulse a 650 nm light emitting diode (LED) with a frequency identical to the scanner actuation frequency. The duty cycle of the pulse is short enough (0.1-0.3%) for the LED to illuminate the scanner only at a specific phase. A delay is applied to the pulse train to fix the illumination at exactly the peak amplitude of scanner rotation. The interferometric measurements of scanner topography is performed with the pulsed LED to obtain the deformation data of the micro-scanner. Fig. 3 plots the measurement results of scanner A at ±4.8° mechanical scan angle; the position-deformation plot on the right shows that large deformations are concentrated at the area 50 μm from the edge of the mirror. 90% of the mirror surface are flat with a δ_{max} of less than ±20 nm.

The rms dynamic deformation versus scan angle of two scanners tested are plotted in Fig. 4. Because deformation is proportional to scan angle, we can obtain the δ_{rms} at ±10° mechanical scan angle by linear extrapolation. Table I lists test results of the two scanners and a controller Scanner C of the same spring design without backside frame. To compare the performance, we use a normalized dynamic deformation factor, which is obtained by dividing the deformation by θf^2 and normalizing the values against the factor of scanner C. It is shown that the dynamic deformation of scanners with backside frame is half of the one without.

4 CONCLUSION

High frequency micro-scanners with backside reinforcement frames are fabricated with a dry-wet combination process. The δ_{rms} of the scanners at ±10° mechanical angle is less than 27 nm, which is much smaller than the 40 μm minimum deformation imposed by optical quality requirements. The mirror is capable of performing 800 pixel scan at ∼30.4 kHz to act as the fast axis of a SVGA display scanner.

REFERENCES

[1] P. J. Brosens, "Dynamic mirror distortions in optical scanning," *Applied Optics*, vol. 11, no. 12, pp. 2987–2989, 1972.

[2] R. A. Conant, J. T. Nee, K. Y. Lau, and R. S. Muller, "A flat high-frequency scanning micromirror," in *Proceedings of Solid-State Sensor, Actuator and Microsystems Workshop*, Hilton Head, SC, June 2000, pp. 6–9.

[3] C.-H. Ji, M. Choi, S.-C. Kim, S.-H. Lee, S.-H. Kim, Y. Yee, and J.-U. Bu, "An electrostatic scanning micromirror with diaphragm mirror plate and diamond-shaped reinforcement frame," *J. Microelectromech. Syst.*, vol. 9, no. 4, pp. 409–418, 2000.

[4] J. T. Nee, R. A. Conant, M. R. Hart, R. S. Muller, and K. Y. Lau, "Stretched-film micromirrors for improved optical flatness," in *The Thirteenth Annual International Conference on Micro Electro Mechanical Systems*, Miyazaki, Japan, Jan 2000, pp. 704–709.

[5] A. Wolter, T. Klose, S.-T. Hsu, H. Schenk, and H. Lakner, "Scanning 2d micromirror with enhanced flatness at high frequency," in *MOEMS Display, Imaging, and Miniaturized Microsystems IV, Proceedings SPIE*, vol. 6114, San Jose, CA, 2006, pp. 207–214.

[6] J. Kiihamaki, H. Kattelus, J. Karttunen, and S. Franssila, "Depth and profile control in plasma etched mems structures," *Sensors and Actuators: Physical*, vol. 82, no. 1, pp. 234–238, May 2000.

[7] M. R. Hart, R. A. Conant, K. Y. Lau, and R. S. Muller, "Stroboscopic interferometer system for dynamic mems characterization," *J. Microelectromech. Syst.*, vol. 9, no. 4, pp. 409–418, 2000.

ThB5
11:45 – 12:00

Combined Device of Optical Microdisplacement Sensor and PZT-Actuated Micromirror

K. Akase,[1] R. Sawada,[1] E. Higurashi,[2] T. Kobayashi,[3] R. Maeda,[3] M. Inokuchi,[1] S. Sanada[1] and I.Ishikawa[1]

[1]Department of Intelligent Machinery and Systems, Kyushu University,
Motooka, Nishi-ku, Fukuoka 819–0395, Japan
[2]Research Center for Advanced Science and Technology, University of Tokyo,
4–6–1 Komaba, Meguro-ku, Tokyo 153–8904, Japan
[3]National Institute of Advanced Industrial Science and Technology (AIST),
1–2–1 Namiki, Tsukuba 305–8564, Japan
E-mail: akase@nano-micro.mech.kyushu-u.ac.jp   Tel / Fax 81 92 802 3817

Abstract

A combined device of a PZT-film-actuated micromirror and microsensor that can detect linear movement and rotation angle of the mirror has been developed[1]. The micromirror is actuated by vertical movement of two PZT cantilevers formed as a unit on the right and left of the movable mirror and directly connected to it via hinges. By detecting reflected light diffused from a VCSEL (vertical-cavity surface-emitting laser) toward the mirror, using two photodiodes (one on each side), displacement and rotation angle are measured with high precision. Since this combined device can feed back the displacement and rotation angle obtained from the sensor, it can compensate for the hysteresis of PZT and therefore enable stable, high-precision optical beam control.

Keywords: Microdisplacement sensor, PZT-Actuated Micromirror, VCSEL, Photodiode.

1 INTRODUCTION

At the previous conference, we reported on a microsensor that made use of a diffused beam from a surface-emitting laser of an integrated type 1.5 mm x 1.5 mm x 1 mm in size and could be used for high-precision measurement of displacement and rotation angle of a small micromirror.

This paper reports on a combined device of a PZT-actuated micromirror and optical microsensor. The PZT micromirror, unlike electrostatic-type micromirrors, has no structure such as an electrode near the rear of the micromirror (the face toward which a beam is not irradiated). Consequently, if the sensor is very small, a sensor for measuring the displacement and rotation angle of the micromirror can be placed on the rear face of the micromirror.

Since a PZT-actuated micromirror has hysteresis, it is essential to feed back the displacement or rotation angle, such as by determining the relationship between the actuation voltage and the displacement or the rotation angle of the micromirror in advance. Even if a micromirror is manufactured and miniaturized with micromachining technology, if the sensor for measuring the displacement of the mirror, etc. is not sufficiently compact the usefulness of the micromirror is greatly impaired. The combined device of a microsensor and a very small micromirror presented here can improve the actuation precision and stability of the mirror without imposing a limiting factor on miniaturization of the micromirror.

2 DEVICE CONFIGURATION AND STRUCTURE

Figure 1 shows the structure of our combined device of a PZT-film-actuated micromirror and microsensor that can detect linear movement and rotation angle of the mirror, and the principle for measurement of the displacement and rotation angle of the micromirror. The hinge, provided at the center of the micromirror, is 1 mm square, is directly connected to two PZT cantilevers (Fig. 2), and allows linear movement or rotation. If the two cantilevers are moved vertically in the same direction, the micromirror moves up or down linearly. If the two cantilevers are moved in opposite directions, the micromirror is tilted. A displacement and rotation-angle sensor is placed on the rear face this movable mirror. Diffused light radiated from the surface-emitting laser is reflected by the movable micromirror. The reflected light is detected by the two photodiodes bonded on the sides of the surface-emitting laser. The linear movement displacement of the micromirror is measured based on the sum of detected output from the two diodes (A+B), and the rotation angle is measured based on the difference (A–B) (Fig. 1-(c)). Figure 3 shows a photo of the combined device. The overall size is 12 mm x 8 mm x 7 mm.

1-4244-0641-2/07/$20.00 ©2007 IEEE

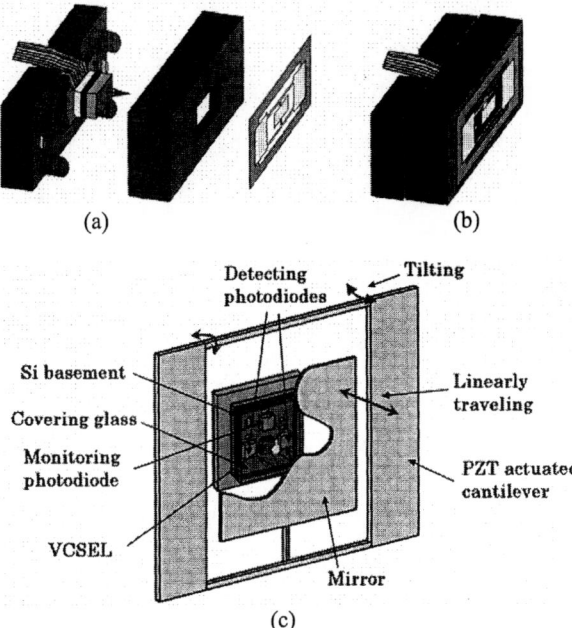

Fig. 1 Structure of our combined device of a PZT-film-actuated micromirror and microsensor that can detect linear movement and rotation angle of the mirror (a) and (b), and the principle of measurement for displacement and rotation angle of the micromirror (c)

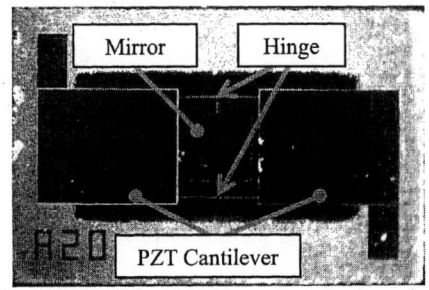

Fig. 2 Photo of the PZT-film-actuated micromirror

Fig. 3 Photo of a combined device of displacement sensor and PZT micromirror

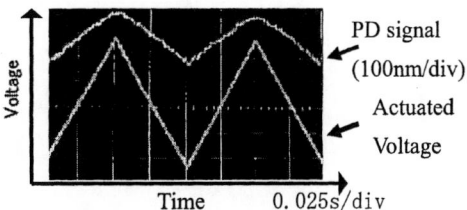

Fig. 4 Output signal from a microdisplacement sensor obtained when a micromirror is moved, with a triangular waveform of voltage that corresponds to 200 nm of movement (PD signal)

3 CHARACTERISTICS

Figure 4 shows the output signal of a micro-displacement sensor obtained when a triangular waveform of voltage that corresponds to 200 nm movement of the micromirror is applied to the PZT cantilevers. It is understood that a minimum displacement of approximately 20 nm can be measured, providing sensor resolution sufficient for compensation of mirror variations based on the hysteresis and temperature changes caused by PZT actuation. In addition, we believe that this resolution satisfies the precision requirements of most applications. Two those PD signals corresponding to triangular-wave actuation volts, which are shown in Figure 4 are different from each other. This seems to be due to the instability, or short-term fluctuation for the mirror movement. Also, this result indicates that the instability can be improved by feeding the output of this accurate sensor signals back.

4 CONCLUSION

We have developed a small device that can perform feedback control of linear displacement of a mirror, by combining a PZT-actuated micromirror and an optical microsensor as a unit. Although this optical displacement sensor is very compact and provides high precision, the measurement range is quite wide, at 1 mm or more. Consequently, the sensor can provide performance that cannot be achieved with any combination of electrical-capacitance sensors and a micromirror, which would necessarily have much larger displacement. In the future, we will acquire data on mirror rotation to check the validity for rotation amount and also reduce electrical signals by combining with a preamplifier (amplifier circuit).

5 ACKNOWLEDGEMENTS

Part of this work was supported by the New Energy and Industrial Technology Development Organization (NEDO).

REFERENCE

[1] R. Sawada, E. Higurashi, S. Sanada, D. Chino and I. Ishikawa, 2006 IEEE/LEOS International Conference on Optical MEMS AND THEIR APPLICATIONS, 2006, pp.52-53.

Author Index

A

Akase, K.	ThB5
Aljasem, K.	MA3
Antoszewski, J.	WA1
Armiger, B.	TuP10
Arnold, M.	WC4
Arora, W. J.	TuB4
Ataka, M.	TuB3
Ataman, C.	MC2
Ayyalasomayajula, P.	TuP17

B

Bakke, T.	WC5
Barbastathis, G.	TuB4, ThA5
Barretto, R. P. J.	MA2
Bergeron, S.	WD2
Blunier, S.	WC4
Borwick, R.	WD1
Brasselet, S.	TuA3
Bräuer, A.	PLE2
Bulgan, E.	TuP20

C

Carberry, D.	TuB1
Catrysse, P. B.	WB6
Chang, C.-M.	TuP6
Chang, P.-H.	TuP22
Chang, T.-L.	MC5
Chauvat, D.	TuA3
Chen, C.-C.	TuP40
Chen, C.-Y.	TuP36, TuP37, TuP38
Chen, C.-N.	TuP12
Chen, H.-W.	TuP33
Chen, S. Y.	TuP35
Chen, T.-Y.	TuP25
Chen, X.	TuP39
Cheng, C.-C.	TuP6
Chiou, P.-Y.	TuA5
Chiou, S.-J.	TuP15
Chiu, Y.	MC5
Cho, Y.-H.	TuB5, WC3
Choo, H.	TuP2
Chung, T.	TuP3
Chyan, J. Y.	TuA4, ThA3
Cocker, E. D.	MA2
Condit, J. C.	MA4
Conedera, V.	ThB1
Conrad, H.	ThB3

D

Dannberg, P.	PLE2
de Rooij, N. F.	TuP17, ThB1
Dell, J. M.	WA1
Denatale, J. F.	WD1
Dual, J.	WC4
Duan, A.	TuP39
Duparré, J.	PLE2

E

Ebisui, A.	MA5
Ebneter, C.	WC4
Eiji, H.	ThB5
Ekekwe, N.	TuP10

F

Fabre, N.	ThB1
Fan, S.	WB6, ThA1

F (cont.)

Fang, W.	MC1, WA4
Faraone, L.	WA1
Felder, F.	WC4
Ferhanoglu, O.	TuP21
Flusberg, B. A.	MA2
Fujita, H.	TuB3, WD3, WD4
Funaki, H.	WD4
Furukawa, R.	TuP26

G

Gacoin, T.	TuA3
Gao, Q.	ThA4
Garmire, D.	TuP2
Ge, Y.	TuP24
Gibson, G.	TuB1
Godbout, N.	WD2
Gordon, R.	TuP7
Gunning, W.	WD1

H

Haga, T.	MA1
Hah, D.	TuP15
Han, W.	TuB5
Hane, K.	MC3, TuP20, TuP23, WB3, ThB2
Hegg, M. C.	MB3
Higo, A.	TuB3, WD3
Hillmer, H. H.	TuP1
Hirabayashi, Y.	TuB3
Hirose, K.	MC4
Ho, F. H.	TuP35
Hoefer, B.	PLE2
Holmstrom, S.	MC2
Horsley, D. A.	TuP32
Hoshino, K.	MA4
Howe, R. T.	TuP32, WB6
Hsieh, C. M.	TuP12
Hsieh, C.-H.	ThA3
Hsieh, J.	MC1
Hsieh, T.-L.	TuP15
Hsu, C. C.	WB5
Hsu, S.-T.	ThB4
Hsu, W.-C.	ThA3
Hu, F.-R.	ThA2
Hu, H. C.	TuP8
Hu, Y.-C.	WB2
Huang, C.-F.	TuP36, TuP37
Huang, J.-J.	TuP36, TuP37

I

In, H. J.	ThA5
Inokuchi, M.	ThB5
Inoue, D.	TuP13
Ishikawa, I.	ThB5
Isikman, S.	MC2
Itaya, K.	WD4

J

Jagadish, C.	ThA4
Jang, W.	TuP5
Jang, Y.-H.	TuP14
Jeon, J.-A.	TuP3
Jin, Y.-H.	WC3
Jo, K.	TuP11, TuP5
Johansen, I.-R.	WC5
John, S.	PLE3
Joyce, H. J.	ThA4
Jung, I. W.	WB1
Jung, J. C.	MA2

201

K

Kalim, S.	TuA5
Kanamori, Y.	TuP20, TuP23
Kant, R.	TuP2, TuP32
Kawashima, H.	TuP26
Keating, A. J.	WA1
Kemp, N. J.	MA4
Kim, M. G.	TuP11, TuP5
Kim, M.	TuP3
Kim, Y.	ThA4
Kim, Y.-K.	TuP14, TuP3
Kiyokura, T.	MA1
Klose, T.	ThB3
Ko, C.-H.	TuP31
Kobayashi, A.	TuP26
Kobayashi, T.	ThB5
Kobrin, P.	WD1
Kodate, K.	WC1
Koga, A.	TuP26
Komai, Y.	WC1
Kumar, K.	MA4
Kwon, H.	TuB3

L

Lacolle, M.	WC5
Lai, N. D.	WB5
Lanzoni, P.	ThB1
Lauxtermann, S.	WD1
Le Xuan, L.	TuA3
Leach, J.	TuB1
Lee, C. B.	WB3
Lee, C.	TuA2, TuP12
Lee, D.	TuP16
Lee, J.-H.	TuP11, TuP5
Lee, R.-K.	MB2, TuP29
Lee, S. K.	WB3
Lee, S.-K.	TuP11
Lee, T.-D.	TuP28
Lim, C. S.	WC2
Lin, H.-Y.	MC1
Lin, J. H.	WB5
Lin, L. Y.	MB3, MB5
Lin, S.-W.	TuP33
Lin, Y.-B.	TuP30
Lin, Y. H.	TuP35
Liotard, A.	ThB1
Liu, A.	WC2
Liu, W. F.	TuP33
Lu, Y.-C.	TuP36, TuP37, TuP38

M

Ma, Y.-F.	TuP34
Maboudian, R.	WB6
Maeda, R.	ThB5
Maheshwari, V.	TuA1
Masson, J.	WD2
Masunishi, K.	TuP26
Mathai, S.	WD5
Matsuyama, N.	TuP23
McElroy, A.	MA4
Milanović, V.	TuP18, TuP19
Miles, M.	TuB1
Milner, T. E.	MA4
Miner, A.	TuP19
Mita, Y.	MC4
Miura, H.	MC3
Moe, S. T.	WC5
Momiuchi, M.	TuP26
Mueller, C.	WA5
Muller, R. S.	TuP2
Murray, K.	TuP10
Musca, C. A.	WA1

N

Nagasaka, Y.	MA5
Nakada, M.	TuB3
Nakano, Y.	WD3
Nguyen, C.	TuA1
Nguyen, T.	WA1
Nichol, A. J.	TuB4, ThA5
Noda, S.	MB1
Noell, W.	TuP17, ThB1

O

Ohtsu, M.	PLE1
Okamoto, K.	WC1

P

Padgett, M. J.	TuB1
Pai, S.-S.	MB4
Park, I.-H.	TuP3
Park, J.-H.	TuP14, TuP3
Park, Y.	TuP5
Peng, C.-L.	TuP6
Perruchas, S.	TuA3
Peter, Y.-A.	WD2
Piyawattanametha, W.	MA2
Poppe, E.	TuP39
Poulin, A.	WD2
Provine, J.	TuP32, WB6

Q

Quack, N.	WC4
Quidant, R.	TuP41

R

Ra, H.	MA2
Rahim, M.	WC4
Righini, M.	TuP41
Roch, J.-F.	TuA3
Roper, C.	WB6

S

Sagberg, H.	WC5
Saito, N.	TuP26
Sakai, S.	MC4
Sanada, S.	ThB5
Sandeau, N.	TuA3
Sandner, T.	ThB3
Saraf, R.	TuA1
Sasaki, M.	MC3, ThB2
Sasaki, T.	MC3
Sawada, R.	ThB5
Schenk, H.	ThB3
Schneider, F.	WA5
Schnitzer, M. J.	MA2
Schreiber, P.	PLE2
Shaik, R. P.	WA2
Shen, J.-T.	ThA1
Shen, K.-C.	TuP36
Shimada, J.	MA1
Silva, K. K. M. B. D.	WA1
Slablab, A.	TuA3
Solgaard, O.	MA2, TuP16, WB1
Song, W. Z.	WC2
Stupar, P.	WD1
Su, J. Y.	TuP35
Sun, C.-W.	TuP15
Suzuki, K.	WD4

T

Tabata, M.	ThB2
Taguchi, Y.	MA5
Takahashi, K.	WD4

Tan, H. H.	ThA4
Tang, T.-Y.	TuP36, TuP37
Tard, C.	TuA3
Tatara, N.	MA1
Teitell, M.	TuA5
Tien, C.-L.	TuP33
Tien, M.-C.	WD5
Ting, T.-L.	TuP30
Toshiyoshi, H.	TuB3, WD3, WD4
Toy, M. F.	TuP21
Tsai, C.	TuP12, TuP6
Tsai, J.-C.	TuP15
Tseng, P.-Y.	TuA5
Tseng, V. F.-G.	MC5

U

Urey, H.	MC2, TuP21

V

Veronis, G.	ThA1
Viereck, V.	TuP1
Villeval, P.	TuA3

W

Waldis, S.	TuP17, ThB1
Wallrabe, U.	WA5
Wang, C.-J.	MB5
Wang, J.-S.	WB1
Wang, M.	TuP24
Wang, W.-S.	TuP30
Wang, Z.	ThA1
Werber, A.	MA3, TuB2
Winchester, K. J.	WA1
Woo, D. K.	WB3
Wu, C. Y.	WB4
Wu, L.	WA3
Wu, M. H.	TuP15, WD5
Wu, M.	MC1
Wu, T.-H.	TuA5

X

Xie, H.	WA3

Y

Yalcinkaya, A. D.	MC2
Yang, C.-C.	TuP36, TuP37, TuP38
Yang, Y.-J.	TuP36
Yao, J.	WD5
Yap, P.	WC2
Ye, J.-S.	TuP23
Yeh, D.-M.	TuP36, TuP37, TuP38
Yeh, J.	TuP12, TuP6, ThA3
Yin, H.-L.	WB2
Yoo, B.-W.	TuP14, TuP3
Yu, C.-S.	TuP9, WB2
Yu, K.	TuP14
Yu, Z.	ThA1

Z

Zamkotsian, F.	TuP17, ThB1
Zappe, H.	MA3, TuB2
Zelenina, A.	TuP41
Zhang, X.	MA4
Zhou, C.	TuA3
Zogg, H.	WC4